Zoonomia;
Or
The Laws Of Organic Life
Vol. 1

by

Erasmus Darwin

Zoonomia; Or The Laws Of Organic Life
Vol. 1
by Erasmus Darwin

ISBN: 978-93-59327-38-9

Published by

DOUBLE 9 BOOKS

2/13-B, Ansari Road
Daryaganj, New Delhi – 110002
info@double9books.com
www.double9books.com
Tel. 011-40042856

ABOUT THE AUTHOR

Charles Hodge (December 27, 1797 – June 19, 1878) was a Reformed Presbyterian theologian and Princeton Theological Seminary president from 1851 until 1878. He was a significant proponent of Princeton Theology, an orthodox Calvinist theological movement that flourished in nineteenth-century America. He strongly defended the Bible's authority as God's Word. In the twentieth century, many of his opinions were shared by Fundamentalists and Evangelicals. Charles Hodge's father, Hugh Hodge, was the son of a Scotsman who emigrated from Northern Ireland in the early eighteenth century. Hugh graduated from Princeton College in 1773, and after serving as a military surgeon during the Revolutionary War, he went on to practice medicine in Philadelphia. He married well-bred Bostonian orphan Mary Blanchard in 1790. The first three sons of the Hodges died in the Yellow Fever Epidemic of 1793 and another in 1795. Hugh Lenox, their first surviving child, was born in 1796. Hugh Lenox went on to become an obstetrics expert, and he remained especially close to Charles, even financially assisting him.

CONTENTS

DEDICATION

To the candid and ingenious Members of the College of Physicians, of the Royal Philosophical Society, of the Two Universities, and to all those, who study the Operations of the Mind as a Science, or who practice Medicine as a Profession, the subsequent Work is, with great respect, inscribed by the Author,

DERBY, May 1, 1794.

PREFACE

The purport of the following pages is an endeavour to reduce the facts belonging to ANIMAL LIFE into classes, orders, genera, and species; and, by comparing them with each other, to unravel the theory of diseases. It happened, perhaps unfortunately for the inquirers into the knowledge of diseases, that other sciences had received improvement previous to their own; whence, instead of comparing the properties belonging to animated nature with each other, they, idly ingenious, busied themselves in attempting to explain the laws of life by those of mechanism and chemistry; they considered the body as an hydraulic machine, and the fluids as passing through a series of chemical changes, forgetting that animation was its essential characteristic.

The great CREATOR of all things has infinitely diversified the works of his hands, but has at the same time stamped a certain similitude on the features of nature, that demonstrates to us, that *the whole is one family of one parent*. On this similitude is founded all rational analogy; which, so long as it is concerned in comparing the essential properties of bodies, leads us to many and important discoveries; but when with licentious activity it links together objects, otherwise discordant, by some fanciful similitude; it may indeed collect ornaments for wit and poetry, but philosophy and truth recoil from its combinations.

The want of a theory, deduced from such strict analogy, to conduct the practice of medicine is lamented by its professors; for, as a great number of unconnected facts are difficult to be acquired, and to be reasoned from, the art of medicine is in many instances less efficacious under the direction of its wisest practitioners; and by that busy crowd, who either boldly wade in darkness, or are led into endless error by the glare of false theory, it is daily practised to the destruction of thousands; add to this the unceasing injury which accrues to the public by the perpetual advertisements of pretended nostrums; the minds of the indolent become superstitiously fearful of diseases, which they do not labour under; and thus become the daily prey of some crafty empyric.

A theory founded upon nature, that should bind together the scattered facts of medical knowledge, and converge into one point of view the laws of organic life, would thus on many accounts contribute to the interest of society. It would capacitate men of moderate abilities to practise the art of healing with real advantage to the public; it would enable every one of literary acquirements to distinguish the genuine disciples of medicine from those of boastful effrontery, or of wily address; and would teach mankind in some important situations the *knowledge of themselves.*

There are some modern practitioners, who declaim against medical theory in general, not considering that to think is to theorize; and that no one can direct a method of cure to a person labouring under disease without thinking, that is, without theorizing; and happy therefore is the patient, whose physician possesses the best theory.

The words idea, perception, sensation, recollection, suggestion, and association, are each of them used in this treatise in a more limited sense than in the writers of metaphysic. The author was in doubt, whether he should rather have substituted new words instead of them; but was at length of opinion, that new definitions of words already in use would be less burthensome to the memory of the reader.

A great part of this work has lain by the writer above twenty years, as some of his friends can testify: he had hoped by frequent revision to have made it more worthy the acceptance of the public; this however his other perpetual occupations have in part prevented, and may continue to prevent, as long as he may be capable of revising it; he therefore begs of the candid reader to accept of it in its present state, and to excuse any inaccuracies of expression, or of conclusion, into which the intricacy of his subject, the general imperfection of language, or the frailty he has in common with other men, may have betrayed him; and from which he has not the vanity to believe this treatise to be exempt.

SECT. I
OF MOTION

The whole of nature may be supposed to consist of two essences or substances; one of which may be termed spirit, and the other matter. The former of these possesses the power to commence or produce motion, and the latter to receive and communicate it. So that motion, considered as a cause, immediately precedes every effect; and, considered as an effect, it immediately succeeds every cause.

The MOTIONS OF MATTER may be divided into two kinds, primary and secondary. The secondary motions are those, which are given to or received from other matter in motion. Their laws have been successfully investigated by philosophers in their treatises on mechanic powers. These motions are distinguished by this circumstance, that the velocity multiplied into the quantity of matter of the body acted upon is equal to the velocity multiplied into the quantity of matter of the acting body.

The primary motions of matter may be divided into three classes, those belonging to gravitation, to chemistry, and to life; and each class has its peculiar laws. Though these three classes include the motions of solid, liquid, and aerial bodies; there is nevertheless a fourth division of motions; I mean those of the supposed ethereal fluids of magnetism, electricity, heat, and light; whose properties are not so well investigated as to be classed with sufficient accuracy.

1st. The gravitating motions include the annual and diurnal rotation of the earth and planets, the flux and reflux of the ocean, the descent of heavy bodies, and other phænomena of gravitation. The unparalleled sagacity of the great NEWTON has deduced the laws of this class of motions from the simple principle of the general attraction of matter. These motions are distinguished by their tendency to or from the centers of the sun or planets.

2d. The chemical class of motions includes all the various appearances of chemistry. Many of the facts, which belong to these branches of science, are nicely ascertained, and elegantly classed; but their laws have not yet been developed from such simple principles as those above-mentioned;

though it is probable, that they depend on the specific attractions belonging to the particles of bodies, or to the difference of the quantity of attraction belonging to the sides and angles of those particles. The chemical motions are distinguished by their being generally attended with an evident decomposition or new combination of the active materials.

3d. The third class includes all the motions of the animal and vegetable world; as well those of the vessels, which circulate their juices, and of the muscles, which perform their locomotion, as those of the organs of sense, which constitute their ideas.

This last class of motion is the subject of the following pages; which, though conscious of their many imperfections, I hope may give some pleasure to the patient reader, and contribute something to the knowledge and to the cure of diseases.

SECT. II
EXPLANATIONS AND DEFINITIONS

I. Outline of the animal economy. — II. 1. *Of the sensorium.* 2. *Of the brain and nervous medulla.* 3. *A nerve.* 4. *A muscular fibre.* 5. *The immediate organs of sense.* 6. *The external organs of sense.* 7. *An idea or sensual motion.* 8. *Perception.* 9. *Sensation.* 10. *Recollection and suggestion.* 11. *Habit, causation, association, catenation.* 12. *Reflexideas.* 13. *Stimulus defined.*

As some explanations and definitions will be necessary in the prosecution of the work, the reader is troubled with them in this place, and is intreated to keep them in his mind as he proceeds, and to take them for granted, till an apt opportunity occurs to evince their truth; to which I shall premise a very short outline of the animal economy.

I.—1. The nervous system has its origin from the brain, and is distributed to every part of the body. Those nerves, which serve the senses, principally arise from that part of the brain, which is lodged in the head; and those, which serve the purposes of muscular motion, principally arise from that part of the brain, which is lodged in the neck and back, and which is erroneously called the spinal marrow. The ultimate fibrils of these nerves terminate in the immediate organs of sense and muscular fibres, and if a ligature be put on any part of their passage from the head or spine, all motion and perception cease in the parts beneath the ligature.

2. The longitudinal muscular fibres compose the locomotive muscles, whose contractions move the bones of the limbs and trunk, to which their extremities are attached. The annular or spiral muscular fibres compose the vascular muscles, which constitute the intestinal canal, the arteries, veins, glands, and absorbent vessels.

3. The immediate organs of sense, as the retina of the eye, probably consist of moving fibrils, with a power of contraction similar to that of the larger muscles above described.

4. The cellular membrane consists of cells, which resemble those of a sponge, communicating with each other, and connecting together all the other parts of the body.

5. The arterial system consists of the aortal and the pulmonary artery, which are attended through their whole course with their correspondent veins. The pulmonary artery receives the blood from the right chamber of the heart, and carries it to the minute extensive ramifications of the lungs, where it is exposed to the action of the air on a surface equal to that of the whole external skin, through the thin moist coats of those vessels, which are spread on the air-cells, which constitute the minute terminal ramifications of the wind-pipe. Here the blood changes its colour from a dark red to a bright scarlet. It is then collected by the branches of the pulmonary vein, and conveyed to the left chamber of the heart.

6. The aorta is another large artery, which receives the blood from the left chamber of the heart, after it has been thus aerated in the lungs, and conveys it by ascending and descending branches to every other part of the system; the extremities of this artery terminate either in glands, as the salivary glands, lacrymal glands, &c. or in capillary vessels, which are probably less involuted glands; in these some fluid, as saliva, tears, perspiration, are separated from the blood; and the remainder of the blood is absorbed or drank up by branches of veins correspondent to the branches of the artery; which are furnished with valves to prevent its return; and is thus carried back, after having again changed its colour to a dark red, to the right chamber of the heart. The circulation of the blood in the liver differs from this general system; for the veins which drink up the refluent blood from those arteries, which are spread on the bowels and mesentery, unite into a trunk in the liver, and form a kind of artery, which is branched into the whole substance of the liver, and is called the vena portarum; and from which the bile is separated by the numerous hepatic glands, which constitute that viscus.

7. The glands may be divided into three systems, the convoluted glands, such as those above described, which separate bile, tears, saliva, &c. Secondly, the glands without convolution, as the capillary vessels, which unite the terminations of the arteries and veins; and separate both the mucus, which lubricates the cellular membrane, and the perspirable matter, which preserves the skin moist and flexible. And thirdly, the whole absorbent system, consisting of the lacteals, which open their mouths into the stomach and intestines, and of the lymphatics, which open their mouths on the external surface of the body, and on the internal linings of all the cells of the cellular membrane, and other cavities of the body.

These lacteal and lymphatic vessels are furnished with numerous valves to prevent the return of the fluids, which they absorb, and terminate in glands, called lymphatic glands, and may hence be considered as long necks or mouths belonging to these glands. To these they convey the chyle and mucus, with a part of the perspirable matter, and atmospheric moisture; all which, after having passed through these glands, and having suffered some change in them, are carried forward into the blood, and supply perpetual nourishment to the system, or replace its hourly waste.

8. The stomach and intestinal canal have a constant vermicular motion, which carries forwards their contents, after the lacteals have drank up the chyle from them; and which is excited into action by the stimulus of the aliment we swallow, but which becomes occasionally inverted or retrograde, as in vomiting, and in the iliac passion.

II. 1. The word *sensorium* in the following pages is designed to express not only the medullary part of the brain, spinal marrow, nerves, organs of sense, and of the muscles; but also at the same time that living principle, or spirit of animation, which resides throughout the body, without being cognizable to our senses, except by its effects. The changes which occasionally take place in the sensorium, as during the exertions of volition, or the sensations of pleasure or pain, are termed *sensorial motions*.

2. The similarity of the texture of the brain to that of the pancreas, and some other glands of the body, has induced the inquirers into this subject to believe, that a fluid, perhaps much more subtile than the electric aura, is separated from the blood by that organ for the purposes of motion and sensation. When we recollect, that the electric fluid itself is actually accumulated and given out voluntarily by the torpedo and the gymnotus electricus, that an electric shock will frequently stimulate into motion a paralytic limb, and lastly that it needs no perceptible tubes to convey it, this opinion seems not without probability; and the singular figure of the brain and nervous system seems well adapted to distribute it over every part of the body.

For the medullary substance of the brain not only occupies the cavities of the head and spine, but passes along the innumerable ramifications of the nerves to the various muscles and organs of sense. In these it lays aside its coverings, and is intermixed with the slender fibres, which constitute those muscles and organs of sense. Thus all these distant ramifications of the sensorium are united at one of their extremities, that is, in the head and spine; and thus these central parts of the sensorium constitute a communication between all the organs of sense and muscles.

3. A *nerve* is a continuation of the medullary substance of the brain from the head or spine towards the other parts of the body, wrapped in its proper membrane.

4. The *muscular fibres* are moving organs intermixed with that medullary substance, which is continued along the nerves, as mentioned above. They are indued with the power of contraction, and are again elongated either by antagonist muscles, by circulating fluids, or by elastic ligaments. So the muscles on one side of the forearm bend the fingers by means of their tendons, and those on the other side of the fore-arm extend them again. The arteries are distended by the circulating blood; and in the necks of quadrupeds there is a strong elastic ligament, which assists the muscles, which elevate the head, to keep it in its horizontal position, and to raise it after it has been depressed.

5. The *immediate organs of sense* consist in like manner of moving fibres enveloped in the medullary substance above mentioned; and are erroneously supposed to be simply an expansion of the nervous medulla, as the retina of the eye, and the rete mucosum of the skin, which are the immediate organs of vision, and of touch. Hence when we speak of the contractions of the fibrous parts of the body, we shall mean both the contractions of the muscles, and those of the immediate organs of sense. These *fibrous motions* are thus distinguished from the *sensorial motions* above mentioned.

6. The *external organs* of sense are the coverings of the immediate organs of sense, and are mechanically adapted for the reception or transmission of peculiar bodies, or of their qualities, as the cornea and humours of the eye, the tympanum of the ear, the cuticle of the fingers and tongue.

7. The word *idea* has various meanings in the writers of metaphysic: it is here used simply for those notions of external things, which our organs of sense bring us acquainted with originally; and is defined a contraction, or motion, or configuration, of the fibres, which constitute the immediate organ of sense; which will be explained at large in another part of the work. Synonymous with the word idea, we shall sometimes use the words *sensual motion* in contradistinction to *muscular motion.*

8. The word *perception* includes both the action of the organ of sense in consequence of the impact of external objects, and our attention to that action; that is, it expresses both the motion of the organ of sense, or idea, and the pain or pleasure that succeeds or accompanies it.

9. The pleasure or pain which necessarily accompanies all those perceptions or ideas which we attend to, either gradually subsides, or is succeeded by other fibrous motions. In the latter case it is termed *sensation,* as explained in Sect. V. 2, and VI. 2.—The reader is intreated to keep this in

his mind, that through all this treatise the word sensation is used to express pleasure or pain only in its active state, by whatever means it is introduced into the system, without any reference to the stimulation of external objects.

10. The vulgar use of the word *memory* is too unlimited for our purpose: those ideas which we voluntarily recall are here termed ideas of *recollection*, as when we will to repeat the alphabet backwards. And those ideas which are suggested to us by preceding ideas are here termed ideas of *suggestion*, as whilst we repeat the alphabet in the usual order; when by habits previously acquired B is suggested by A, and C by B, without any effort of deliberation.

11. The word *association* properly signifies a society or convention of things in some respects similar to each other. We never say in common language, that the effect is associated with the cause, though they necessarily accompany or succeed each other. Thus the contractions of our muscles and organs of sense may be said to be associated together, but cannot with propriety be said to be associated with irritations, or with volition, or with sensation; because they are caused by them, as mentioned in Sect. IV. When fibrous contractions succeed other fibrous contractions, the connection is termed *association*; when fibrous contractions succeed sensorial motions, the connection is termed *causation*; when fibrous and sensorial motions reciprocally introduce each other in progressive trains or tribes, it is termed *catenation* of animal motions. All these connections are said to be produced by *habit*; that is, by frequent repetition.

12. It may be proper to observe, that by the unavoidable idiom of our language the ideas of perception, of recollection, or of imagination, in the plural number signify the ideas belonging to perception, to recollection, or to imagination; whilst the idea of perception, of recollection, or of imagination, in the singular number is used for what is termed "a reflex idea of any of those operations of the sensorium."

13. By the word *stimulus* is not only meant the application of external bodies to our organs of sense and muscular fibres, which excites into action the sensorial power termed irritation; but also pleasure or pain, when they excite into action the sensorial power termed sensation; and desire or aversion, when they excite into action the power of volition; and lastly, the fibrous contractions which precede association; as is further explained in Sect. XII. 2. 1.

SECT. III
THE MOTIONS OF THE RETINA
DEMONSTRATED BY EXPERIMENTS

I. Of animal motions and of ideas. II. The fibrous structure of the retina. III. The activity of the retina in vision. 1. Rays of light have no momentum. 2. Objects long viewed become fainter. 3. Spectra of black objects become luminous. 4. Varying spectra from gyration. 5. From long inspection of various colours. IV. Motions of the organs of sense constitute ideas. 1. Light from pressing the eye-ball, and sound from the pulsation of the carotid artery. 2. Ideas in sleep mistaken for perceptions. 3. Ideas of imagination produce pain and sickness like sensations. 4. When the organ of sense is destroyed, the ideas belonging to that sense perish. V. Analogy between muscular motions and sensual motions, or ideas. 1. They are both originally excited by irritations. 2. And associated together in the same manner. 3. Both act in nearly the same times. 4. Are alike strengthened or fatigued by exercise. 5. Are alike painful from inflammation. 6. Are alike benumbed by compression. 7. Are alike liable to paralysis. 8. To convulsion. 9. To the influence of old age.—VI. Objections answered. 1. Why we cannot invent new ideas. 2. If ideas resemble external objects. 3. Of the imagined sensation in an amputated limb. 4. Abstract ideas.— VII. What are ideas, if they are not animal motions?

Before the great variety of animal motions can be duly arranged into natural classes and orders, it is necessary to smooth the way to this yet unconquered field of science, by removing some obstacles which thwart our passage. I. To demonstrate that the retina and other immediate organs of sense possess a power of motion, and that these motions constitute our ideas, according to the fifth and seventh of the preceding assertions, claims our first attention.

Animal motions are distinguished from the communicated motions, mentioned in the first section, as they have no mechanical proportion to their cause; for the goad of a spur on the skin of a horse shall induce him to move a load of hay. They differ from the gravitating motions there mentioned as they are exerted with equal facility in all directions, and they differ from the chemical class of motions, because no apparent decompositions or new combinations are produced in the moving materials.

Hence, when we say animal motion is excited by irritation, we do not mean that the motion bears any proportion to the mechanical impulse of the stimulus; nor that it is affected by the general gravitation of the two bodies; nor by their chemical properties, but solely that certain animal fibres are excited into action by something external to the moving organ.

In this sense the stimulus of the blood produces the contractions of the heart; and the substances we take into our stomach and bowels stimulate them to perform their necessary functions. The rays of light excite the retina into animal motion by their stimulus; at the same time that those rays of light themselves are physically converged to a focus by the inactive humours of the eye. The vibrations of the air stimulate the auditory nerve into animal action; while it is probable that the tympanum of the ear at the same time undergoes a mechanical vibration.

To render this circumstance more easy to be comprehended, *motion may be defined to be a variation of figure*; for the whole universe may be considered as one thing possessing a certain figure; the motions of any of its parts are a variation of this figure of the whole: this definition of motion will be further explained in Section XIV. 2. 2. on the production of ideas.

Now the motions of an organ of sense are a succession of configurations of that organ; these configurations succeed each other quicker or slower; and whatever configuration of this organ of sense, that is, whatever portion of the motion of it is, or has usually been, attended to, constitutes an idea. Hence the configuration is not to be considered as an effect of the motion of the organ, but rather as a part or temporary termination of it; and that, whether a pause succeeds it, or a new configuration immediately takes place. Thus when a succession of moving objects are presented to our view, the ideas of trumpets, horns, lords and ladies, trains and canopies, are configurations, that is, parts or links of the successive motions of the organ of vision.

Plate I.

These motions or configurations of the organs of sense differ from the sensorial motions to be described hereafter, as they appear to be simply contractions of the fibrous extremities of those organs, and in that respect exactly resemble the motions or contractions of the larger muscles, as appears from the following experiment. Place a circular piece of red silk about an inch in diameter on a sheet of white paper in a strong light, as in Plate I.—look for a minute on this area, or till the eye becomes somewhat fatigued, and then, gently closing your eyes, and shading them with your hand, a circular green area of the same apparent diameter becomes visible in the closed eye. This green area is the colour reverse to the red area, which had been previously inspected, as explained in the experiments on ocular spectra at the end of the work, and in Botanical Garden, P. 1. additional note, No. 1. Hence it appears, that a part of the retina, which had been fatigued by contraction in one direction, relieves itself by exerting the antagonist fibres, and producing a contraction in an opposite direction, as is common in the exertions of our muscles. Thus when we are tired with long action of our arms in one direction, as in holding a bridle on a journey, we occasionally throw them into an opposite position to relieve the fatigued muscles.

Mr. Locke has defined an idea to be "whatever is present to the mind;" but this would include the exertions of volition, and the sensations of pleasure and pain, as well as those operations of our system, which acquaint us with external objects; and is therefore too unlimited for our purpose. Mr. Lock seems to have fallen into a further error, by conceiving, that the mind could form a general or abstract idea by its own operation, which was the copy of no particular perception; as of a triangle in general, that was neither acute, obtuse, nor right angled. The ingenious Dr. Berkley and Mr. Hume have demonstrated, that such general ideas have no existence in nature, not even in the mind of their celebrated inventor. We shall therefore take for granted at present, that our recollection or imagination of external objects consists of a partial repetition of the perceptions, which were excited by those external objects, at the time we became acquainted with them; and that our reflex ideas of the operations of our minds are partial repetitions of those operations.

II. The following article evinces that the organ of vision consists of a fibrous part as well as of the nervous medulla, like other white muscles; and hence, as it resembles the muscular parts of the body in its structure, we may conclude, that it must resemble them in possessing a power of being excited into animal motion.—The subsequent experiments on the optic nerve, and on the colours remaining in the eye, are copied from a paper on ocular spectra published in the seventy-sixth volume of the Philos. Trans. by Dr. R. Darwin of Shrewsbury; which, as I shall have frequent occasion to refer to, is reprinted in this work, Sect. XL. The retina of an ox's eye was suspended in a glass of warm water, and forcibly torn in a few places; the edges of these parts appeared jagged and hairy, and did not contract and become smooth like simple mucus, when it is distended till it breaks; which evinced that it consisted of fibres. This fibrous construction became still more distinct to the light by adding some caustic alcali to the water; as the adhering mucus was first eroded, and the hair-like fibres remained floating in the vessel. Nor does the degree of transparency of the retina invalidate this evidence of its fibrous structure, since Leeuwenhoek has shewn, that the crystalline humour itself consists of fibres. Arc. Nat. V. I. 70.

Hence it appears, that as the muscles consist of larger fibres intermixed with a smaller quantity of nervous medulla, the organ of vision consists of a greater quantity of nervous medulla intermixed with smaller fibres. It is probable that the locomotive muscles of microscopic animals may have greater tenuity than these of the retina; and there is reason to conclude from analogy, that the other immediate organs of sense, as the portio mollis of the auditory nerve, and the rete mucosum of the skin, possess a similarity of structure with the retina, and a similar power of being excited into animal motion.

III. The subsequent articles shew, that neither mechanical impressions, nor chemical combinations of light, but that the animal activity of the retina constitutes vision.

1. Much has been conjectured by philosophers about the momentum of the rays of light; to subject this to experiment a very light horizontal balance was constructed by Mr. Michel, with about an inch square of thin leaf-copper suspended at each end of it, as described in Dr. Priestley's History of Light and Colours. The focus of a very large convex mirror was thrown by Dr. Powel, in his lectures on experimental philosophy, in my presence, on one wing of this delicate balance, and it receded from the light; thrown on

the other wing, it approached towards the light, and this repeatedly; so that no sensible impulse could be observed, but what might well be ascribed to the ascent of heated air.

Whence it is reasonable to conclude, that the light of the day must be much too weak in its dilute state to make any mechanical impression on so tenacious a substance as the retina of the eye.—Add to this, that as the retina is nearly transparent, it could therefore make less resistance to the mechanical impulse of light; which, according, to the observations related by Mr. Melvil in the Edinburgh Literary Essays, only communicates heat, and should therefore only communicate momentum, where it is obstructed, reflected, or refracted.—From whence also may be collected the final cause of this degree of transparency of the retina, viz. left by the focus of stronger lights, heat and pain should have been produced in the retina, instead of that stimulus which excites it into animal motion.

2. On looking long on an area of scarlet silk of about an inch in diameter laid on white paper, as in Plate I. the scarlet colour becomes fainter, till at length it entirely vanishes, though the eye is kept uniformly and steadily upon it. Now if the change or motion of the retina was a mechanical impression, or a chemical tinge of coloured light, the perception would every minute become stronger and stronger,—whereas in this experiment it becomes every instant weaker and weaker. The same circumstance obtains in the continued application of sound, or of sapid bodies, or of odorous ones, or of tangible ones, to their adapted organs of sense.

Plate II.

Thus when a circular coin, as a shilling, is pressed on the palm of the hand, the sense of touch is mechanically compressed; but it is the stimulus of this pressure that excites the organ of touch into animal action, which constitutes the perception of hardness and of figure; for in some minutes the perception ceases, though the mechanical pressure of the object remains.

3. Make with ink on white paper a very black spot about half an inch in diameter, with a tail about an inch in length, so as to resemble a tadpole, as in Plate II.; look steadfastly for a minute on the center of this spot, and, on moving the eye a little, the figure of the tadpole will be seen on the white part of the paper; which figure of the tadpole will appear more luminous than the other part of the white paper; which can only be explained by supposing that a part of the retina, on which the tadpole was delineated, to have become more sensible to light than the other parts of it, which were exposed to the white paper; and not from any idea of mechanical impression or chemical combination of light with the retina.

4. When any one turns round rapidly, till he becomes dizzy, and falls upon the ground, the spectra of the ambient objects continue to present themselves in rotation, and he seems to behold the objects still in motion. Now if these spectra were impressions on a passive organ, they either must continue as they were received last, or not continue at all.

5. Place a piece of red silk about an inch in diameter on a sheet of white paper in a strong light, as in Plate I; look steadily upon it from the distance of about half a yard for a minute; then closing your eye-lids, cover them with your hands and handkerchief, and a green spectrum will be seen in your eyes resembling in form the piece of red silk. After some seconds of time the spectrum will disappear, and in a few more seconds will reappear; and thus alternately three or four times, if the experiment be well made, till at length it vanishes entirely.

Plate III.

6. Place a circular piece of white paper, about four inches in diameter, in the sunshine, cover the center of this with a circular piece of black silk, about three inches in diameter; and the center of the black silk with a circle of pink silk, about two inches in diameter; and the center of the pink silk with a circle of yellow silk, about one inch in diameter; and the center of this with a circle of blue silk, about half an inch in diameter; make a small spot with ink in the center of the blue silk, as in Plate III.; look steadily for a minute on this central spot, and then closing your eyes, and applying your hand at about an inch distance before them, so as to prevent too much or too little light from passing through the eye-lids, and you will see the most beautiful circles of colours that imagination can conceive; which are most resembled by the colours occasioned by pouring a drop or two of oil on a still lake in a bright day. But these circular irises of colours are not only different from the colours of the silks above mentioned, but are at the same time perpetually changing as long as they exist.

From all these experiments it appears, that these spectra in the eye are not owing to the mechanical impulse of light impressed on the retina; nor to its chemical combination with that organ; nor to the absorption and emission of light, as is supposed, perhaps erroneously, to take place in calcined shells and other phosphorescent bodies, after having been exposed to the light: for in all these cases the spectra in the eye should either remain of the same colour, or gradually decay, when the object is withdrawn; and neither their evanescence during the presence of their object, as in the second experiment, nor their change from dark to luminous, as in the third experiment, nor their rotation, as in the fourth experiment, nor the alternate presence and evanescence of them, as in the fifth experiment, nor the perpetual change of colours of them, as in the last experiment, could exist.

IV. The subsequent articles shew, that these animal motions or configurations of our organs of sense constitute our ideas.

1. If any one in the dark presses the ball of his eye, by applying his finger to the external corner of it, a luminous appearance is observed; and by a smart stroke on the eye great slashes of fire are perceived. (Newton's Optics.) So that when the arteries, that are near the auditory nerve, make stronger pulsations than usual, as in some fevers, an undulating sound is excited in the ears. Hence it is not the presence of the light and sound, but the motions of the organ, that are immediately necessary to constitute the perception or idea of light and sound.

2. During the time of sleep, or in delirium, the ideas of imagination are mistaken for the perceptions of external objects; whence it appears, that

these ideas of imagination, are no other than a reiteration of those motions of the organs of sense, which were originally excited by the stimulus of external objects: and in our waking hours the simple ideas, that we call up by recollection or by imagination, as the colour of red, or the smell of a rose, are exact resemblances of the same simple ideas from perception; and in consequence must be a repetition of those very motions.

3. The disagreeable sensation called the tooth-edge is originally excited by the painful jarring of the teeth in biting the edge of the glass, or porcelain cup, in which our food was given us in our infancy, as is further explained in the Section XVI. 10, on Instinct. — This disagreeable sensation is afterwards excitable not only by a repetition of the sound, that was then produced, but by imagination alone, as I have myself frequently experienced; in this case the idea of biting a china cup, when I imagine it very distinctly, or when I see another person bite a cup or glass, excites an actual pain in the nerves of my teeth. So that this idea and pain seem to be nothing more than the reiterated motions of those nerves, that were formerly so disagreeably affected.

Other ideas that are excited by imagination or recollection in many instances produce similar effects on the constitution, as our perceptions had formerly produced, and are therefore undoubtedly a repetition of the same motions. A story which the celebrated Baron Van Swieton relates of himself is to this purpose. He was present when the putrid carcase of a dead dog exploded with prodigious stench; and some years afterwards, accidentally riding along the same road, he was thrown into the same sickness and vomiting by the idea of the stench, as he had before experienced from the perception of it.

4. Where the organ of sense is totally destroyed, the ideas which were received by that organ seem to perish along with it, as well as the power of perception. Of this a satisfactory instance has fallen under my observation. A gentleman about sixty years of age had been totally deaf for near thirty years: he appeared to be a man of good understanding, and amused himself with reading, and by conversing either by the use of the pen, or by signs made with his fingers, to represent letters. I observed that he had so far forgot the pronunciation of the language, that when he attempted to speak, none of his words had distinct articulation, though his relations could sometimes understand his meaning. But, which is much to the point, he assured me, that in his dreams he always imagined that people conversed with him by signs or writing, and never that he heard any one speak to him. From hence it appears, that with the perceptions of sounds he has also lost

the ideas of them; though the organs of speech still retain somewhat of their usual habits of articulation.

This observation may throw some light on the medical treatment of deaf people; as it may be learnt from their dreams whether the auditory nerve be paralytic, or their deafness be owing to some defect of the external organ.

It rarely happens that the immediate organ of vision is perfectly destroyed. The most frequent causes of blindness are occasioned by defects of the external organ, as in cataracts and obfuscations of the cornea. But I have had the opportunity of conversing with two men, who had been some years blind; one of them had a complete gutta serena, and the other had lost the whole substance of his eyes. They both told me that they did not remember to have ever dreamt of visible objects, since the total loss of their sight.

V. Another method of discovering that our ideas are animal motions of the organs of sense, is from considering the great analogy they bear to the motions of the larger muscles of the body. In the following articles it will appear that they are originally excited into action by the irritation of external objects like our muscles; are associated together like our muscular motions; act in similar time with them; are fatigued by continued exertion like them; and that the organs of sense are subject to inflammation, numbness, palsy, convulsion, and the defects of old age, in the same manner as the muscular fibres.

1. All our perceptions or ideas of external objects are universally allowed to have been originally excited by the stimulus of those external objects; and it will be shewn in a succeeding section, that it is probable that all our muscular motions, as well those that are become voluntary as those of the heart and glandular system, were originally in like manner excited by the stimulus of something external to the organ of motion.

2. Our ideas are also associated together after their production precisely in the same manner as our muscular motions; which will likewise be fully explained in the succeeding section.

3. The time taken up in performing an idea is likewise much the same as that taken up in performing a muscular motion. A musician can press the keys of an harpsichord with his fingers in the order of a tune he has been accustomed to play, in as little time as he can run over those notes in his mind. So we many times in an hour cover our eye-balls with our eye-lids without perceiving that we are in the dark; hence the perception or idea of light is not changed for that of darkness in so small a time as the twinkling

of an eye; so that in this case the muscular motion of the eye-lid is performed quicker than the perception of light can be changed for that of darkness. — So if a fire-stick be whirled round in the dark, a luminous circle appears to the observer; if it be whirled somewhat slower, this circle becomes interrupted in one part; and then the time taken up in such a revolution of the stick is the same that the observer uses in changing his ideas: thus the δολικοσκοτον εγκος of Homer, the long shadow of the flying javelin, is elegantly designed to give us an idea of its velocity, and not of its length.

4. The fatigue that follows a continued attention of the mind to one object is relieved by changing the subject of our thoughts; as the continued movement of one limb is relieved by moving another in its stead. Whereas a due exercise of the faculties of the mind strengthens and improves those faculties, whether of imagination or recollection; as the exercise of our limbs in dancing or fencing increases the strength and agility of the muscles thus employed.

5. If the muscles of any limb are inflamed, they do not move without pain; so when the retina is inflamed, its motions also are painful. Hence light is as intolerable in this kind of ophthalmia, as pressure is to the finger in the paronychia. In this disease the patients frequently dream of having their eyes painfully dazzled; hence the idea of strong light is painful as well as the reality. The first of these facts evinces that our perceptions are motions of the organs of sense; and the latter, that our imaginations are also motions of the same organs.

6. The organs of sense, like the moving muscles, are liable to become benumbed, or less sensible, from compression. Thus, if any person on a light day looks on a white wall, he may perceive the ramifications of the optic artery, at every pulsation of it, represented by darker branches on the white wall; which is evidently owing to its compressing the retina during the diastole of the artery. Savage Nosolog.

7. The organs of sense and the moving muscles are alike liable to be affected with palsy, as in the gutta serena, and in some cases of deafness; and one side of the face has sometimes lost its power of sensation, but retained its power of motion; other parts of the body have lost their motions but retained their sensation, as in the common hemiplagia; and in other instances both these powers have perished together.

8. In some convulsive diseases a delirium or insanity supervenes, and the convulsions cease; and conversely the convulsions shall supervene, and the delirium cease. Of this I have been a witness many times in a day in the paroxysms of violent epilepsies; which evinces that one kind of delirium is

a convulsion of the organs of sense, and that our ideas are the motions of these organs: the subsequent cases will illustrate this observation.

Miss G— —, a fair young lady, with light eyes and hair, was seized with most violent convulsions of her limbs, with outrageous hiccough, and most vehement efforts to vomit: after near an hour was elapsed this tragedy ceased, and a calm talkative delirium supervened for about another hour; and these relieved each other at intervals during the greatest part of three or four days. After having carefully considered this disease, I thought the convulsions of her ideas less dangerous than those of her muscles; and having in vain attempted to make any opiate continue in her stomach, an ounce of laudanum was rubbed along the spine of her back, and a dram of it was used as an enema; by this medicine a kind of drunken delirium was continued many hours; and when it ceased the convulsions did not return; and the lady continued well many years, except some lighter relapses, which were relieved in the same manner.

Miss H— —, an accomplished young lady, with light eyes and hair, was seized with convulsions of her limbs, with hiccough, and efforts to vomit, more violent than words can express; these continued near an hour, and were succeeded with a cataleptic spasm of one arm, with the hand applied to her head; and after about twenty minutes these spasms ceased, and a talkative reverie supervened for near an other hour, from which no violence, which it was proper to use, could awaken her. These periods of convulsions, first of the muscles, and then of the ideas, returned twice a day for several weeks; and were at length removed by great doses of opium, after a great variety of other medicines and applications had been in vain experienced. This lady was subject to frequent relapses, once or twice a year for many years, and was as frequently relieved by the same method.

Miss W— —, an elegant young lady, with black eyes and hair, had sometimes a violent pain of her side, at other times a most painful strangury, which were every day succeeded by delirium; which gave a temporary relief to the painful spasms. After the vain exhibition of variety of medicines and applications by different physicians, for more than a twelvemonth, she was directed to take some doses of opium, which were gradually increased, by which a drunken delirium was kept up for a day or two, and the pains prevented from returning. A flesh diet, with a little wine or beer, instead of the low regimen she had previously used, in a few weeks completely established her health; which, except a few relapses, has continued for many years.

9. Lastly, as we advance in life all the parts of the body become more rigid, and are rendered less susceptible of new habits of motion, though they retain

those that were before established. This is sensibly observed by those who apply themselves late in life to music, fencing, or any of the mechanic arts. In the same manner many elderly people retain the ideas they had learned early in life, but find great difficulty in acquiring new trains of memory; insomuch that in extreme old age we frequently see a forgetfulness of the business of yesterday, and at the same time a circumstantial remembrance of the amusements of their youth; till at length the ideas of recollection and activity of the body gradually cease together,—such is the condition of humanity!—and nothing remains but the vital motions and sensations.

VI. 1. In opposition to this doctrine of the production of our ideas, it may be asked, if some of our ideas, like other animal motions, are voluntary, why can we not invent new ones, that have not been received by perception? The answer will be better understood after having perused the succeeding section, where it will be explained, that the muscular motions likewise are originally excited by the stimulus of bodies external to the moving organ; and that the will has only the power of repeating the motions thus excited.

2. Another objector may ask, Can the motion of an organ of sense resemble an odour or a colour? To which I can only answer, that it has not been demonstrated that any of our ideas resemble the objects that excite them; it has generally been believed that they do not; but this shall be discussed at large in Sect. XIV.

3. There is another objection that at first view would seem less easy to surmount. After the amputation, of a foot or a finger, it has frequently happened, that an injury being offered to the stump of the amputated limb, whether from cold air, too great pressure, or other accidents, the patient has complained, of a sensation of pain in the foot or finger, that was cut off. Does not this evince that all our ideas are excited in the brain, and not in the organs of sense? This objection is answered, by observing that our ideas of the shape, place, and solidity of our limbs, are acquired by our organs of touch and of sight, which are situated in our fingers and eyes, and not by any sensations in the limb itself.

In this case the pain or sensation, which formerly has arisen in the foot or toes, and been propagated along the nerves to the central part of the sensorium, was at the same time accompanied with a visible idea of the shape and place, and with a tangible idea of the solidity of the affected limb: now when these nerves are afterwards affected by any injury done to the remaining stump with a similar degree or kind of pain, the ideas of the shape, place, or solidity of the lost limb, return by association; as these ideas belong to the organs of sight and touch, on which they were first excited.

4. If you wonder what organs of sense can be excited into motion, when you call up the ideas of wisdom or benevolence, which Mr. Locke has termed abstracted ideas; I ask you by what organs of sense you first became acquainted with these ideas? And the answer will be reciprocal; for it is certain that all our ideas were originally acquired by our organs of sense; for whatever excites our perception must be external to the organ that perceives it, and we have no other inlets to knowledge but by our perceptions: as will be further explained in Section XIV. and XV. on the Productions and Classes of Ideas.

VII. If our recollection or imagination be not a repetition of animal movements, I ask, in my turn, What is it? You tell me it consists of images or pictures of things. Where is this extensive canvas hung up? or where are the numerous receptacles in which those are deposited? or to what else in the animal system have they any similitude?

That pleasing picture of objects, represented in miniature on the retina of the eye, seems to have given rise to this illusive oratory! It was forgot that this representation belongs rather to the laws of light, than to those of life; and may with equal elegance be seen in the camera obscura as in the eye; and that the picture vanishes for ever, when the object is withdrawn.

SECT. IV
LAWS OF ANIMAL CAUSATION

I. The fibres, which constitute the muscles and organs of sense, possess a power of contraction. The circumstances attending the exertion of this power of CONTRACTION constitute the laws of animal motion, as the circumstances attending the exertion of the power of ATTRACTION constitute the laws of motion of inanimate matter.

II. The spirit of animation is the immediate cause of the contraction of animal fibres, it resides in the brain and nerves, and is liable to general or partial diminution or accumulation.

III. The stimulus of bodies external to the moving organ is the remote cause of the original contractions of animal fibres.

IV. A certain quantity of stimulus produces irritation, which is an exertion of the spirit of animation exciting the fibres into contraction.

V. A certain quantity of contraction of animal fibres, if it be perceived at all, produces pleasure; a greater or less quantity of contraction, if it be perceived at all, produces pain; these constitute sensation.

VI. A certain quantity of sensation produces desire or aversion; these constitute volition.

VII. All animal motions which have occurred at the same time, or in immediate succession, become so connected, that when one of them is reproduced, the other has a tendency to accompany or succeed it. When fibrous contractions succeed or accompany other fibrous contractions, the connection is termed association; when fibrous contractions succeed sensorial motions, the connexion is termed causation; when fibrous and sensorial motions reciprocally introduce each other, it is termed catenation of animal motions. All these connections are said to be produced by habit, that is, by frequent repetition. These laws of animal causation will be evinced by numerous facts, which occur in our daily exertions; and will afterwards be employed to explain the more recondite phænomena of the production, growth, diseases, and decay of the animal system.

SECT. V
OF THE FOUR FACULTIES OR
MOTIONS OF THE SENSORIUM

1. Four sensorial powers. 2. Irritation, sensation, volition, association defined. 3. Sensorial motions distinguished from fibrous motions.

1. The spirit of animation has four different modes of action, or in other words the animal sensorium possesses four different faculties, which are occasionally exerted, and cause all the contractions of the fibrous parts of the body. These are the faculty of causing fibrous contractions in consequence of the irritations excited by external bodies, in consequence of the sensations of pleasure or pain, in consequence of volition, and in consequence of the associations of fibrous contractions with other fibrous contractions, which precede or accompany them.

These four faculties of the sensorium during their inactive state are termed irritability, sensibility, voluntarity, and associability; in their active state they are termed as above, irritation, sensation, volition, association.

2. IRRITATION is an exertion or change of some extreme part of the sensorium residing in the muscles or organs of sense, in consequence of the appulses of external bodies.

SENSATION is an exertion or change of the central parts of the sensorium, or of the whole of it, *beginning* at some of those extreme parts of it, which reside in the muscles or organs of sense.

VOLITION is an exertion or change of the central parts of the sensorium, or of the whole of it, *terminating* in some of those extreme parts of it, which reside in the muscles or organs of sense.

ASSOCIATION is an exertion or change of some extreme part of the sensorium residing in the muscles or organs of sense, in consequence of some antecedent or attendant fibrous contractions.

3. These four faculties of the animal sensorium may at the time of their exertions be termed motions without impropriety of language; for we cannot pass from a state of insensibility or inaction to a state of sensibility or of exertion without some change of the sensorium, and every change includes motion. We shall therefore sometimes term the above described faculties *sensorial motions* to distinguish them from *fibrous motions*; which latter expression includes the motions of the muscles and organs of sense.

The active motions of the fibres, whether those of the muscles or organs of sense, are probably simple contractions; the fibres being again elongated by antagonist muscles, by circulating fluids, or sometimes by elastic ligaments, as in the necks of quadrupeds. The sensorial motions, which constitute the sensations of pleasure or pain, and which constitute volition, and which cause the fibrous contractions in consequence of irritation or of association, are not here supposed to be fluctuations or refluctuations of the spirit of animation; nor are they supposed to be vibrations or revibrations, nor condensations or equilibrations of it; but to be changes or motions of it peculiar to life.

SECT. VI
OF THE FOUR CLASSES OF FIBROUS MOTIONS

I. Origin of fibrous contractions. II. Distribution of them into four classes, irritative motions, sensitive motions, voluntary motions, and associate motions, defined.

I. All the fibrous contractions of animal bodies originate from the sensorium, and resolve themselves into four classes, correspondent with the four powers or motions of the sensorium above described, and from which they have their causation.

1. These fibrous contractions were originally caused by the irritations excited by objects, which are external to the moving organ. As the pulsations of the heart are owing to the irritations excited by the stimulus of the blood; and the ideas of perception are owing to the irritations excited by external bodies.

2. But as painful or pleasurable sensations frequently accompanied those irritations, by habit these fibrous contractions became causeable by the sensations, and the irritations ceased to be necessary to their production. As the secretion of tears in grief is caused by the sensation of pain; and the ideas of imagination, as in dreams or delirium, are excited by the pleasure or pain, with which they were formerly accompanied.

3. But as the efforts of the will frequently accompanied these painful or pleasureable sensations, by habit the fibrous contractions became causable by volition; and both the irritations and sensations ceased to be necessary to their production. As the deliberate locomotions of the body, and the ideas of recollection, as when we will to repeat the alphabet backwards.

4. But as many of these fibrous contractions frequently accompanied other fibrous contractions, by habit they became causable by their associations with them; and the irritations, sensations, and volition, ceased to be necessary to their production. As the actions of the muscles of the lower limbs in fencing are associated with those of the arms; and the ideas of suggestion are associated with other ideas, which precede or accompany

them; as in repeating carelessly the alphabet in its usual order after having began it.

II. We shall give the following names to these four classes of fibrous motions, and subjoin their definitions.

1. Irritative motions. That exertion or change of the sensorium, which is caused by the appulses of external bodies, either simply subsides, or is succeeded by sensation, or it produces fibrous motions; it is termed irritation, and irritative motions are those contractions of the muscular fibres, or of the organs of sense, that are immediately consequent to this exertion or change of the sensorium.

2. Sensitive motions. That exertion or change of the sensorium, which constitutes pleasure or pain, either simply subsides, or is succeeded by volition, or it produces fibrous motions; it is termed sensation, and the sensitive motions are those contractions of the muscular fibres, or of the organs of sense, that are immediately consequent to this exertion or change of the sensorium.

3. Voluntary motions. That exertion or change of the sensorium, which constitutes desire or aversion, either simply subsides, or is succeeded by fibrous motions; it is then termed volition, and voluntary motions are those contractions of the muscular fibres, or of the organs of sense, that are immediately consequent to this exertion or change of the sensorium.

4. Associate motions. That exertion or change of the sensorium, which accompanies fibrous motions, either simply subsides, or is succeeded by sensation or volition, or it produces other fibrous motions; it is then termed association, and the associate motions are those contractions of the muscular fibres, or of the organs of sense, that are immediately consequent to this exertion or change of the sensorium.

SECT. VI
OF IRRITATIVE MOTIONS

I. 1. *Some muscular motions are excited by perpetual irritations. 2. Others more frequently by sensations. 3. Others by volition. Case of involuntary stretchings in paralytic limbs. 4. Some sensual motions are excited by perpetual irritations. 5. Others more frequently by sensation or volition.*

II. 1. *Muscular motions excited by perpetual irritations occasionally become obedient sensation and to volition. 2. And the sensual motions.*

III. 1. *Other muscular motions are associated with the irritative ones. 2. And other ideas with irritative ones. Of letters, language, hieroglyphics. Irritative ideas exist without our attention to them.*

I. 1. Many of our muscular motions are excited by perpetual irritations, as those of the heart and arterial system by the circumfluent blood. Many other of them are excited by intermitted irritations, as those of the stomach and bowels by the aliment we swallow; of the bile-ducts by the bile; of the kidneys, pancreas, and many other glands, by the peculiar fluids they separate from the blood; and those of the lacteal and other absorbent vessels by the chyle, lymph, and moisture of the atmosphere. These motions are accelerated or retarded, as their correspondent irritations are increased or diminished, without our attention or consciousness, in the same manner as the various secretions of fruit, gum, resin, wax, and, honey, are produced in the vegetable world, and as the juices of the earth and the moisture of the atmosphere are absorbed by their roots and foliage.

2. Other muscular motions, that are most frequently connected with our sensations, as those of the sphincters of the bladder and anus, and the musculi erectores penis, were originally excited into motion by irritation, for young children make water, and have other evacuations without attention to these circumstances; "et primis etiam ab incunabulis tenduntur sæpius puerorum penes, amore nondum expergefacto." So the nipples of young women are liable to become turgid by irritation, long before they are in a situation to be excited by the pleasure of giving milk to the lips of a child.

3. The contractions of the larger muscles of our bodies, that are most frequently connected with volition, were originally excited into action by internal irritations: as appears from the stretching or yawning of all animals after long sleep. In the beginning of some fevers this irritation of the muscles produces perpetual stretching and yawning; in other periods of fever an universal restlessness arises from the same cause, the patient changing the attitude of his body every minute. The repeated struggles of the fœtus in the uterus must be owing to this internal irritation: for the fœtus can have no other inducement to move its limbs but the tædium or irksomeness of a continued posture.

The following case evinces, that the motions of stretching the limbs after a continued attitude are not always owing to the power of the will. Mr. Dean, a mason, of Austry in Leicestershire, had the spine of the third vertebra of the back enlarged; in some weeks his lower extremities became feeble, and at length quite paralytic: neither the pain of blisters, the heat of fomentations, nor the utmost efforts of the will could produce the least motion in these limbs; yet twice or thrice a day for many months his feet, legs, and thighs, were affected for many minutes with forceable stretchings, attended with the sensation of fatigue; and he at length recovered the use of his limbs, though the spine continued protuberant. The same circumstance is frequently seen in a less degree in the common hemiplagia; and when this happens, I have believed repeated and strong shocks of electricity to have been of great advantage.

4. In like manner the various organs of sense are originally excited into motion by various external stimuli adapted to this purpose, which motions are termed perceptions or ideas; and many of these motions during our waking hours are excited by perpetual irritation, as those of the organs of hearing and of touch. The former by the constant low indistinct noises that murmur around us, and the latter by the weight of our bodies on the parts which support them; and by the unceasing variations of the heat, moisture, and pressure of the atmosphere; and these sensual motions, precisely as the muscular ones above mentioned, obey their correspondent irritations without our attention or consciousness.

5. Other classes of our ideas are more frequently excited by our sensations of pleasure or pain, and others by volition: but that these have all been originally excited by stimuli from external objects, and only vary in their combinations or reparations, has been fully evinced by Mr. Locke: and are by him termed the ideas of perception in contradistinction to those, which he calls the ideas of reflection.

II. 1. These muscular motions, that are excited by perpetual irritation, are nevertheless occasionally excitable by the sensations of pleasure or pain, or by volition; as appears by the palpitation of the heart from fear, the increased secretion of saliva at the sight of agreeable food, and the glow on the skin of those who are ashamed. There is an instance told in the Philosophical Transactions of a man, who could for a time stop the motion of his heart when he pleased; and Mr. D. has often told me, be could so far increase the peristaltic motion of his bowels by voluntary efforts, as to produce an evacuation by stool at any time in half an hour.

2. In like manner the sensual motions, or ideas, that are excited by perpetual irritation, are nevertheless occasionally excited by sensation or volition; as in the night, when we listen under the influence of fear, or from voluntary attention, the motions excited in the organ of hearing by the whispering of the air in our room, the pulsation of our own arteries, or the faint beating of a distant watch, become objects of perception.

III. 1. Innumerable trains or tribes of other motions are associated with these muscular motions which are excited by irritation; as by the stimulus of the blood in the right chamber of the heart, the lungs are induced to expand themselves; and the pectoral and intercostal muscles, and the diaphragm, act at the same time by their associations with them. And when the pharinx is irritated by agreeable food, the muscles of deglutition are brought into action by association. Thus when a greater light falls on the eye, the iris is brought into action without our attention; and the ciliary process, when the focus is formed before or behind the retina, by their associations with the increased irritative motions of the organ of vision. Many common actions of life are produced in a similar manner. If a fly settle on my forehead, whilst I am intent on my present occupation, I dislodge it with my finger, without exciting my attention or breaking the train of my ideas.

2. In like manner the irritative ideas suggest to us many other trains or tribes of ideas that are associated with them. On this kind of connection, language, letters, hieroglyphics, and every kind of symbol, depend. The symbols themselves produce irritative ideas, or sensual motions, which we do not attend to; and other ideas, that are succeeded by sensation, are excited by their association with them. And as these irritative ideas make up a part of the chain of our waking thoughts, introducing other ideas that engage our attention, though themselves are unattended to, we find it very difficult to investigate by what steps many of our hourly trains of ideas gain their admittance.

It may appear paradoxical, that ideas can exist, and not be attended to; but all our perceptions are ideas excited by irritation, and succeeded by sensation. Now when these ideas excited by irritation give us neither pleasure nor pain, we cease to attend to them. Thus whilst I am walking through that grove before my window, I do not run against the trees or the benches, though my thoughts are strenuously exerted on some other object. This leads us to a distinct knowledge of irritative ideas, for the idea of the tree or bench, which I avoid, exists on my retina, and induces by association the action of certain locomotive muscles; though neither itself nor the actions of those muscles engage my attention.

Thus whilst we are conversing on this subject, the tone, note, and articulation of every individual word forms its correspondent irritative idea on the organ of hearing; but we only attend to the associated ideas, that are attached by habit to these irritative ones, and are succeeded by sensation; thus when we read the words "PRINTING-PRESS" we do not attend to the shape, size, or existence of the letters which compose these words, though each of them excites a correspondent irritative motion of our organ of vision, but they introduce by association our idea of the most useful of modern inventions; the capacious reservoir of human knowledge, whose branching streams diffuse sciences, arts, and morality, through all nations and all ages.

SECT. VIII
OF SENSITIVE MOTIONS

I. 1. Sensitive muscular motions were originally excited into action by irritation. 2. And sensitive sensual motions, ideas of imagination, dreams. II. 1. Sensitive muscular motions are occasionally obedient to volition. 2. And sensitive sensual motions. III. 1. Other muscular motions are associated with the sensitive ones. 2. And other sensual motions.

I. 1. Many of the motions of our muscles, that are excited into action by irritation, are at the same time accompanied with painful or pleasurable sensations; and at length become by habit causable by the sensations. Thus the motions of the sphincters of the bladder and anus were originally excited into action by irritation; for young children give no attention to these evacuations; but as soon as they become sensible of the inconvenience of obeying these irritations, they suffer the water or excrement to accumulate, till it disagreeably affects them; and the action of those sphincters is then in consequence of this disagreeable sensation. So the secretion of saliva, which in young children is copiously produced by irritation, and drops from their mouths, is frequently attended with the agreeable sensation produced by the mastication of tasteful food;, till at length the sight of such food to a hungry person excites into action these salival glands; as is seen in the slavering of hungry dogs.

The motions of those muscles, which are affected by lascivious ideas, and those which are exerted in smiling, weeping, starting from fear, and winking at the approach of danger to the eye, and at times the actions of every large muscle of the body become causable by our sensations. And all these motions are performed with strength and velocity in proportion to the energy of the sensation that excites them, and the quantity of sensorial power.

2. Many of the motions of our organs of sense, or ideas, that were originally excited into action by irritation, become in like manner more frequently causable by our sensations of pleasure or pain. These motions are then termed the ideas of imagination, and make up all the scenery and

transactions of our dreams. Thus when any painful or pleasurable sensations possess us, as of love, anger, fear; whether in our sleep or waking hours, the ideas, that have been formerly excited by the objects of these sensations, now vividly recur before us by their connection with these sensations themselves. So the fair smiling virgin, that excited your love by her presence, whenever that sensation recurs, rises before you in imagination; and that with all the pleasing circumstances, that had before engaged your attention. And in sleep, when you dream under the influence of fear, all the robbers, fires, and precipices, that you formerly have seen or heard of, arise before you with terrible vivacity. All these sensual motions, like the muscular ones above mentioned, are performed with strength and velocity in proportion to the energy of the sensation of pleasure or pain, which excites them, and the quantity of sensorial power.

II. 1. Many of these muscular motions above described, that are most frequently excited by our sensations, are nevertheless occasionally causable by volition; for we can smile or frown spontaneously, can make water before the quantity or acrimony of the urine produces a disagreeable sensation, and can voluntarily masticate a nauseous drug, or swallow a bitter draught, though our sensation would strongly dissuade us.

2. In like manner the sensual motions, or ideas, that are most frequently excited by our sensations, are nevertheless occasionally causeable by volition, as we can spontaneously call up our last night's dream before us, tracing it industriously step by step through all its variety of scenery and transaction; or can voluntarily examine or repeat the ideas, that have been excited by out disgust or admiration.

III. 1. Innumerable trains or tribes of motions are associated with these sensitive muscular motions above mentioned; as when a drop of water falling into the wind-pipe disagreeably affects the air-vessels of the lungs, they are excited into violent action; and with these sensitive motions are associated the actions of the pectoral and intercostal muscles, and the diaphragm; till by their united and repeated succussions the drop is returned through the larinx. The same occurs when any thing disagreeably affects the nostrils, or the stomach, or the uterus; variety of muscles are excited by association into forcible action, not to be suppressed by the utmost efforts of the will; as in sneezing, vomiting, and parturition.

2. In like manner with these sensitive sensual motions, or ideas of imagination, are associated many other trains or tribes of ideas, which by some writers of metaphysics have been classed under the terms of resemblance, causation, and contiguity; and will be more fully treated of hereafter.

SECT. IX
OF VOLUNTARY MOTIONS

I. 1. Voluntary muscular motions are originally excited by irritations. 2. And voluntary ideas. Of reason. II. 1. Voluntary muscular motions are occasionally causable by sensations. 2. And voluntary ideas. III. 1. Voluntary muscular motions are occasionally obedient to irritations. 2. And voluntary ideas. IV. 1. Voluntary muscular motions are associated with other muscular motions. 2. And voluntary ideas.

When pleasure or pain affect the animal system, many of its motions both muscular and sensual are brought into action; as was shewn in the preceding section, and were called sensitive motions. The general tendency of these motions is to arrest and to possess the pleasure, or to dislodge or avoid the pain: but if this cannot immediately be accomplished, desire or aversion are produced, and the motions in consequence of this new faculty of the sensorium are called voluntary.

I. 1. Those muscles of the body that are attached to bones, have in general their principal connections with volition, as I move my pen or raise my body. These motions were originally excited by irritation, as was explained in the section on that subject, afterwards the sensations of pleasure or pain, that accompanied the motions thus excited, induced a repetition of them; and at length many of them were voluntarily practised in succession or in combination for the common purposes of life, as in learning to walk, or to speak; and are performed with strength and velocity in proportion to the energy of the volition, that excites them, and the quantity of sensorial power.

2. Another great class of voluntary motions consists of the ideas of recollection. We will to repeat a certain train of ideas, as of the alphabet backwards; and if any ideas, that do not belong to this intended train, intrude themselves by other connections, we will to reject them, and voluntarily persist in the determined train. So at my approach to a house which I have but once visited, and that at the distance of many months, I will to recollect

the names of the numerous family I expect to see there, and I do recollect them.

On this voluntary recollection of ideas our faculty of reason depends, as it enables us to acquire an idea of the dissimilitude of any two ideas. Thus if you voluntarily produce the idea of a right-angled triangle, and then of a square; and after having excited these ideas repeatedly, you excite the idea of their difference, which is that of another right-angled triangle inverted over the former; you are said to reason upon this subject, or to compare your ideas.

These ideas of recollection, like the muscular motions above mentioned, were originally excited by the irritation of external bodies, and were termed ideas of perception: afterwards the pleasure or pain, that accompanied these motions, induced a repetition of them in the absence of the external body, by which they were first excited; and then they were termed ideas of imagination. At length they become voluntarily practised in succession or in combination for the common purposes of life; as when we make ourselves masters of the history of mankind, or of the sciences they have investigated; and are then called ideas of recollection; and are performed with strength and velocity in proportion to the energy of the volition that excites them, and the quantity of sensorial power.

II. 1. The muscular motions above described, that are most frequently obedient to the will are nevertheless occasionally causable by painful or pleasurable sensation, as in the starting from fear, and the contraction of the calf of the leg in the cramp.

2. In like manner the sensual motions, or ideas, that are most frequently connected with volition, are nevertheless occasionally causable by painful or pleasurable sensation. As the histories of men, or the description of places, which we have voluntarily taken pains to remember, sometimes occur to us in our dreams.

III. 1. The muscular motions that are generally subservient to volition, are also occasionally causable by irritation, as in stretching the limbs after sleep, and yawning. In this manner a contraction of the arm is produced by passing the electric fluid from the Leyden phial along its muscles; and that even though the limb is paralytic. The sudden motion of the arm produces a disagreeable sensation in the joint, but the muscles seem to be brought into action simply by irritation.

2. The ideas, that are generally subservient to the will, are in like manner occasionally excited by irritation; as when we view again an object, we have before well studied, and often recollected.

IV. 1. Innumerable trains or tribes of motions are associated with these voluntary muscular motions above mentioned; as when I will to extend my arm to a distant object, some other muscles are brought into action, and preserve the balance of my body. And when I wish to perform any steady exertion, as in threading a needle, or chopping with an ax, the pectoral muscles are at the same time brought into action to preserve the trunk of the body motionless, and we cease to respire for a time.

2. In like manner the voluntary sensual motions, or ideas of recollection, are associated with many other trains or tribes of ideas. As when I voluntarily recollect a gothic window, that I saw some time ago, the whole front of the cathedral occurs to me at the same time.

SECT. X
OF ASSOCIATE MOTIONS

I. 1. Many muscular motions excited by irritations in trains or tribes become associated. 2. And many ideas. II. 1. Many sensitive muscular motions become associated. 2. And many sensitive ideas. III. 1. Many voluntary muscular motions become associated. 2. And then become obedient to sensation or irritation. 3. And many voluntary ideas become associated.

All the fibrous motions, whether muscular or sensual, which are frequently brought into action together, either in combined tribes, or in successive trains, become so connected by habit, that when one of them is reproduced the others have a tendency to succeed or accompany it.

I. 1. Many of our muscular motions were originally excited in successive trains, as the contractions of the auricles and of the ventricles of the heart; and others in combined tribes, as the various divisions of the muscles which compose the calf of the leg, which were originally irritated into synchronous action by the tædium or irksomeness of a continued posture. By frequent repetitions these motions acquire associations, which continue during our lives, and even after the destruction of the greatest part of the sensorium; for the heart of a viper or frog will continue to pulsate long after it is taken from the body; and when it has entirely ceased to move, if any part of it is goaded with a pin, the whole heart will again renew its pulsations. This kind of connection we shall term irritative association, to distinguish it from sensitive and voluntary associations.

2. In like manner many of our ideas are originally excited in tribes; as all the objects of sight, after we become so well acquainted with the laws of vision, as to distinguish figure and distance as well as colour; or in trains, as while we pass along the objects that surround us. The tribes thus received by irritation become associated by habit, and have been termed complex ideas by the writers of metaphysics, as this book, or that orange. The trains have received no particular name, but these are alike associations of ideas, and frequently continue during our lives. So the taste of a pine-apple, though we

eat it blindfold, recalls the colour and shape of it; and we can scarcely think on solidity without figure.

II. 1. By the various efforts of our sensations to acquire or avoid their objects, many muscles are daily brought into successive or synchronous actions; these become associated by habit, and are then excited together with great facility, and in many instances gain indissoluble connections. So the play of puppies and kittens is a representation of their mode of fighting or of taking their prey; and the motions of the muscles necessary for those purposes become associated by habit, and gain a great adroitness of action by these early repetitions: so the motions of the abdominal muscles, which were originally brought into concurrent action, with the protrusive motion of the rectum or bladder by sensation, become so conjoined with them by habit, that they not only easily obey these sensations occasioned by the stimulus of the excrement and urine, but are brought into violent and unrestrainable action in the strangury and tenesmus. This kind of connection we shall term sensitive association.

2. So many of our ideas, that have been excited together or in succession by our sensations, gain synchronous or successive associations, that are sometimes indissoluble but with life. Hence the idea of an inhuman or dishonourable action perpetually calls up before us the idea of the wretch that was guilty of it. And hence those unconquerable antipathies are formed, which some people have to the sight of peculiar kinds of food, of which in their infancy they have eaten to excess or by constraint.

III. 1. In learning any mechanic art, as music, dancing, or the use of the sword, we teach many of our muscles to act together or in succession by repeated voluntary efforts; which by habit become formed into tribes or trains of association, and serve all our purposes with great facility, and in some instances acquire an indissoluble union. These motions are gradually formed into a habit of acting together by a multitude of repetitions, whilst they are yet separately causable by the will, as is evident from the long time that is taken up by children in learning to walk and to speak; and is experienced by every one, when he first attempts to skate upon the ice or to swim: these we shall term voluntary associations.

2. All these muscular movements, when they are thus associated into tribes or trains, become afterwards not only obedient to volition, but to the sensations and irritations; and the same movement composes a part of many different tribes or trains of motion. Thus a single muscle, when it acts in consort with its neighbours on one side, assists to move the limb in one direction; and in another, when, it acts with those in its neighbourhood on the other side; and in other directions, when it acts separately or jointly with

those that lie immediately under or above it; and all these with equal facility after their associations have been well established.

The facility, with which each muscle changes from one associated tribe to another, and that either backwards or forwards, is well observable in the muscles of the arm in moving the windlass of an air-pump; and the slowness of those muscular movements, that have not been associated by habit, may be experienced by any one, who shall attempt to saw the air quick perpendicularly with one hand, and horizontally with the other at the same time.

3. In learning every kind of science we voluntarily associate many tribes and trains of ideas, which afterwards are ready for all the purposes either of volition, sensation, or irritation; and in some instances acquire indissoluble habits of acting together, so as to affect our reasoning, and influence our actions. Hence the necessity of a good education.

These associate ideas are gradually formed into habits of acting together by frequent repetition, while they are yet separately obedient to the will; as is evident from the difficulty we experience in gaining so exact an idea of the front of St. Paul's church, as to be able to delineate it with accuracy, or in recollecting a poem of a few pages.

And these ideas, thus associated into tribes, not only make up the parts of the trains of volition, sensation, and irritation; but the same idea composes a part of many different tribes and trains of ideas. So the simple idea of whiteness composes a part of the complex idea of snow, milk, ivory; and the complex idea of the letter A composes a part of the several associated trains of ideas that make up the variety of words, in which this letter enters.

The numerous trains of these associated ideas are divided by Mr. Hume into three classes, which he has termed contiguity, causation, and resemblance. Nor should we wonder to find them thus connected together, since it is the business of our lives to dispose them into those three classes; and we become valuable to ourselves and our friends, as we succeed in it. Those who have combined an extensive class of ideas by the contiguity of time or place, are men learned in the history of mankind, and of the sciences they have cultivated. Those who have connected a great class of ideas of resemblances, possess the source of the ornaments of poetry and oratory, and of all rational analogy. While those who have connected great classes of ideas of causation, are furnished with the powers of producing effects. These are the men of active wisdom, who lead armies to victory, and kingdoms to prosperity; or discover and improve the sciences, which meliorate and adorn the condition of humanity.

SECT. XI
ADDITIONAL OBSERVATIONS ON THE SENSORIAL POWERS

I. Stimulation is of various kinds adapted to the organs of sense, to the muscles, to hollow membranes, and glands. Some objects irritate our senses by repeated impulses. II. 1. Sensation and volition frequently affect the whole sensorium. 2. Emotions, passions, appetites. 3. Origin of desire and aversion. Criterion of voluntary actions, difference of brutes and men. 4. Sensibility and voluntarity. III. Associations formed before nativity, irritative motions mistaken for officiated ones.

Irritation.

I. The various organs of sense require various kinds of stimulation to excite them into action; the particles of light penetrate the cornea and humours of the eye, and then irritate the naked retina; rapid particles, dissolved or diffused in water or saliva, and odorous ones, mixed or combined with the air, irritate the extremities of the nerves of taste and smell; which either penetrate, or are expanded on the membranes of the tongue and nostrils; the auditory nerves are stimulated by the vibrations of the atmosphere communicated by means of the tympanum and of the fluid, whether of air or of water, behind it; and the nerves of touch by the hardness of surrounding bodies, though the cuticle is interposed between these bodies and the medulla of the nerve.

As the nerves of the senses have each their appropriated objects, which stimulate them into activity; so the muscular fibres, which are the terminations of other sets of nerves, have their peculiar objects, which excite them into action; the longitudinal muscles are stimulated into contraction by extension, whence the stretching or pandiculation after a long continued posture, during which they have been kept in a state of extension; and the hollow muscles are excited into action by distention, as those of the rectum and bladder are induced to protrude their contents from their sense of the distention rather than of the acrimony of those contents.

There are other objects adapted to stimulate the nerves, which terminate in variety of membranes, and those especially which form the terminations of canals; thus the preparations of mercury particularly affect the salivary glands, ipecacuanha the stomach, aloe the sphincter of the anus, cantharides that of the bladder, and lastly every gland of the body appears to be indued with a kind of taste, by which it selects or forms each its peculiar fluid from the blood; and by which it is irritated into activity.

Many of these external properties of bodies, which stimulate our organs of sense, do not seem to effect this by a single impulse, but by repeated impulses; as the nerve of the ear is probably not excitable by a single vibration of air, nor the optic nerve by a single particle of light; which circumstance produces some analogy between those two senses, at the same time the solidity of bodies is perceived by a single application of a solid body to the nerves of touch, and that even through the cuticle; and we are probably possessed of a peculiar sense to distinguish the nice degrees of heat and cold.

The senses of touch and of hearing acquaint us with the mechanical impact and vibration of bodies, those of smell and taste seem to acquaint us with some of their chemical properties, while the sense of vision and of heat acquaint us with the existence of their peculiar fluids.

Sensation and Volition.

II. Many motions are produced by pleasure or pain, and that even in contradiction to the power of volition, as in laughing, or in the strangury; but as no name has been given to pleasure or pain, at the time it is exerted so as to cause fibrous motions, we have used the term sensation for this purpose; and mean it to bear the same analogy to pleasure and pain, that the word volition does to desire and aversion.

1. It was mentioned in the fifth Section, that, what we have termed sensation is a motion of the central parts, or of the whole sensorium, *beginning* at some of the extremities of it. This appears first, because our pains and pleasures are always caused by our ideas or muscular motions, which are the motions of the extremities of the sensorium. And, secondly, because the sensation of pleasure or pain frequently continues some time after the ideas or muscular motions which excited it have ceased: for we often feel a glow of pleasure from an agreeable reverie, for many minutes after the ideas, that were the subject of it, have escaped our memory; and frequently experience a dejection of spirits without being able to assign the cause of it but by much recollection.

When the sensorial faculty of desire or aversion is exerted so as to cause fibrous motions, it is termed volition; which is said in Sect. V. to be a motion of the central parts, or of the whole sensorium, *terminating* in some of the extremities of it. This appears, first, because our desires and aversions always terminate in recollecting and comparing our ideas, or in exerting our muscles; which are the motions of the extremities of the sensorium. And, secondly, because desire or aversion begins, and frequently continues for a time in the central parts of the sensorium, before it is peculiarly exerted at the extremities of it; for we sometimes feel desire or aversion without immediately knowing their objects, and in consequence without immediately exerting any of our muscular or sensual motions to attain them: as in the beginning of the passion of love, and perhaps of hunger, or in the ennui of indolent people.

Though sensation and volition begin or terminate at the extremities or central parts of the sensorium, yet the whole of it is frequently influenced by the exertion of these faculties, as appears from their effects on the external habit: for the whole skin is reddened by shame, and an universal trembling is produced by fear: and every muscle of the body is agitated in angry people by the desire of revenge.

There is another very curious circumstance, which shews that sensation and volition are movements of the sensorium in contrary directions; that is, that volition begins at the central parts of it, and proceeds to the extremities; and that sensation begins at the extremities, and proceeds to the central parts: I mean that these two sensorial faculties cannot be strongly exerted at the same time; for when we exert our volition strongly, we do not attend to pleasure or pain; and conversely, when we are strongly affected with the sensation of pleasure or pain, we use no volition. As will be further explained in Section XVIII. on sleep, and Section XXXIV. on volition.

2. All our emotions and passions seem to arise out of the exertions of these two faculties of the animal sensorium. Pride, hope, joy, are the names of particular pleasures: shame, despair, sorrow, are the names of peculiar pains: and love, ambition, avarice, of particular desires: hatred, disgust, fear, anxiety, of particular aversions. Whilst the passion of anger includes the pain from a recent injury, and the aversion to the adversary that occasioned it. And compassion is the pain we experience at the sight of misery, and the desire of relieving it.

There is another tribe of desires, which are commonly termed appetites, and are the immediate consequences of the absence of some irritative motions. Those, which arise from defect of internal irritations, have proper names conferred upon them, as hunger, thirst, lust, and the desire of air, when our respiration is impaired by noxious vapours; and of warmth, when we are exposed to too great a degree of cold. But those, whose stimuli are external to the body, are named from the objects, which are by nature constituted to excite them; these desires originate from our past experience of the pleasurable sensations they occasion, as the smell of an hyacinth, or the taste of a pine-apple.

Whence it appears, that our pleasures and pains are at least as various and as numerous as our irritations; and that our desires and aversions must be as numerous as our pleasures and pains. And that as sensation is here used as a general term for our numerous pleasures and pains, when they produce the contractions of our fibres; so volition is the general name for our desires and aversions, when they produce fibrous contractions. Thus when a motion of the central parts, or of the whole sensorium, terminates in the exertion of our muscles, it is generally called voluntary action; when it terminates in the exertion of our ideas, it is termed recollection, reasoning, determining.

3. As the sensations of pleasure and pain are originally introduced by the irritations of external objects: so our desires and aversions are originally introduced by those sensations; for when the objects of our pleasures or pains are at a distance, and we cannot instantaneously possess the one, or avoid the other, then desire or aversion is produced, and a voluntary exertion of our ideas or muscles succeeds.

The pain of hunger excites you to look out for food, the tree, that shades you, presents its odoriferous fruit before your eyes, you approach, pluck, and eat.

The various movements of walking to the tree, gathering the fruit, and masticating it, are associated motions introduced by their connection with sensation; but if from the uncommon height of the tree, the fruit be inaccessible, and you are prevented from quickly possessing the intended pleasure, desire is produced. The consequence of this desire is, first, a deliberation about the means to gain the object of pleasure in process of time, as it cannot be procured immediately; and, secondly, the muscular action necessary for this purpose.

You voluntarily call up all your ideas of causation, that are related to the effect you desire, and voluntarily examine and compare them, and at length determine whether to ascend the tree, or to gather stones from the neighbouring brook, is easier to practise, or more promising of success; and, finally, you gather the stones, and repeatedly fling them to dislodge the fruit.

Hence then we gain a criterion to distinguish voluntary acts or thoughts from those caused by sensation. As the former are always employed about the *means* to acquire pleasurable objects, or the *means* to avoid painful ones;

while the latter are employed in the possession of those, which are already in our power.

Hence the activity of this power of volition produces the great difference between the human and the brute creation. The ideas and the actions of brutes are almost perpetually employed about their present pleasures, or their present pains; and, except in the few instances which are mentioned in Section XVI, on instinct, they seldom busy themselves about the means of procuring future bliss, or of avoiding future misery; so that the acquiring of languages, the making of tools, and labouring for money, which are all only the means to procure pleasures; and the praying to the Deity, as another means to procure happiness, are characteristic of human nature.

4. As there are many diseases produced by the quantity of the sensation of pain or pleasure being too great or too little; so are there diseases produced by the susceptibility of the constitution to motions causable by these sensations being too dull or too vivid. This susceptibility of the system to sensitive motions is termed sensibility, to distinguish it from sensation, which is the actual existence or exertion of pain or pleasure.

Other classes of diseases are owing to the excessive promptitude, or sluggishness of the constitution to voluntary exertions, as well as to the quantity of desire or of aversion. This susceptibility of the system to voluntary motions is termed voluntarity, to distinguish it from volition, which is the exertion of desire or aversion; these diseases will be treated of at length in the progress of the work.

Association.

III. 1. It is not easy to assign a cause, why those animal movements, that have once occurred in succession, or in combination, should afterwards have a tendency to succeed or accompany each other. It is a property of animation, and distinguishes this order of being from the other productions of nature.

When a child first wrote the word man, it was distinguished in his mind into three letters, and those letters into many parts of letters; but by repeated use the word man becomes to his hand in writing it, as to his organs of speech in pronouncing it, but one movement without any deliberation, or sensation, or irritation, interposed between the parts of it. And as many separate motions of our muscles thus become united, and form, as it were, one motion; so each separate motion before such union may be conceived to consist of many parts or spaces moved through; and perhaps even the individual fibres of our muscles have thus gradually been brought to act in concert, which habits began to be acquired as early as the very formation of the moving organs, long before the nativity of the animal; as explained in the Section XVI. 2. on instinct.

2. There are many motions of the body, belonging to the irritative class, which might by a hasty observer be mistaken for associated ones; as the peristaltic motion of the stomach and intestines, and the contractions of the heart and arteries, might be supposed to be associated with the irritative motions of their nerves of sense, rather than to be excited by the irritation of their muscular fibres by the distention, acrimony, or momentum of the blood. So the distention or elongation of muscles by objects external to them irritates them into contraction, though the cuticle or other parts may intervene between the stimulating body and the contracting muscle. Thus a horse voids his excrement when its weight or bulk irritates the rectum or sphincter ani. These muscles act from the irritation of distention, when he excludes his excrement, but the muscles of the abdomen and diaphragm are brought into motion by association with those of the sphincter and rectum.

SECT. XII
OF STIMULUS, SENSORIAL EXERTION, AND FIBROUS CONTRACTION

I. Of fibrous contraction. 1. *Two particles of a fibre cannot approach without the intervention of something, as in magnetism, electricity, elasticity. Spirit of life is not electric ether. Galvani's experiments. 2. Contraction of a fibre. 3. Relaxation succeeds. 4. Successive contractions, with intervals. Quick pulse from debility, from paucity of blood. Weak contractions performed in less time, and with shorter intervals. 5. Last situation of the fibres continues after contraction. 6. Contraction greater than usual induces pleasure or pain. 7. Mobility of the fibres uniform. Quantity of sensorial power fluctuates. Constitutes excitability.* II. Of sensorial exertion. 1. *Animal motion includes stimulus, sensorial power, and contractile fibres. The sensorial faculties act separately or conjointly. Stimulus of four kinds. Strength and weakness defined. Sensorial power perpetually exhausted and renewed. Weakness from defect of stimulus. From defect of sensorial power, the direct and indirect debility of Dr. Brown. Why we become warm in Buxton bath after a time, and see well after a time in a darkish room. Fibres may act violently, or with their whole force, and yet feebly. Great exertion in inflammation explained. Great muscular force of some insane people. 2. Occasional accumulation of sensorial power in muscles subject to constant stimulus. In animals sleeping in winter. In eggs, seeds, schirrous tumours, tendons, bones. 3. Great exertion introduces pleasure or pain. Inflammation. Libration of the system between torpor and activity. Fever-fits. 4. Desire and aversion introduced. Excess of volition cures fevers.* III. Of repeated stimulus. 1. *A stimulus repeated too frequently looses effect. As opium, wine, grief. Hence old age. Opium and aloes in small doses. 2. A stimulus not repeated too frequently does not lose effect. Perpetual movement of the vital organs. 3. A stimulus repeated at uniform times produces greater effect. Irritation combined with association. 4. A stimulus repeated frequently and uniformly may be withdrawn, and the action of the organ will continue. Hence the bark cures agues, and strengthens weak constitutions. 5. Defect of stimulus repeated at certain intervals causes fever-fits. 6. Stimulus long applied ceases to act a second time. 7. If a stimulus excites sensation in an organ not usually excited into sensation, inflammation is produced.* IV. Of stimulus greater than natural. 1. *A stimulus greater than natural diminishes the quantity of sensorial power in general. 2. In particular organs. 3. Induces the organ into spasmodic actions. 4. Induces the antagonist fibres into action. 5. Induces the organ into convulsive or fixed spasms. 6. Produces paralysis of the organ.* V. Of stimulus less than natural. 1. *Stimulus less than natural occasions accumulation of sensorial power in general. 2. In particular organs, flushing of the face in a frosty morning. In fibres subject to perpetual stimulus only. Quantity of sensorial power inversely as the stimulus. 3. Induces pain. As of cold, hunger, head-ach. 4. Induces more feeble and frequent contraction. As in low fevers. Which are frequently owing to deficiency of sensorial power rather than to deficiency of stimulus. 5. Inverts successive trains of motion. Inverts ideas. 6. Induces paralysis and death.* VI. Cure of increased exertion. 1. *Natural cure of exhaustion of sensorial power. 2. Decrease the irritations. Venesection. Cold. Abstinence. 3. Prevent the previous cold fit. Opium. Bark. Warmth. Anger. Surprise. 4. Excite some other part of the system. Opium and warm bath relieve pains both from defect and from excess of stimulus. 5. First increase the stimulus above, and then decrease it beneath the natural quantity.* VII. Cure of decreased exertion. 1. *Natural cure by accumulation of sensorial power. Ague-fits. Syncope. 2. Increase the stimulation, by wine, opium, given so as not to intoxicate. Cheerful ideas. 3. Change the kinds of stimulus. 4. Stimulate the associated organs. Blisters of use in heart-burn, and cold extremities. 5. Decrease the stimulation for a time, cold bath. 6. Decrease the stimulation below natural, and then increase it above natural. Bark after emetics. Opium after venesection. Practice of Sydenham in chlorosis. 7. Prevent unnecessary expenditure of sensorial power. Decumbent posture, silence, darkness. Pulse quickened by rising out of bed. 8. To the greatest degree of quiescence apply the least stimulus. Otherwise paralysis or inflammation of the organ ensues. Gin, wine, blisters, destroy by too great stimulation in fevers with debility. Intoxication in the slightest degree succeeded by debility. Golden rule for determining the best degree of stimulus in low*

I. *Of fibrous contraction.*

1. If two particles of iron lie near each other without motion, and afterwards approach each other; it is reasonable to conclude that something besides the iron particles is the cause of their approximation; this invisible something is termed magnetism. In the same manner, if the particles, which compose an animal muscle, do not touch each other in the relaxed state of the muscle, and are brought into contact during the contraction of the muscle, it is reasonable to conclude, that some other agent is the cause of this new approximation. For nothing can act, where it does not exist; for to act includes to exist; and therefore the particles of the muscular fibre (which in its state of relaxation are supposed not to touch) cannot affect each other without the influence of some intermediate agent; this agent is here termed the spirit of animation, or sensorial power, but may with equal propriety be termed the power, which causes contraction; or may be called by any other name, which the reader may choose to affix to it.

The contraction of a muscular fibre may be compared to the following electric experiment, which is here mentioned not as a philosophical analogy, but as an illustration or simile to facilitate the conception of a difficult subject. Let twenty very small Leyden phials properly coated be hung in a row by fine silk threads at a small distance from each other; let the internal charge of one phial be positive, and of the other negative alternately, if a communication be made from the internal surface of the first to the external surface of the last in the row, they will all of them instantly approach each other, and thus shorten a line that might connect them like a muscular fibre. See Botanic Garden, p. 1. Canto I. 1. 202, note on Gymnotus.

The attractions of electricity or of magnetism do not apply philosophically to the illustration of the contraction of animal fibres, since the force of those attractions increases in some proportion inversely as the distance, but in muscular motion there appears no difference in velocity or strength during the beginning or end of the contraction, but what may be clearly ascribed to the varying mechanic advantage in the approximation of one bone to another. Nor can muscular motion be assimilated with greater plausibility to the attraction of cohesion or elasticity; for in bending a steel spring, as a small sword, a less force is required to bend it the first inch than the second; and the second than the third; the particles of steel on the convex side of the bent spring endeavouring to restore themselves more powerfully the further they are drawn from each other. See Botanic Garden, P. I. addit. Note XVIII.

I am aware that this may be explained another way, by supposing the elasticity of the spring to depend more on the compression of the particles on the concave side than on the extension of them on the convex side; and by supposing the elasticity of the elastic gum to depend more on the resistance to the lateral compression of its particles than to the longitudinal extension of them. Nevertheless in muscular contraction, as above observed, there appears no difference in the velocity or force of it at its commencement or at its termination; from whence we must conclude that animal contraction is governed by laws of its own, and not by those of mechanics, chemistry, magnetism, or electricity.

On these accounts I do not think the experiments conclusive, which were lately published by Galvani, Volta, and others, to shew a similitude between the spirit of animation, which contracts the muscular fibres, and the electric fluid. Since the electric fluid may act only as a more potent stimulus exciting the muscular fibres into action, and not by supplying them with a new quantity of the spirit of life. Thus in a recent hemiplegia I have frequently observed, when the patient yawned and stretched himself, that the paralytic limbs moved also, though they were totally disobedient to the will. And when he was electrified by passing shocks from the affected hand to the affected foot, a motion of the paralytic limbs was also produced. Now as in the act of yawning the muscles of the paralytic limbs were excited into action by the stimulus of the irksomeness of a continued posture, and not by any additional quantity of the spirit of life; so we may conclude, that the passage of the electric fluid, which produced a similar effect, acted only as a stimulus, and not by supplying any addition of sensorial power.

If nevertheless this theory should ever become established, a stimulus must be called an eductor of vital ether; which stimulus may consist of sensation or volition, as in the electric eel, as well as in the appulses of external bodies; and by drawing off the charges of vital fluid may occasion the contraction or motions of the muscular fibres, and organs of sense.

2. The immediate effect of the action of the spirit of animation or sensorial power on the fibrous parts of the body, whether it acts in the mode of irritation, sensation, volition, or association, is a contraction of the animal fibre, according to the second law of animal causation. Sect. IV. Thus the stimulus of the blood induces the contraction of the heart; the agreeable taste of a strawberry produces the contraction of the muscles of deglutition; the effort of the will contracts the muscles, which move the limbs in walking; and by association other muscles of the trunk are brought into contraction to preserve the balance of the body. The fibrous extremities of the organs of sense have been shewn, by the ocular spectra in Sect. III. to

suffer similar contraction by each of the above modes of excitation; and by their configurations to constitute our ideas.

3. After animal fibres have for some time been excited into contraction, a relaxation succeeds, even though the exciting cause continues to act. In respect to the irritative motions this is exemplified in the peristaltic contractions of the bowels; which cease and are renewed alternately, though the stimulus of the aliment continues to be uniformly applied; in the sensitive motions, as in strangury, tenesmus, and parturition, the alternate contractions and relaxations of the muscles exist, though the stimulus is perpetual. In our voluntary exertions it is experienced, as no one can hang long by the hands, however vehemently he wills so to do; and in the associate motions the constant change of our attitudes evinces the necessity of relaxation to those muscles, which have been long in action.

This relaxation of a muscle after its contraction, even though the stimulus continues to be applied, appears to arise from the expenditure or diminution of the spirit of animation previously resident in the muscle, according to the second law of animal causation in Sect. IV. In those constitutions, which are termed weak, the spirit of animation becomes sooner exhausted, and tremulous motions are produced, as in the hands of infirm people, when they lift a cup to their mouths. This quicker exhaustion of the spirit of animation is probably owing to a less quantity of it residing in the acting fibres, which therefore more frequently require a supply from the nerves, which belong to them.

4. If the sensorial power continues to act, whether it acts in the mode of irritation, sensation, volition, or association, a new contraction of the animal fibre succeeds after a certain interval; which interval is of shorter continuance in weak people than in strong ones. This is exemplified in the shaking of the hands of weak people, when they attempt to write. In a manuscript epistle of one of my correspondents, which is written in a small hand, I observed from four to six zigzags in the perpendicular stroke of every letter, which shews that both the contractions of the fingers, and intervals between them, must have been performed in very short periods of time.

The times of contraction of the muscles of enfeebled people being less, and the intervals between those contractions being less also, accounts for the quick pulse in fevers with debility, and in dying animals. The shortness of the intervals between one contraction and another in weak constitutions, is probably owing to the general deficiency of the quantity of the spirit of animation, and that therefore there is a less quantity of it to be received at each interval of the activity of the fibres. Hence in repeated motions, as of the fingers in performing on the harpsichord, it would at first sight appear, that swiftness and strength were incompatible; nevertheless the single contraction of a muscle is performed with greater velocity as well as with greater force by vigorous constitutions, as in throwing a javelin.

There is however another circumstance, which may often contribute to cause the quickness of the pulse in nervous fevers, as in animals bleeding to death in the slaughter-house; which is the deficient quantity of blood; whence the heart is but half distended, and in consequencc sooner contracts. See Sect. XXXII. 2. 1.

For we must not confound frequency of repetition with quickness of motion, or the number of pulsations with the velocity, with which the fibres, which constitute the coats of the arteries, contract themselves. For where the frequency of the pulsations is but seventy-five in a minute, as in health; the contracting fibres, which constitute the sides of the arteries, may move through a greater space in a given time, than where the frequency of pulsation is one hundred and fifty in a minute, as in some fevers with great debility. For if in those fevers the arteries do not expand themselves in their diastole to more than half the usual diameter of their diastole in health, the fibres which constitute their coats, will move through a less space in a minute than in health, though they make two pulsations for one.

Suppose the diameter of the artery during its systole to be one line, and that the diameter of the same artery during its diastole is in health is four lines, and in a fever with, great debility only two lines. It follows, that the arterial fibres contract in health from a circle of twelve lines in circumference to a circle of three lines in circumference, that is they move through a space of nine lines in length. While the arterial fibres in the fever with debility would twice contract from a circle of six lines to a circle of three lines; that is while they move through a space equal to six lines. Hence though the frequency of pulsation in fever be greater as two to one, yet the velocity of contraction in health is greater as nine to six, or as three to two.

On the contrary in inflammatory diseases with strength, as in the pleurisy, the velocity of the contracting sides of the arteries is much greater than in health, for if we suppose the number of pulsations in a pleurisy to be half as much more than in health, that is as one hundred and twenty to eighty, (which is about what generally happens in inflammatory diseases) and if the diameter of the artery in diastole be one third greater than in health, which I believe is near the truth, the result will be, that the velocity of the contractile sides of the arteries will be in a pleurisy as two and a half to one, compared to the velocity of their contraction in a state of health, for if the circumference of the systole of the artery be three lines, and the diastole in health be twelve lines in circumference, and in a pleurisy eighteen lines;

and secondly, if the artery pulsates thrice in the diseased state for twice in the healthy one, it follows, that the velocity of contraction in the diseased state to that in the healthy state will be forty-five to eighteen, or as two and a half to one.

From hence it would appear, that if we had a criterion to determine the velocity of the arterial contractions, it would at the same time give us their strength, and thus be of more service in distinguishing diseases, than the knowledge of their frequency. As such a criterion cannot be had, the frequency of pulsation, the age of the patient being allowed for, will in some measure assist us to distinguish arterial strength from arterial debility, since in inflammatory diseases with strength the frequency seldom exceeds one hundred and eighteen or one hundred and twenty pulsations in a minute; unless under peculiar circumstance, as the great additional stimuli of wine or of external heat.

5. After a muscle or organ of sense has been excited into contraction, and the sensorial power ceases to act, the last situation or configuration of it continues; unless it be disturbed by the action of some antagonist fibres, or other extraneous power. Thus in weak or languid people, wherever they throw their limbs on their bed or sofa, there they lie, till another exertion changes their attitude; hence one kind of ocular spectra seems to be produced after looking at bright objects; thus when a fire-stick is whirled round in the night, there appears in the eye a complete circle of fire; the action or configuration of one part of the retina not ceasing before the return of the whirling fire.

Thus if any one looks at the setting sun for a short time, and then covers his closed eyes with his hand, he will for many seconds of time perceive the image of the sun on his retina. A similar image of all other bodies would remain some time in the eye, but is effaced by the eternal change of the motions of the extremity of this nerve in our attention to other objects. See Sect. XVIII. 5. on Sleep. Hence the dark spots, and other ocular spectra, are more frequently attended to, and remain longer in the eyes of weak people, as after violent exercise, intoxication, or want of sleep.

6. A contraction of the fibres somewhat greater than usual introduces pleasurable sensation into the system, according to the fourth law of animal causation. Hence the pleasure in the beginning of drunkenness is owing to the increased action of the system from the stimulus of vinous spirit or of opium. If the contractions be still greater in energy or duration, painful sensations are introduced, as in consequence of great heat, or caustic applications, or fatigue.

If any part of the system, which is used to perpetual activity, as the stomach, or heart, or the fine vessels of the skin, acts for a time with less energy, another kind of painful sensation ensues, which is called hunger, or faintness, or cold. This occurs in a less degree in the locomotive muscles, and is called wearysomeness. In the two former kinds of sensation there is an expenditure of sensorial power, in these latter there is an accumulation of it.

7. We have used the words exertion of sensorial power as a general term to express either irritation, sensation, volition, or association; that is, to express the activity or motion of the spirit of animation, at the time it produces the contractions of the fibrous parts of the system. It may be supposed that there may exist a greater or less mobility of the fibrous parts of our system, or a propensity to be stimulated into contraction by the greater or less quantity or energy of the spirit of animation; and that hence if the exertion of the sensorial power be in its natural state, and the mobility of the fibres be increased, the same quantity of fibrous contraction will be caused, as if the mobility of the fibres continues in its natural state, and the sensorial exertion be increased.

Thus it may be conceived, that in diseases accompanied with strength, as in inflammatory fevers with arterial strength, that the cause of greater fibrous contraction, may exist in the increased mobility of the fibres, whose contractions are thence both more forceable and more frequent. And that in diseases attended with debility, as in nervous fevers, where the fibrous contractions are weaker, and more frequent, it may be conceived that the cause consists in a decrease of mobility of the fibres; and that those weak constitutions, which are attended with cold extremities and large pupils of the eyes, may possess less mobility of the contractile fibres, as well as less quantity of exertion of the spirit of animation.

In answer to this mode of reasoning it may be sufficient to observe, that the contractile fibres consist of inert matter, and when the sensorial power is withdrawn, as in death, they possess no power of motion at all, but remain in their last state, whether of contraction or relaxation, and must thence derive the whole of this property from the spirit of animation. At the same time it is not improbable, that the moving fibres of strong people may possess a capability of receiving or containing a greater quantity of the spirit of animation than those of weak people.

In every contraction of a fibre there is an expenditure of the sensorial power, or spirit of animation; and where the exertion of this sensorial power has been for some time increased, and the muscles or organs of sense have in consequence acted with greater energy, its propensity to activity is proportionally lessened; which is to be ascribed to the exhaustion or diminution of its quantity. On the contrary, where there has been less fibrous contraction than usual for a certain time, the sensorial power or spirit of animation becomes accumulated in the inactive part of the system. Hence vigour succeeds rest, and hence the propensity to action of all our organs of sense and muscles is in a state of perpetual fluctuation. The irritability for instance of the retina, that is, its quantity of sensorial power, varies every moment according to the brightness or obscurity of the object last beheld compared with the present one. The same occurs to our sense of heat, and to every part of our system, which is capable of being excited into action.

When this variation of the exertion of the sensorial power becomes much and permanently above or beneath the natural quantity, it becomes a disease. If the irritative motions be too great or too little, it shews that the stimulus of external things affect this sensorial power too violently or too inertly. If the sensitive motions be too great or too little, the cause arises from the deficient or exuberant quantity of sensation produced in consequence of the motions of the muscular fibres or organs of sense; if the voluntary actions are diseased the cause is to be looked for in the quantity of volition produced in consequence of the desire or aversion occasioned by the painful or pleasurable sensations above mentioned. And the diseases of associations probably depend on the greater or less quantity of the other three sensorial powers by which they were formed.

From whence it appears that the propensity to action, whether it be called irritability, sensibility, voluntarity, or associability, is only another mode of expression for the quantity of sensorial power residing in the organ to be excited. And that on the contrary the words inirritability and insensibility, together with inaptitude to voluntary and associate motions, are synonymous with deficiency of the quantity of sensorial power, or of the spirit of animation, residing in the organs to be excited.

II. *Of sensorial Exertion.*

1. There are three circumstances to be attended to in the production of animal motions, 1st. The stimulus. 2d. The sensorial power. 3d. The contractile fibre. 1st. A stimulus, external to the organ, originally induces into action the sensorial faculty termed irritation; this produces the contraction of the fibres, which, if it be perceived at all, introduces pleasure or pain; which in their active state are termed sensation; which is another sensorial faculty, and occasionally produces contraction of the fibres; this pleasure or pain is therefore to be considered as another stimulus, which may either act alone or in conjunction with the former faculty of the sensorium termed irritation.

This new stimulus of pleasure or pain either induces into action the sensorial faculty termed sensation, which then produces the contraction of the fibres; or it introduces desire or aversion, which excite into action another sensorial faculty, termed volition, and may therefore be considered as another stimulus, which either alone or in conjunction with one or both of the two former faculties of the sensorium produces the contraction of animal fibres. There is another sensorial power, that of association, which perpetually, in conjunction with one or more of the above, and frequently singly, produces the contraction of animal fibres, and which is itself excited into action by the previous motions of contracting fibres.

Now as the sensorial power, termed irritation, residing in any particular fibres, is excited into exertion by the stimulus of external bodies acting on those fibres; the sensorial power, termed sensation, residing in any particular fibres is excited into exertion by the stimulus of pleasure or pain acting on those fibres; the sensorial power, termed volition, residing in any particular fibres is excited into exertion by the stimulus of desire or aversion; and the sensorial power, termed association, residing in any particular fibres, is excited into action by the stimulus of other fibrous motions, which had frequently preceded them. The word stimulus may therefore be used without impropriety of language, for any of these four causes, which excite the four sensorial powers into exertion. For though the immediate cause of volition has generally been termed *a motive*; and that of irritation only has generally obtained the name of *stimulus*; yet as the immediate cause, which excites the sensorial powers of sensation, or of association into exertion, have obtained no general name, we shall use the word stimulus for them all.

Hence the quantity of motion produced in any particular part of the animal system will be as the quantity of stimulus and the quantity of sensorial power, or spirit of animation, residing in the contracting fibres. Where both these quantities are great, *strength* is produced, when that word is applied to the motions of animal bodies. Where either of them is deficient, *weakness* is produced, as applied to the motions of animal bodies.

Now as the sensorial power, or spirit of animation, is perpetually exhausted by the expenditure of it in fibrous contractions, and is perpetually renewed by the secretion or production of it in the brain and spinal marrow, the quantity of animal strength must be in a perpetual state of fluctuation on this account; and if to this be added the unceasing variation of all the four kinds of stimulus above described, which produce the exertions of the sensorial powers, the ceaseless vicissitude of animal strength becomes easily comprehended.

If the quantity of sensorial power remains the same, and the quantity of stimulus be lessened, a weakness of the fibrous contractions ensues, which may be denominated *debility from defect of stimulus*. If the quantity of stimulus remains the same, and the quantity of sensorial power be lessened, another kind of weakness ensues, which may be termed *debility from defect of sensorial power*; the former of these is called by Dr. Brown, in his Elements of Medicine, direct debility, and the latter indirect debility. The coincidence of some parts of this work with correspondent deductions in the Brunonian Elementa Medicina, a work (with some exceptions) of great genius, must be considered as confirmations of the truth of the theory, as they were probably arrived at by different trains of reasoning.

Thus in those who have been exposed to cold and hunger there is a deficiency of stimulus. While in nervous fever there is a deficiency of sensorial power. And in habitual drunkards, in a morning before their usual potation, there is a deficiency both of stimulus and of sensorial power. While, on the other hand, in the beginning of intoxication there is an excess of stimulus; in the hot-ach, after the hands have been immersed in snow, there is a redundancy of sensorial power; and in inflammatory diseases with arterial strength, there is an excess of both.

Hence if the sensorial power be lessened, while the quantity of stimulus remains the same as in nervous fever, the frequency of repetition of the arterial contractions may continue, but their force in respect to removing obstacles, as in promoting the circulation of the blood, or the velocity of each contraction, will be diminished, that is, the animal strength will be lessened. And secondly, if the quantity of sensorial power be lessened, and the stimulus be increased to a certain degree, as in giving opium in nervous fevers, the arterial contractions may be performed more frequently than natural, yet with less strength.

And thirdly, if the sensorial power continues the same in respect to quantity, and the stimulus be somewhat diminished, as in going into a darkish room, or into a coldish bath, suppose of about eighty degrees of heat, as Buxton-bath, a temporary weakness of the affected fibres is induced, till an accumulation of sensorial power gradually succeeds, and counterbalances the deficiency of stimulus, and then the bath ceases to feel cold, and the room ceases to appear dark; because the fibres of the subcutaneous vessels, or of the organs of sense, act with their usual energy.

A set of muscular fibres may thus be stimulated into violent exertion, that is, they may act frequently, and with their whole sensorial power, but may nevertheless not act strongly; because the quantity of their sensorial power was originally small, or was previously exhausted. Hence a stimulus may be great, and the irritation in consequence act with its full force, as in the hot paroxysms of nervous fever; but if the sensorial power, termed irritation, be small in quantity, the force of the fibrous contractions, and the times of their continuance in their contracted state, will be proportionally small.

In the same manner in the hot paroxysm of putrid fevers, which are shewn in Sect. XXXIII. to be inflammatory fevers with arterial debility, the sensorial power termed sensation is exerted with great activity, yet the fibrous contractions, which produce the circulation of the blood, are performed without strength, because the quantity of sensorial power then residing in that part of the system is small.

Thus in irritative fever with arterial strength, that is, with excess of spirit of animation, the quantity of exertion during the hot part of the paroxysm is to be estimated from the quantity of stimulus, and the quantity of sensorial power. While in sensitive (or inflammatory) fever with arterial strength, that is, with excess of spirit of animation, the violent and forcible actions of the vascular system during the hot part of the paroxysm are induced by the exertions of two sensorial powers, which are excited by two kinds of stimulus. These are the sensorial power of irritation excited by the stimulus of bodies external to the moving fibres, and the sensorial power of sensation excited by the pain in consequence of the increased contractions of those moving fibres.

And in insane people in some cases the force of their muscular actions will be in proportion to the quantity of sensorial power, which they possess, and the quantity of the stimulus of desire or aversion, which excites their volition into action. At the same time in other cases the stimulus of pain or pleasure, and the stimulus of external bodies, may excite into action the sensorial powers of sensation and irritation, and thus add greater force to their muscular actions.

2. The application of the stimulus, whether that stimulus be some quality of external bodies, or pleasure or pain, or desire or aversion, or a link of association, excites the correspondent sensorial power into action, and this causes the contraction of the fibre. On the contraction of the fibre a part of the spirit of animation becomes expended, and the fibre ceases to contract, though the stimulus continues to be applied; till in a certain time the fibre having received a supply of sensorial power is ready to contract again, if the stimulus continues to be applied. If the stimulus on the contrary be withdrawn, the same quantity of quiescent sensorial power becomes resident in the fibre as before its contraction; as appears from the readiness

for action of the large locomotive muscles of the body in a short time after common exertion.

But in those muscular fibres, which are subject to constant stimulus, as the arteries, glands, and capillary vessels, another phenomenon occurs, if their accustomed stimulus be withdrawn; which is, that the sensorial power becomes accumulated in the contractile fibres, owing to the want of its being perpetually expended, or carried away, by their usual unremitted contractions. And on this account those muscular fibres become afterwards excitable into their natural actions by a much weaker stimulus; or into unnatural violence of action by their accustomed stimulus, as is seen in the hot fits of intermittent fevers, which are in consequence of the previous cold ones. Thus the minute vessels of the skin are constantly stimulated by the fluid matter of heat; if the quantity of this stimulus of heat be a while diminished, as in covering the hands with snow, the vessels cease to act, as appears from the paleness of the skin; if this cold application of snow be continued but a short time, the sensorial power, which had habitually been supplied to the fibres, becomes now accumulated in them, owing to the want of its being expended by their accustomed contractions. And thence a less stimulus of heat will now excite them into violent contractions.

If the quiescence of fibres, which had previously been subject to perpetual stimulus, continues a longer time; or their accustomed stimulus be more completely withdrawn; the accumulation of sensorial power becomes still greater, as in those exposed to cold and hunger; pain is produced, and the organ gradually dies from the chemical changes, which take place in it; or it is at a great distance of time restored to action by stimulus applied with great caution in small quantity, as happens to some larger animals and to many insects, which during the winter months lie benumbed with cold, and are said to sleep, and to persons apparently drowned, or apparently frozen to death. Snails have been said to revive by throwing them into water after having been many years shut up in the cabinets of the curious; and eggs and seeds in general are restored to life after many months of torpor by the stimulus of warmth and moisture.

The inflammation of schirrous tumours, which have long existed in a state of inaction, is a process of this kind; as well as the sensibility acquired by inflamed tendons and bones, which had at their formation a similar sensibility, which had so long lain dormant in their uninflamed state.

3. If after long quiescence from defect of stimulus the fibres, which had previously been habituated to perpetual stimulus, are again exposed to but their usual quantity of it; as in those who have suffered the extremes of cold

or hunger; a violent exertion of the affected organ commences, owing, as above explained, to the great accumulation of sensorial power. This violent exertion not only diminishes the accumulated spirit of animation, but at the same time induces pleasure or pain into the system, which, whether it be succeeded by inflammation or not, becomes an additional stimulus, and acting along with the former one, produces still greater exertions; and thus reduces the sensorial power in the contracting fibres beneath its natural quantity.

When the spirit of animation is thus exhausted by useless exertions, the organ becomes torpid or unexcitable into action, and a second fit of quiescence succeeds that of abundant activity. During this second fit of quiescence the sensorial power becomes again accumulated, and another fit of exertion follows in train. These vicissitudes of exertion and inertion of the arterial system constitute the paroxysms of remittent fevers; or intermittent ones, when there is an interval of the natural action of the arteries between the exacerbations.

In these paroxysms of fevers, which consist of the libration of the arterial system between the extremes of exertion and quiescence, either the fits become less and less violent from the contractile fibres becoming coming less excitable to the stimulus by habit, that is, by becoming accustomed to it, as explained below XII. 3. 1. or the whole sensorial power becomes exhausted, and the arteries cease to beat, and the patient dies in the cold part of the paroxysm. Or secondly, so much pain is introduced into the system by the violent contractions of the fibres, that inflammation arises, which prevents future cold fits by expending a part of the sensorial power in the extension of old vessels or the production of new ones; and thus preventing the too great accumulation or exertion of it in other parts of the system; or which by the great increase of stimulus excites into great action the whole glandular system as well as the arterial, and thence a greater quantity of sensorial power is produced in the brain, and thus its exhaustion in any peculiar part of the system ceases to be affected.

4. Or thirdly, in consequence of the painful or pleasurable sensation above mentioned, desire and aversion are introduced, and inordinate volition succeeds; which by its own exertions expends so much of the spirit of animation, that the two other sensorial faculties, or irritation and sensation, act so much more feebly; that the paroxysms of fever, or that libration between the extremes of exertion and inactivity of the arterial system, gradually subsides. On this account a temporary insanity is a favourable sign in fevers, as I have had some opportunities of observing.

III. *Of repeated Stimulus.*

1. When a stimulus is repeated more frequently than the expenditure of sensorial power can be renewed in the acting organ, the effect of the stimulus becomes gradually diminished. Thus if two grains of opium be swallowed by a person unused to so strong a stimulus, all the vascular systems in the body act with greater energy, all the secretions and the absorption from those secreted fluids are increased in quantity; and pleasure or pain are introduced into the system, which adds an additional stimulus to that already too great. After some hours the sensorial power becomes diminished in quantity, expended by the great activity of the system; and thence, when the stimulus of the opium is withdrawn, the fibres will not obey their usual degree of natural stimulus, and a consequent torpor or quiescence succeeds, as is experienced by drunkards, who on the day after a great excess of spirituous potation feel indigestion, head-ach, and general debility.

In this fit of torpor or quiescence of a part or of the whole of the system, an accumulation of the sensorial power in the affected fibres is formed, and occasions a second paroxysm of exertion by the application only of the natural stimulus, and thus a libration of the sensorial exertion between one excess and the other continues for two or three days, where the stimulus was violent in degree; and for weeks in some fevers, from the stimulus of contagious matter.

But if a second dose of opium be exhibited before the fibres have regained their natural quantity of sensorial power, its effect will be much less than the former, because the spirit of animation or sensorial power is in part exhausted by the previous excess of exertion. Hence all medicines repeated too frequently gradually lose their effect, as opium and wine. Many things of disagreeable taste at first cease to be disagreeable by frequent repetition, as tobacco; grief and pain gradually diminish, and at length cease altogether, and hence life itself becomes tolerable.

Besides the temporary diminution of the spirit of animation or sensorial power, which is naturally stationary or resident in every living fibre, by a single exhibition of a powerful stimulus, the contractile fibres themselves, by the perpetual application of a new quantity of stimulus, before they have regained their natural quantity of sensorial power, appear to suffer in their capability of receiving so much as the natural quantity of sensorial power; and hence a permanent deficiency of spirit of animation takes place, however long the stimulus may have been withdrawn. On this cause depends the permanent debility of those, who have been addicted to intoxication, the general weakness of old age, and the natural debility or inirritability of those, who have pale skins and large pupils of their eyes.

There is a curious phenomenon belongs to this place, which has always appeared difficult of solution; and that is, that opium or aloes may be exhibited in small doses at first, and gradually increased to very large ones without producing stupor or diarrhœa. In this case, though the opium and aloes are given in such small doses as not to produce intoxication or catharsis, yet they are exhibited in quantities sufficient in some degree to exhaust the sensorial power, and hence a stronger and a stronger dose is required; otherwise the medicine would soon cease to act at all.

On the contrary, if the opium or aloes be exhibited in a large dose at first, so as to produce intoxication or diarrhœa; after a few repetitions the quantity of either of them may be diminished, and they will still produce this effect. For the more powerful stimulus dissevers the progressive catenations of animal motions, described in Sect. XVII. and introduces a new link between them; whence every repetition strengthens this new association or catenation, and the stimulus may be gradually decreased, or be nearly withdrawn, and yet the effect shall continue; because the sensorial power of association or catenation being united with the stimulus, increases in energy with every repetition of the catenated circle; and it is by these means that all the irritative associations of motions are originally produced.

2. When a stimulus is repeated at such distant intervals of time, that the natural quantity of sensorial power becomes completely restored in the acting fibres, it will act with the same energy as when first applied. Hence those who have lately accustomed themselves to large doses of opium by beginning with small ones, and gradually increasing them, and repeating them frequently, as mentioned in the preceding paragraph; if they intermit the use of it for a few days only, must begin again with as small doses as they took at first, otherwise they will experience the inconveniences of intoxication.

On this circumstance depend the constant unfailing effects of the various kinds of stimulus, which excite into action all the vascular systems in the body; the arterial, venous, absorbent, and glandular vessels, are brought into perpetual unwearied action by the fluids, which are adapted to stimulate them; but these have the sensorial power of association added to that of irritation, and even in some degree that of sensation, and even of volition, as will be spoken of in their places; and life itself is thus carried on by the production of sensorial power being equal to its waste or expenditure in the perpetual movement of the vascular organization.

3. When a stimulus is repeated at uniform intervals of time with such distances between them, that the expenditure of sensorial power in the acting fibres becomes completely renewed, the effect is produced with greater facility or energy. For the sensorial power of association is combined with the sensorial power of irritation, or, in common language, the acquired habit assists the power of the stimulus.

This circumstance not only obtains in the annual and diurnal catenations of animal motions explained in Sect. XXXVI. but in every less circle of actions or ideas, as in the burthen of a song, or the iterations of a dance; and constitutes the pleasure we receive from repetition and imitation; as treated of in Sect. XXII. 2.

4. When a stimulus has been many times repeated at uniform intervals, so as to produce the complete action of the organ, it may then be gradually diminished, or totally withdrawn, and the action of the organ will continue. For the sensorial power of association becomes united with that of irritation, and by frequent repetition becomes at length of sufficient energy to carry on the new link in the circle of actions, without the irritation which at first introduced it.

Hence, when the bark is given at stated intervals for the cure of intermittent fevers, if sixty grains of it be given every three hours for the twenty-four hours preceding the expected paroxysm, so as to stimulate the defective part of the system into action, and by that means to prevent the torpor or quiescence of the fibres, which constitutes the cold fit; much less than half the quantity, given before the time at which another paroxysm of quiescence would have taken place, will be sufficient to prevent it; because now the sensorial power, termed association, acts in a twofold manner. First, in respect to the period of the catenation in which the cold fit was produced, which is now dissevered by the stronger stimulus of the first doses of the bark; and, secondly, because each dose of bark being repeated at periodical times, has its effect increased by the sensorial faculty of association being combined with that of irritation.

Now, when sixty grains of Peruvian bark are taken twice a day, suppose at ten o'clock and at six, for a fortnight, the irritation excited by this additional stimulus becomes a part of the diurnal circle of actions, and will at length carry on the increased action of the system without the assistance of the stimulus of the bark. On this theory the bitter medicines, chalybeates, and opiates in appropriated doses, exhibited for a fortnight, give permanent strength to pale feeble children, and other weak constitutions.

5. When a defect of stimulus, as of heat, recurs at certain diurnal intervals, which induces some torpor or quiescence of a part of the system,

the diurnal catenation of actions becomes disordered, and a new association with this link of torpid action is formed; on the next period the quantity of quiescence will be increased, suppose the same defect of stimulus to recur, because now the new association conspires with the defective irritation in introducing the torpid action of this part of the diurnal catenation. In this manner many fever-fits commence, where the patient is for some days indisposed at certain hours, before the cold paroxysm of fever is completely formed. See Sect. XVII. 3. 3. on Catenation of Animal Motions.

6. If a stimulus, which at first excited the affected organ into so great exertion as to produce sensation, be continued for a certain time, it will cease to produce sensation both then and when repeated, though the irritative motions in consequence of it may continue or be re-excited.

Many catenations of irritative motions were at first succeeded by sensation, as the apparent motions of objects when we walk past them, and probably the vital motions themselves in the early state of our existence. But as those sensations were followed by no movements of the system in consequence of them, they gradually ceased to be produced, not being joined to any succeeding link of catenation. Hence contagious matter, which has for some weeks stimulated the system into great and permanent sensation, ceases afterwards to produce general sensation, or inflammation, though it may still induce topical irritations. See Sect. XXXIII. 2. 8. XIX. 9.

Our absorbent system then seems to receive those contagious matters, which it has before experienced, in the same manner as it imbibes common moisture or other fluids; that is, without being thrown into so violent action as to produce sensation; the consequence of which is an increase of daily energy or activity, till inflammation and its consequences succeed.

7. If a stimulus excites an organ into such violent contractions as to produce sensation, the motions of which organ had not usually produced sensation, this new sensorial power, added to the irritation occasioned by the stimulus, increases the activity of the organ. And if this activity be catenated with the diurnal circle of actions, an increasing inflammation is produced; as in the evening paroxysms of small-pox, and other fevers with inflammation. And hence schirrous tumours, tendons and membranes, and probably the arteries themselves become inflamed, when they are strongly stimulated.

IV. Of Stimulus greater than natural.

1. A quantity of stimulus greater than natural, producing an increased exertion of sensorial power, whether that exertion be in the mode of irritation, sensation, volition, or association, diminishes the general quantity of it. This fact is observable in the progress of intoxication, as the increased

quantity or energy of the irritative motions, owing to the stimulus of vinous spirit, introduces much pleasurable sensation into the system, and much exertion of muscular or sensual motions in consequence of this increased sensation; the voluntary motions, and even the associate ones, become much impaired or diminished; and delirium and staggering succeed. See Sect. XXI. on Drunkenness. And hence the great prostration of the strength of the locomotive muscles in some fevers, is owing to the exhaustion of sensorial power by the increased action of the arterial system.

In like manner a stimulus greater than natural, applied to a part of the system, increases the exertion of sensorial power in that part, and diminishes it in some other part. As in the commencement of scarlet fever, it is usual to see great redness and heat on the faces and breasts of children, while at the same time their feet are colder than natural; partial heats are observable in other fevers with debility, and are generally attended with torpor or quiescence of some other part of the system. But these partial exertions of sensorial power are sometimes attended with increased partial exertions in other parts of the system, which sympathize with them, as the flushing of the face after a full meal. Both these therefore are to be ascribed to sympathetic associations, explained in Sect. XXXV. and not to general exhaustion or accumulation of sensorial power.

2. A quantity of stimulus greater than natural, producing an increased exertion of sensorial power in any particular organ, diminishes the quantity of it in that organ. This appears from the contractions of animal fibres being not so easily excited by a less stimulus after the organ has been subjected to a greater. Thus after looking at any luminous object of a small size, as at the setting sun, for a short time, so as not much to fatigue the eye, this part of the retina becomes less sensible to smaller quantities of light; hence when the eyes are turned on other less luminous parts of the sky, a dark spot is seen resembling the shape of the sun, or other luminous object which we last behold. See Sect. XL. No. 2.

Thus we are some time before we can distinguish objects in an obscure room after coming from bright day-light, though the iris presently contracts itself. We are not able to hear weak sounds after loud ones. And the stomachs of those who have been much habituated to the stronger stimulus of fermented or spirituous liquors, are not excited into due action by weaker ones.

3. A quantity of stimulus something greater than the last mentioned, or longer continued, induces the organ into spasmodic action, which ceases and recurs alternately. Thus on looking for a time on the setting sun, so as not greatly to fatigue the sight, a yellow spectrum is seen when the eyes are

closed and covered, which continues for a time, and then disappears and recurs repeatedly before it entirely vanishes. See Sect. XL. No. 5. Thus the action of vomiting ceases and is renewed by intervals, although the emetic drug is thrown up with the first effort. A tenesmus continues by intervals some time after the exclusion of acrid excrement; and the pulsations of the heart of a viper are said to continue some time after it is cleared from its blood.

In these cases the violent contractions of the fibres produce pain according to law 4; and this pain constitutes an additional kind or quantity of excitement, which again induces the fibres into contraction, and which painful excitement is again renewed, and again induces contractions of the fibres with gradually diminishing effect.

4. A quantity of stimulus greater than that last mentioned, or longer continued, induces the antagonist muscles into spasmodic action. This is beautifully illustrated by the ocular spectra described in Sect. XL. No. 6. to which the reader is referred. From those experiments there is reason to conclude that the fatigued part of the retina throws itself into a contrary mode of action like oscitation or pandiculation, as soon as the stimulus, which has fatigued it, is withdrawn; but that it still remains liable to be excited into action by any other colours except the colour with which it has been fatigued. Thus the yawning and stretching the limbs after a continued action or attitude seems occasioned by the antagonist muscles being stimulated by their extension during the contractions of those in action, or in the situation in which that action last left them.

5. A quantity of stimulus greater than the last, or longer continued, induces variety of convulsions or fixed spasms either of the affected organ or of the moving fibres in the other parts of the body. In respect to the spectra in the eye, this is well illustrated in No. 7 and 8, of Sect. XL. Epileptic convulsions, as the emprosthotonos and opisthotonos, with the cramp of the calf of the leg, locked jaw, and other cataleptic fits, appear to originate from pain, as some of these patients scream aloud before the convulsion takes place; which seems at first to be an effort to relieve painful sensation, and afterwards an effort to prevent it.

In these cases the violent contractions of the fibres produce so much pain, as to constitute a perpetual excitement; and that in so great a degree as to allow but small intervals of relaxation of the contracting fibres as in convulsions, or no intervals at all as in fixed spasms.

6. A quantity of stimulus greater than the last, or longer continued, produces a paralysis of the organ. In many cases this paralysis is only a temporary effect, as on looking long on a small area of bright red silk

placed on a sheet of white paper on the floor in a strong light, the red silk gradually becomes paler, and at length disappears; which evinces that a part of the retina, by being violently excited, becomes for a time unaffected by the stimulus of that colour. Thus cathartic medicines, opiates, poisons, contagious matter, cease to influence our system after it has been habituated to the use of them, except by the exhibition of increased quantities of them; our fibres not only become unaffected by stimuli, by which they have previously been violently irritated, as by the matter of the small-pox or measles; but they also become unaffected by sensation, where the violent exertions, which disabled them, were in consequence of too great quantity of sensation. And lastly the fibres, which become disobedient to volition, are probably disabled by their too violent exertions in consequence of too great a quantity of volition.

After every exertion of our fibres a temporary paralysis succeeds, whence the intervals of all muscular contractions, as mentioned in No. 3 and 4 of this Section; the immediate cause of these more permanent kinds of paralysis is probably owing in the same manner to the too great exhaustion of the spirit of animation in the affected part; so that a stronger stimulus is required, or one of a different kind from that, which occasioned those too violent contractions, to again excite the affected organ into activity; and if a stronger stimulus could be applied, it must again induce paralysis.

For these powerful stimuli excite pain at the same time, that they produce irritation; and this pain not only excites fibrous motions by its stimulus, but it also produces volition; and thus all these stimuli acting at the same time, and sometimes with the addition of their associations, produce so great exertion as to expend the whole of the sensorial power in the affected fibres.

V. Of Stimulus less than natural.

1. A quantity of stimulus less than natural, producing a decreased exertion of sensorial power, occasions an accumulation of the general quantity of it. This circumstance is observable in the hemiplagia, in which the patients are perpetually moving the muscles, which are unaffected. On this account we awake with greater vigour after sleep, because during so many hours, the great usual expenditure of sensorial power in the performance of voluntary actions, and in the exertions of our organs of sense, in consequence of the irritations occasioned by external objects had been suspended, and a consequent accumulation had taken place.

In like manner the exertion of the sensorial power less than natural in one part of the system, is liable to produce an increase of the exertion of it in some other part. Thus by the action of vomiting, in which the natural exertion of the motions of the stomach are destroyed or diminished, an increased absorption of the pulmonary and cellular lymphatics is produced, as is known by the increased absorption of the fluid deposited in them in dropsical cases. But these partial quiescences of sensorial power are also sometimes attended with other partial quiescences, which sympathize with them, as cold and pale extremities from hunger. These therefore are to be ascribed to the associations of sympathy explained in Sect. XXXV. and not to the general accumulation of sensorial power.

2. A quantity of stimulus less than natural, applied to fibres previously accustomed to perpetual stimulus, is succeeded by accumulation of sensorial power in the affected organ. The truth of this proposition is evinced, because a stimulus less than natural, if it be somewhat greater than that above mentioned, will excite the organ so circumstanced into violent activity. Thus on a frosty day with wind, the face of a person exposed to the wind is at first pale and shrunk; but on turning the face from the wind, it becomes soon of a glow with warmth and flushing. The glow of the skin in emerging from the cold-bath is owing to the same cause.

It does not appear, that an accumulation of sensorial power above the natural quantity is acquired by those muscles, which are not subject to perpetual stimulus, as the locomotive muscles: these, after the greatest fatigue, only acquire by rest their usual aptitude to motion; whereas the vascular system, as the heart and arteries, after a short quiescence, are thrown into violent action by their natural quantity of stimulus.

Nevertheless by this accumulation of sensorial power during the application of decreased stimulus, and by the exhaustion of it during the action of increased stimulus, it is wisely provided, that the actions of the vascular muscles and organs of sense are not much deranged by small variations of stimulus; as the quantity of sensorial power becomes in some measure inversely as the quantity of stimulus.

3. A quantity of stimulus less than that mentioned above, and continued for some time, induces pain in the affected organ, as the pain of cold in the hands, when they are immersed in snow, is owing to a deficiency of the stimulation of heat. Hunger is a pain from the deficiency of the stimulation of food. Pain in the back at the commencement of ague-fits, and the head-achs which attend feeble people, are pains from defect of stimulus, and are hence relieved by opium, essential oils, spirit of wine.

As the pains, which originate from defect of stimulus, only occur in those parts of the system, which have been previously subjected to perpetual stimulus; and as an accumulation of sensorial power is produced in the quiescent organ along with the pain, as in cold or hunger, there is reason to believe, that the pain is owing to the accumulation of sensorial power. For, in the locomotive muscles, in the retina of the eye, and other organs of senses, no pain occurs from the absence of stimulus, nor any great accumulation of sensorial power beyond their natural quantity, since these organs have not been used to a perpetual supply of it. There is indeed a greater accumulation occurs in the organ of vision after its quiescence, because it is subject to more constant stimulus.

4. A certain quantity of stimulus less than natural induces the moving organ into feebler and more frequent contractions, as mentioned in No. I. 4. of this Section. For each contraction moving through a less space, or with less force, that is, with less expenditure of the spirit of animation, is sooner relaxed, and the spirit of animation derived at each interval into the acting fibres being less, these intervals likewise become shorter. Hence the tremours of the hands of people accustomed to vinous spirit, till they take their usual stimulus; hence the quick pulse in fevers attended with debility, which is greater than in fevers attended with strength; in the latter the pulse seldom beats above 120 times in a minute, in the former it frequently exceeds 140.

It must be observed, that in this and the two following articles the decreased action of the system is probably more frequently occasioned by deficiency in the quantity of sensorial power, than in the quantity of stimulus. Thus those feeble constitutions which have large pupils of their eyes, and all who labour under nervous fevers, seem to owe their want of natural quantity of activity in the system to the deficiency of sensorial power; since, as far as can be seen, they frequently possess the natural quantity of stimulus.

5. A certain quantity of stimulus, less than that above mentioned, inverts the order of successive fibrous contractions; as in vomiting the vermicular motions of the stomach and duodenum are inverted, and their contents ejected, which is probably owing to the exhaustion of the spirit of animation in the acting muscles by a previous excessive stimulus, as by the root of ipecacuanha, and the consequent defect of sensorial power. The same retrograde motions affect the whole intestinal canal in ileus; and the œsophagus in globus hystericus. See this further explained in Sect. XXIX. No. 11. on Retrograde Motions.

I must observe, also, that something similar happens in the production of our ideas, or sensual motions, when they are too weakly excited; when any one is thinking intensely about one thing, and carelessly conversing about another, he is liable to use the word of a contrary meaning to that which he designed, as cold weather for hot weather, summer for winter.

6. A certain quantity of stimulus, less than that above mentioned, is succeeded by paralysis, first of the voluntary and sensitive motions, and afterwards of those of irritation, and of association, which constitutes death.

VI. Cure of increased Exertion.

1. The cure, which nature has provided for the increased exertion of any part of the system, consists in the consequent expenditure of the sensorial power. But as a greater torpor follows this exhaustion of sensorial power, as explained in the next paragraph, and a greater exertion succeeds this torpor, the constitution frequently sinks under these increasing librations between exertion and quiescence; till at length complete quiescence, that is, death, closes the scene.

For, during the great exertion of the system in the hot fit of fever, an increase of stimulus is produced from the greater momentum of the blood, the greater distention of the heart and arteries, and the increased production of heat, by the violent actions of the system occasioned by this augmentation of stimulus, the sensorial power becomes diminished in a few hours much beneath its natural quantity, the vessels at length cease to obey even these great degrees of stimulus, as shewn in Sect. XL. 9. 1. and a torpor of the whole or of a part of the system ensues.

Now as this second cold fit commences with a greater deficiency of sensorial power, it is also attended with a greater deficiency of stimulus than in the preceding cold fit, that is, with less momentum of blood, less distention of the heart. On this account the second cold fit becomes more violent and of longer duration than the first; and as a greater accumulation of sensorial power must be produced before the system of vessels will again obey the diminished stimulus, it follows, that the second hot fit of fever will be more violent than the former one. And that unless some other causes counteract either the violent exertions in the hot fit, or the great torpor in the cold fit, life will at length be extinguished by the expenditure of the whole of the sensorial power. And from hence it appears, that the true means of curing fevers must be such as decrease the action of the system in the hot fit, and increase it in the cold fit; that is, such as prevent the too great diminution of sensorial power in the hot fit, and the too great accumulation of it in the cold one.

2. Where the exertion of the sensorial powers is much increased, as in the hot fits of fever or inflammation, the following are the usual means of relieving it. Decrease the irritations by blood-letting, and other evacuations; by cold water taken into the stomach, or injected as an enema, or used externally; by cold air breathed into the lungs, and diffused over the skin; with food of less stimulus than the patient has been accustomed to.

3. As a cold fit, or paroxysm of inactivity of some parts of the system, generally precedes the hot fit, or paroxysm of exertion, by which the sensorial power becomes accumulated, this cold paroxysm should be prevented by stimulant medicines and diet, as wine, opium, bark, warmth, cheerfulness, anger, surprise.

4. Excite into greater action some other part of the system, by which means the spirit of animation may be in part expended, and thence the inordinate actions of the diseased part may be lessened. Hence when a part of the skin acts violently, as of the face in the eruption of the small-pox, if the feet be cold they should be covered. Hence the use of a blister applied near a topical inflammation. Hence opium and warm bath relieve pains both from excess and defect of stimulus.

5. First increase the general stimulation above its natural quantity, which may in some degree exhaust the spirit of animation, and then decrease the stimulation beneath its natural quantity. Hence after sudorific medicines and warm air, the application of refrigerants may have greater effect, if they could be administered without danger of producing too great torpor of some part of the system; as frequently happens to people in health from coming out of a warm room into the cold air, by which a topical inflammation in consequence of torpor of the mucous membrane of the nostril is produced, and is termed a cold in the head.

VII. *Cure of decreased Exertion.*

1. Where the exertion of the sensorial powers is much decreased, as in the cold fits of fever, a gradual accumulation of the spirit of animation takes place; as occurs in all cases where inactivity or torpor of a part of the system exists; this accumulation of sensorial power increases, till stimuli less than natural are sufficient to throw it into action, then the cold fit ceases; and from the action of the natural stimuli a hot one succeeds with increased activity of the whole system.

So in fainting fits, or syncope, there is a temporary deficiency of sensorial exertion, and a consequent quiescence of a great part of the system. This quiescence continues, till the sensorial power becomes again accumulated in the torpid organs; and then the usual diurnal stimuli excite the revivescent parts again into action; but as this kind of quiescence continues but a short time compared to the cold paroxysm of an ague, and less affects the circulatory system, a less superabundancy of exertion succeeds in the organs previously torpid, and a less excess of arterial activity. See Sect. XXXIV. 1. 6.

2. In the diseases occasioned by a defect of sensorial exertion, as in cold fits of ague, hysteric complaint, and nervous fever, the following means are those commonly used. 1. Increase the stimulation above its natural quantity

for some weeks, till a new habit of more energetic contraction of the fibres is established. This is to be done by wine, opium, bark, steel, given at exact periods, and in appropriate quantities; for if these medicines be given in such quantity, as to induce the least degree of intoxication, a debility succeeds from the useless exhaustion of spirit of animation in consequence of too great exertion of the muscles or organs of sense. To these irritative stimuli should be added the sensitive ones of cheerful ideas, hope, affection.

3. Change the kinds of stimulus. The habits acquired by the constitution depend on such nice circumstances, that when one kind of stimulus ceases to excite the sensorial power into the quantity of exertion necessary to health, it is often sufficient to change the stimulus for another apparently similar in quantity and quality. Thus when wine ceases to stimulate the constitution, opium in appropriate doses supplies the defect; and the contrary. This is also observed in the effects of cathartic medicines, when one loses its power, another, apparently less efficacious, will succeed. Hence a change of diet, drink, and stimulating medicines, is often advantageous in diseases of debility.

4. Stimulate the organs, whose motions are associated with the torpid parts of the system. The actions of the minute vessels of the various parts of the external skin are not only associated with each other, but are strongly associated with those of some of the internal membranes, and particularly of the stomach. Hence when the exertion of the stomach is less than natural, and indigestion and heartburn succeed, nothing so certainly removes these symptoms as the stimulus of a blister on the back. The coldness of the extremities, as of the nose, ears, or fingers, are hence the best indication for the successful application of blisters.

5. Decrease the stimulus for a time. By lessening the quantity of heat for a minute or two by going into the cold bath, a great accumulation of sensorial power is produced; for not only the minute vessels of the whole external skin for a time become inactive, as appears by their paleness; but the minute vessels of the lungs lose much of their activity also by concert with those of the skin, as appears from the difficulty of breathing at first going into cold water. On emerging from the bath the sensorial power is thrown into great exertion by the stimulus of the common degree of the warmth of the atmosphere, and a great production of animal heat is the consequence. The longer a person continues in the cold bath the greater must be the present inertion of a great part of the system, and in consequence a greater accumulation of sensorial power. Whence M. Pomè recommends some melancholy patients to be kept from two to six hours in spring-water, and in baths still colder.

6. Decrease the stimulus for a time below the natural, and then increase it above natural. The effect of this process, improperly used, is seen in giving much food, or applying much warmth, to those who have been previously exposed to great hunger, or to great cold. The accumulated sensorial power is thrown into so violent exertion, that inflammations and mortifications supervene, and death closes the catastrophe. In many diseases this method is the most successful; hence the bark in agues produces more certain effect after the previous exhibition of emetics. In diseases attended with violent pain, opium has double the effect, if venesection and a cathartic have been previously used. On this seems to have been founded the successful practice of Sydenham, who used venesection and a cathartic in chlorosis before the exhibition of the bark, steel, and opiates.

7. Prevent any unnecessary expenditure of sensorial power. Hence in fevers with debility, a decumbent posture is preferred, with silence, little light, and such a quantity of heat as may prevent any chill sensation, or any coldness of the extremities. The pulse of patients in fevers with debility increases in frequency above ten pulsations in a minute on their rising out of bed. For the expenditure of sensorial power to preserve an erect posture of the body adds to the general deficiency of it, and thus affects the circulation.

8. The longer in time and the greater in degree the quiescence or inertion of an organ has been, so that it still retains life or excitability, the less stimulus should at first be applied to it. The quantity of stimulation is a matter of great nicety to determine, where the torpor or quiescence of the fibres has been experienced in a great degree, or for a considerable time, as in cold fits of the ague, in continued fevers with great debility, or in people famished at sea, or perishing with cold. In the two last cases, very minute quantities of food should be first supplied, and very few additional degrees of heat. In the two former cases, but little stimulus of wine or medicine, above what they had been lately accustomed to, should be exhibited, and this at frequent and stated intervals, so that the effect of one quantity may be observed before the exhibition of another.

If these circumstances are not attended to, as the sensorial power becomes accumulated in the quiescent fibres, an inordinate exertion takes place by the increase of stimulus acting on the accumulated quantity of sensorial power, and either the paralysis, or death of the contractile fibres ensues, from the total expenditure of the sensorial power in the affected organ, owing to this increase of exertion, like the debility after intoxication. Or, secondly, the violent exertions above mentioned produce painful sensation, which becomes a new stimulus, and by thus producing inflammation, and increasing the activity of the fibres already too great, sooner exhausts the whole of the sensorial power in the acting organ, and mortification, that is, the death of the part, supervenes.

Hence there have been many instances of people, whose limbs have been long benumbed by exposure to cold, who have lost them by mortification on their being too hastily brought to the fire; and of others, who were nearly famished at sea, who have died soon after having taken not more than an usual meal of food. I have heard of two well-attested instances of patients in the cold fit of ague, who have died from the exhibition of gin and vinegar, by the inflammation which ensued. And in many fevers attended with debility, the unlimited use of wine, and the wanton application of blisters, I believe, has destroyed numbers by the debility consequent to too great stimulation, that is, by the exhaustion of the sensorial power by its inordinate exertion.

Wherever the least degree of intoxication exists, a proportional debility is the consequence; but there is a golden rule by which the necessary and useful quantity of stimulus in fevers with debility may be ascertained. When wine or beer are exhibited either alone or diluted with water, if the pulse becomes slower the stimulus is of a proper quantity; and should be repeated every two or three hours, or when the pulse again becomes quicker.

In the chronical debility brought on by drinking spirituous or fermented liquors, there is another golden rule by which I have successfully directed the quantity of spirit which they may safely lessen, for there is no other means by which they can recover their health. It should be premised, that where the power of digestion in these patients is totally destroyed, there is not much reason to expect a return to healthful vigour.

I have directed several of these patients to omit one fourth part of the quantity of vinous spirit they have been lately accustomed to, and if in a fortnight their appetite increases, they are advised to omit another fourth part; but if they perceive that their digestion becomes impaired from the want of this quantity of spirituous potation, they are advised to continue as they are, and rather bear the ills they have, than risk the encounter of greater. At the same time flesh-meat with or without spice is recommended, with Peruvian bark and steel in small quantities between their meals, and half a grain of opium or a grain, with five or eight grains of rhubarb at night.

SECT. XIII
OF VEGETABLE ANIMATION

I. 1. Vegetables are irritable; mimosa, dionæa muscipula. Vegetable secretions. 2. Vegetable buds are inferior animals, are liable to greater or less irritability. II. Stamens and pistils of plants shew marks of sensibility. III. Vegetables possess some degree of volition. IV. Motions of plants are associated like those of animals. V. 1. Vegetable structure like that of animals, their anthers and stigmas are living creatures. Male-flowers of Vallisneria. 2. Whether vegetables, possess ideas? They have organs of sense as of touch and smell, and ideas of external things?

I. 1. The fibres of the vegetable world, as well as those of the animal, are excitable into a variety of motion by irritations of external objects. This appears particularly in the mimosa or sensitive plant, whose leaves contract on the slightest injury; the dionæa muscipula, which was lately brought over from the marshes of America, presents us with another curious instance of vegetable irritability; its leaves are armed with spines on their upper edge, and are spread on the ground around the stem; when an insect creeps on any of them in its passage to the flower or seed, the leaf shuts up like a steel rat-trap, and destroys its enemy. See Botanic Garden, Part II. note on Silene.

The various secretions of vegetables, as of odour, fruit, gum, resin, wax, honey, seem brought about in the same manner as in the glands of animals; the tasteless moisture of the earth is converted by the hop-plant into a bitter juice; as by the caterpillar in the nut-shell the sweet kernel is converted into a bitter powder. While the power of absorption in the roots and barks of vegetables is excited into action by the fluids applied to their mouths like the lacteals and lymphatics of animals.

2. The individuals of the vegetable world may be considered as inferior or less perfect animals; a tree is a congeries of many living buds, and in this respect resembles the branches of coralline, which are a congeries of a multitude of animals. Each of these buds of a tree has its proper leaves or petals for lungs, produces its viviparous or its oviparous offspring in buds or seeds; has its own roots, which extending down the stem of the tree are

interwoven with the roots of the other buds, and form the bark, which is the only living part of the stem, is annually renewed, and is superinduced upon the former bark, which then dies, and with its stagnated juices gradually hardening into wood forms the concentric circles, which we see in blocks of timber.

The following circumstances evince the individuality of the buds of trees. First, there are many trees, whose whole internal wood is perished, and yet the branches are vegete and healthy. Secondly, the fibres of the barks of trees are chiefly longitudinal, resembling roots, as is beautifully seen in those prepared barks, that were lately brought from Otaheita. Thirdly, in horizontal wounds of the bark of trees, the fibres of the upper lip are always elongated downwards like roots, but those of the lower lip do not approach to meet them. Fourthly, if you wrap wet moss round any joint of a vine, or cover it with moist earth, roots will shoot out from it. Fifthly, by the inoculation or engrafting of trees many fruits are produced from one stem. Sixthly, a new tree is produced from a branch plucked from an old one, and set in the ground. Whence it appears that the buds of deciduous trees are so many annual plants, that the bark is a contexture of the roots of each individual bud; and that the internal wood is of no other use but to support them in the air, and that thus they resemble the animal world in their individuality.

The irritability of plants, like that of animals, appears liable to be increased or decreased by habit; for those trees or shrubs, which are brought from a colder climate to a warmer, put out their leaves and blossoms a fortnight sooner than the indigenous ones.

Professor Kalm, in his Travels in New York, observes that the apple-trees brought from England blossom a fortnight sooner than the native ones. In our country the shrubs, that are brought a degree or two from the north, are observed to flourish better than those, which come from the south. The Siberian barley and cabbage are said to grow larger in this climate than the similar more southern vegetables. And our hoards of roots, as of potatoes and onions, germinate with less heat in spring, after they have been accustomed to the winter's cold, than in autumn after the summer's heat.

II. The stamens and pistils of flowers shew evident marks of sensibility, not only from many of the stamens and some pistils approaching towards each other at the season of impregnation, but from many of them closing their petals and calyxes during the cold parts of the day. For this cannot be ascribed to irritation, because cold means a defect of the stimulus of heat; but as the want of accustomed stimuli produces pain, as in coldness, hunger, and thirst of animals, these motions of vegetables in closing up their flowers must be ascribed to the disgreeable sensation, and not to the irritation of cold. Others close up their leaves during darkness, which, like the former, cannot be owing to irritation, as the irritating material is withdrawn.

The approach of the anthers in many flowers to the stigmas, and of the pistils of some flowers to the anthers, must be ascribed to the passion of love, and hence belongs to sensation, not to irritation.

III. That the vegetable world possesses some degree of voluntary powers, appears from their necessity to sleep, which we have shewn in Sect. XVIII. to consist in the temporary abolition of voluntary power. This voluntary power seems to be exerted in the circular movement of the tendrils of vines, and other climbing vegetables; or in the efforts to turn the upper surface of their leaves, or their flowers to the light.

IV. The associations of fibrous motions are observable in the vegetable world, as well as in the animal. The divisions of the leaves of the sensitive plant have been accustomed to contract at the same time from the absence of light; hence if by any other circumstance, as a slight stroke or injury, one division is irritated into contraction, the neighbouring ones contract also, from their motions being associated with those of the irritated part. So the various stamina of the class of syngenesia have been accustomed to contract together in the evening, and thence if you stimulate one of them with a pin, according to the experiment of M. Colvolo, they all contract from their acquired associations.

To evince that the collapsing of the sensitive plant is not owing to any mechanical vibrations propagated along the whole branch, when a single leaf is struck with the finger, a leaf of it was slit with sharp scissors, and some seconds of time passed before the plant seemed sensible of the injury; and then the whole branch collapsed as far as the principal stem: this experiment was repeated several times with the least possible impulse to the plant.

V. 1. For the numerous circumstances in which vegetable buds are analogous to animals, the reader is referred to the additional notes at the end of the Botanic Garden, Part I. It is there shewn, that the roots of vegetables resemble the lacteal system of animals; the sap-vessels in the early spring, before their leaves expand, are analogous to the placental vessels of the fœtus; that the leaves of land-plants resemble lungs, and those of aquatic plants the gills of fish; that there are other systems of vessels resembling the vena portarum of quadrupeds, or the aorta of fish; that the digestive power of vegetables is similar to that of animals converting the fluids, which they absorb, into sugar; that their seeds resemble the eggs of animals, and their buds and bulbs their viviparous offspring. And, lastly, that the anthers and stigmas are real animals, attached indeed to their parent tree like polypi or coral insects, but capable of spontaneous motion; that they are affected with the passion of love, and furnished with powers of reproducing their species, and are fed with honey like the moths and butterflies, which plunder their nectaries. See Botanic Garden, Part I. add. note XXXIX.

The male flowers of vallisneria approach still nearer to apparent animality, as they detach themselves from the parent plant, and float on the surface of the water to the female ones. Botanic Garden, Part II. Art. Vallisneria. Other flowers of the classes of monecia and diecia, and polygamia, discharge the fecundating farina, which floating in the air is carried to the stigma of the female flowers, and that at considerable distances. Can this be effected by any specific attraction? or, like the diffusion of the odorous particles of flowers, is it left to the currents of winds, and the accidental miscarriages of it counteracted by the quantity of its production?

2. This leads us to a curious enquiry, whether vegetables have ideas of external things? As all our ideas are originally received by our senses, the question may be changed to, whether vegetables possess any organs of sense? Certain it is, that they possess a sense of heat and cold, another of moisture and dryness, and another of light and darkness; for they close their petals occasionally from the presence of cold, moisture, or darkness. And it has been already shewn, that these actions cannot be performed simply from irritation, because cold and darkness are negative quantities, and on that account sensation or volition are implied, and in consequence a sensorium or union of their nerves. So when we go into the light, we contract the iris; not from any stimulus of the light on the fine muscles of the iris, but from its motions being associated with the sensation of too much light on the retina: which could not take place without a sensorium or center of union of the nerves of the iris with those of vision. See Botanic Garden, Part I. Canto 3. l. 440. note.

Besides these organs of sense, which distinguish cold, moisture, and darkness, the leaves of mimosa, and of dionæa, and of drosera, and the stamens of many flowers, as of the berbery, and the numerous class of syngenesia, are sensible to mechanic impact, that is, they possess a sense of touch, as well as a common sensorium; by the medium of which their muscles are excited into action. Lastly, in many flowers the anthers, when mature, approach the stigma, in others the female organ approaches to the male. In a plant of collinsonia, a branch of which is now before me, the two yellow stamens are about three eights of an inch high, and diverge from each other, at an angle of about fifteen degrees, the purple style is half an inch high, and in some flowers is now applied to the stamen on the right hand, and in others to that of the left; and will, I suppose, change place to-morrow in those, where the anthers have not yet effused their powder.

I ask, by what means are the anthers in many flowers, and stigmas in other flowers, directed to find their paramours? How do either of them know, that the other exists in their vicinity? Is this curious kind of storge produced by mechanic attraction, or by the sensation of love? The latter opinion is supported by the strongest analogy, because a reproduction of the species is the consequence; and then another organ of sense must be wanted to direct these vegetable amourettes to find each other, one probably analogous to our sense of smell, which in the animal world directs the new-born infant to its source of nourishment, and they may thus possess a faculty of perceiving as well as of producing odours.

Thus, besides a kind of taste at the extremities of their roots, similar to that of the extremities of our lacteal vessels, for the purpose of selecting their proper food: and besides different kinds of irritability residing in the various glands, which separate honey, wax, resin, and other juices from their blood; vegetable life seems to possess an organ of sense to distinguish the variations of heat, another to distinguish the varying degrees of moisture, another of light, another of touch, and probably another analogous to our sense of smell. To these must be added the indubitable evidence of their passion of love, and I think we may truly conclude, that they are furnished with a common sensorium belonging to each bud and that they must occasionally repeat those perceptions either in their dreams or waking hours, and consequently possess ideas of so many of the properties of the external world, and of their own existence.

SECT. XIV
OF THE PRODUCTION OF IDEAS

I. *Of material and immaterial beings. Doctrine of St. Paul. II. 1. Of the sense of touch. Of solidity. 2. Of figure. Motion. Time. Place. Space. Number. 3. Of the penetrability of matter. 4. Spirit of animation possesses solidity, figure, visibility, &c. Of Spirits and angels. 5. The existence of external things. III. Of vision. IV. Of hearing. V. Of smell and taste. VI. Of the organ of sense by which we perceive heat and cold, not by the sense of touch. VII. Of the sense of extension, the whole of the locomotive muscles may be considered as one organ of sense. VIII. Of the senses of hunger, thirst, want of fresh air, suckling children, and lust. IX. Of many other organs of sense belonging to the glands. Of painful sensations from the excess of light, pressure, heat, itching, caustics, and electricity.*

I. Philosophers have been much perplexed to understand, in what manner we become acquainted with the external world; insomuch that Dr. Berkly even doubted its existence, from having observed (as he thought) that none of our ideas resemble their correspondent objects. Mr. Hume asserts, that our belief depends on the greater distinctness or energy of our ideas from perception; and Mr. Reid has lately contended, that our belief of external objects is an innate principle necessarily joined with our perceptions.

So true is the observation of the famous Malbranch, "that our senses are not given us to discover the essences of things, but to acquaint us with the means of preserving our existence," (L. I. ch. v.) a melancholy reflection to philosophers!

Some philosophers have divided all created beings into material and immaterial: the former including all that part of being, which obeys the mechanic laws of action and reaction, but which can begin no motion of itself; the other is the cause of all motion, and is either termed the power of gravity, or of specific attraction, or the spirit of animation. This immaterial agent is supposed to exist in or with matter, but to be quite distinct from it, and to be equally capable of existence, after the matter, which now possesses it, is decomposed.

Nor is this theory ill supported by analogy, since heat, electricity, and magnetism, can be given to or taken from a piece of iron; and must therefore exist, whether separated from the metal, or combined with it. From a parity of reasoning, the spirit of animation, would appear to be capable of existing as well separately from the body as with it.

I beg to be understood, that I do not wish to dispute about words, and am ready to allow, that the powers of gravity, specific attraction, electricity, magnetism, and even the spirit of animation, may consist of matter of a finer kind; and to believe, with St. Paul and Malbranch, that the ultimate cause only of all motion is immaterial, that is God. St. Paul says, "in him we live and move, and have our being;" and, in the 15th chapter to the Corinthians, distinguishes between the psyche or living spirit, and the pneuma or reviving spirit. By the words spirit of animation or sensorial power, I mean only that animal life, which mankind possesses in common with brutes, and in some degree even with vegetables, and leave the consideration of the immortal part of us, which is the object of religion, to those who treat of revelation.

II. 1. Of the Sense of Touch.

The first idea we become acquainted with, are those of the sense of touch; for the fœtus must experience some varieties of agitation, and exert some muscular action, in the womb; and may with great probability be supposed thus to gain some ideas of its own figure, of that of the uterus, and of the tenacity of the fluid, that surrounds it, (as appears from the facts mentioned in the succeeding Section upon Instinct.)

Many of the organs of sense are confined to a small part of the body, as the nostrils, ear, or eye, whilst the sense of touch is diffused over the whole skin, but exists with a more exquisite degree of delicacy at the extremities of the fingers and thumbs, and in the lips. The sense of touch is thus very commodiously disposed for the purpose of encompassing smaller bodies, and for adapting itself to the inequalities of larger ones. The figure of small bodies seems to be learnt by children by their lips as much as by their fingers; on which account they put every new object to their mouths, when they are satiated with food, as well as when they are hungry. And puppies seem to learn their ideas of figure principally by the lips in their mode of play.

We acquire our tangible ideas of objects either by the simple pressure of this organ of touch against a solid body, or by moving our organ of touch along the surface of it. In the former case we learn the length and breadth of the object by the quantity of our organ of touch, that is impressed by it: in the latter case we learn the length and breadth of objects by the continuance of their pressure on our moving organ of touch.

It is hence, that we are very slow in acquiring our tangible ideas, and very slow in recollecting them; for if I now think of the tangible idea of a cube, that is, if I think of its figure, and of the solidity of every part of that figure, I must conceive myself as passing my fingers over it, and seem in some measure to feel the idea, as I formerly did the impression, at the ends of them, and am thus very slow in distinctly recollecting it.

When a body compresses any part of our sense of touch, what happens? First, this part of our sensorium undergoes a mechanical compression, which is termed a stimulus; secondly, an idea, or contraction of a part of the organ of sense is excited; thirdly, a motion of the central parts, or of the whole sensorium, which is termed sensation, is produced; and these three constitute the perception of solidity.

2. *Of Figure, Motion, Time, Place, Space, Number.*

No one will deny, that the medulla of the brain and nerves has a certain figure; which, as it is diffused through nearly the whole of the body, must have nearly the figure of that body. Now it follows, that the spirit of animation, or living principle, as it occupies this medulla, and no other part, (which is evinced by a great variety of cruel experiments on living animals,) it follows, that this spirit of animation has also the same figure as the medulla above described. I appeal to common sense! the spirit of animation acts, Where does it act? It acts wherever there is the medulla above mentioned; and that whether the limb is yet joined to a living animal, or whether it be recently detached from it; as the heart of a viper or frog will renew its contractions, when pricked with a pin, for many minutes of time after its exsection from the body.—Does it act any where else?—No; then it certainly exists in this part of space, and no where else; that is, it hath figure; namely, the figure of the nervous system, which is nearly the figure of the body. When the idea of solidity is excited, as above explained, a part of the extensive organ of touch is compressed by some external body, and this part of the sensorium so compressed exactly resembles *in figure* the figure of the body that compressed it. Hence, when we acquire the idea of solidity, we acquire at the same time the idea of FIGURE; and this idea of figure, or motion of *a part* of the organ of touch, exactly resembles *in its figure* the figure of the body that occasions it; and thus exactly acquaints us with this property of the external world.

Now, as the whole universe with all its parts possesses a certain form or figure, if any part of it moves, that form or figure of the whole is varied:

hence, as MOTION is no other than a perpetual variation of figure, our idea of motion is also a real resemblance of the motion that produced it.

It may be said in objection to this definition of motion, that an ivory globe may revolve on its axis, and that here will be a motion without change of figure. But the figure of the particle *x* on one side of this globe is not the *same* figure as the figure of *y* on the other side, any more than the particles themselves are the same, though they are *similar* figures; and hence they cannot change place with each other without disturbing or changing the figure of the whole.

Our idea of TIME is from the same source, but is more abstracted, as it includes only the comparative velocities of these variations of figure; hence if it be asked, How long was this book in printing? it may be answered, Whilst the sun was passing through Aries.

Our idea of PLACE includes only the figure of a group of bodies, not the figures of the bodies themselves. If it be asked where is Nottinghamshire, the answer is, it is surrounded by Derbyshire, Lincolnshire and Leicestershire; hence place is our idea of the figure of one body surrounded by the figures of other bodies.

The idea of SPACE is a more abstracted idea of place excluding the group of bodies.

The idea of NUMBER includes only the particular arrangements, or distributions of a group of bodies, and is therefore only a more abstracted idea of the parts of the figure of the group of bodies; thus when I say England is divided into forty counties, I only speak of certain divisions of its figure.

Hence arises the certainty of the mathematical sciences, as they explain these properties of bodies, which are exactly resembled by our ideas of them, whilst we are obliged to collect almost all our other knowledge from experiment; that is, by observing the effects exerted by one body upon another.

3. *Of the Penetrability of Matter.*

The impossibility of two bodies existing together in the same space cannot be deduced from our idea of solidity, or of figure. As soon as we perceive the motions of objects that surround us, and learn that we possess a power to move our own bodies, we experience, that those objects, which excite in us the idea of solidity and of figure, oppose this voluntary movement of our own organs; as whilst I endeavour to compress between my hands an ivory ball into a spheroid. And we are hence taught by experience, that our own body and those, which we touch, cannot exist in the same part of space.

But this by no means demonstrates, that no two bodies can exist together in the same part of space. Galilæo in the preface to his works seems to be of opinion, that matter is not impenetrable; Mr. Michel, and Mr. Boscowich in his Theoria. Philos. Natur. have espoused this hypothesis: which has been lately published by Dr. Priestley, to whom the world is much indebted for so many important discoveries in science. (Hist. of Light and Colours, p. 391.) The uninterrupted passage of light through transparent bodies, of the electric æther through metallic and aqueous bodies, and of the magnetic effluvia through all bodies, would seem to give some probability to this opinion. Hence it appears, that beings may exist without possessing the property of solidity, as well as they can exist without possessing the properties, which excite our smell or taste, and can thence occupy space without detruding other bodies from it; but we cannot become acquainted with such beings by our sense of touch, any more than we can with odours or flavours without our senses of smell and taste.

But that any being can exist without existing in space, is to my ideas utterly incomprehensible. My appeal is to common sense. *To be* implies a when and a where; the one is comparing it with the motions of other beings, and the other with their situations.

If there was but one object, as the whole creation may be considered as one object, then I cannot ask where it exists? for there are no other objects to compare its situation with. Hence if any one denies, that a being exists in space, he denies, that there are any other beings but that one; for to answer the question, "Where does it exist?" is only to mention the situation of the objects that surround it.

In the same manner if it be asked—"When does a being exist?" The answer only specifies the successive motions either of itself, or of other bodies; hence to say, a body exists not in time, is to say, that there is, or was, no motion in the world.

4. *Of the Spirit of Animation.*

But though there may exist beings in the universe, that have not the property of solidity; that is, which can possess any part of space, at the same time that it is occupied by other bodies; yet there may be other beings, that can assume this property of solidity, or disrobe themselves of it occasionally, as we are taught of spirits, and of angels; and it would seem, that THE SPIRIT OF ANIMATION must be endued with this property, otherwise how could it occasionally give motion to the limbs of animals?—or be itself stimulated into motion by the obtrusions of surrounding bodies, as of light, or odour?

If the spirit of animation was always necessarily penetrable, it could not influence or be influenced by the solidity of common matter; they would exist together, but could not detrude each other from the part of space, where they exist; that is, they could not communicate motion to each other. *No two things can influence or affect each other, which have not some property common to both of them*; for to influence or affect another body is to give or communicate some property to it, that it had not before; but how can one body give that to another, which it does not possess itself?—The words imply, that they must agree in having the power or faculty of possessing some common property. Thus if one body removes another from the part of space, that it possesses, it must have the power of occupying that space itself: and if one body communicates heat or motion to another, it follows, that they have alike the property of possessing heat or motion.

Hence the spirit of animation at the time it communicates or receives motion from solid bodies, must itself possess some property of solidity. And in consequence at the time it receives other kinds of motion from light, it must possess that property, which light possesses, to communicate that kind of motion; and for which no language has a name, unless it may be termed Visibility. And at the time it is stimulated into other kinds of animal motion by the particles of sapid and odorous bodies affecting the senses of taste and smell, it must resemble these particles of flavour, and of odour, in possessing some similar or correspondent property; and for which language has no name, unless we may use the words Saporosity and Odorosity for those common properties, which are possessed by our organs of taste and smell, and by the particles of sapid and odorous bodies; as the words Tangibility and Audibility may express the common property possessed by our organs of touch, and of hearing, and by the solid bodies, or their vibrations, which affect those organs.

5. Finally, though the figures of bodies are in truth resembled by the figure of the part of the organ of touch, which is stimulated into motion; and that organ resembles the solid body, which stimulates it, in its property of solidity; and though the sense of hearing resembles the vibrations of external bodies in its capability of being stimulated into motion by those vibrations; and though our other organs of sense resemble the bodies, that stimulate them, in their capability of being stimulated by them; and we hence become acquainted with these properties of the external world; yet as we can repeat all these motions of our organs of sense by the efforts of volition, or in consequence of the sensation of pleasure or pain, or by their association with other fibrous motions, as happens in our reveries or in sleep, there would still appear to be some difficulty in demonstrating the existence of any thing external to us.

In our dreams we cannot determine this circumstance, because our power of volition is suspended, and the stimuli of external objects are excluded; but in our waking hours we can compare our ideas belonging to one sense with those belonging to another, and can thus distinguish the ideas occasioned by irritation from those excited by sensation, volition, or association. Thus if the idea of the sweetness of sugar should be excited in our dreams, the whiteness and hardness of it occur at the same time by association; and we believe a material lump of sugar present before us. But if, in our waking hours, the idea of the sweetness of sugar occurs to us, the stimuli of surrounding objects, as the edge of the table, on which we press, or green colour of the grass, on which we tread, prevent the other ideas of the hardness and whiteness of the sugar from being exerted by association. Or if they should occur, we voluntarily compare them with the irritative ideas of the table or grass above mentioned, and detect their fallacy. We can thus distinguish the ideas caused by the stimuli of external objects from those, which are introduced by association, sensation, or volition; and during our waking hours can thus acquire a knowledge of the external world. Which nevertheless we cannot do in our dreams, because we have neither perceptions of external bodies, nor the power of volition to enable us to compare them with the ideas of imagination.

III. *Of Vision.*

Our eyes observe a difference of colour, or of shade, in the prominences and depressions of objects, and that those shades uniformly vary, when the sense of touch observes any variation. Hence when the retina becomes stimulated by colours or shades of light in a certain form, as in a circular spot; we know by experience, that this is a sign, that a tangible body is before us; and that its figure is resembled by the miniature figure of the part of the organ of vision, that is thus stimulated.

Here whilst the stimulated part of the retina resembles exactly the visible figure of the whole in miniature, the various kinds of stimuli from different colours mark the visible figures of the minuter parts; and by habit we instantly recall the tangible figures.

Thus when a tree is the object of sight, a part of the retina resembling a flat branching figure is stimulated by various shades of colours; but it is by suggestion, that the gibbosity of the tree, and the moss, that fringes its trunk, appear before us. These are ideas of suggestion, which we feel or attend to, associated with the motions of the retina, or irritative ideas, which we do not attend to.

So that though our visible ideas resemble in miniature the outline of the figure of coloured bodies, in other respects they serve only as a language, which by acquired associations introduce the tangible ideas of bodies. Hence it is, that this sense is so readily deceived by the art of the painter to our amusement and instruction. The reader will find much very curious knowledge on this subject in Bishop Berkley's Essay on Vision, a work of great ingenuity.

The immediate object however of the sense of vision is light; this fluid, though its velocity is so great, appears to have no perceptible mechanical impulse, as was mentioned in the third Section, but seems to stimulate the retina into animal motion by its transmission through this part of the sensorium: for though the eyes of cats or other animals appear luminous in obscure places; yet it is probable, that none of the light, which falls on the retina, is reflected from it, but adheres to or enters into combination with the choroide coat behind it.

The combination of the particles of light with opake bodies, and therefore with the choroide coat of the eye, is evinced from the heat, which is given out, as in other chemical combinations. For the sunbeams communicate no heat in their passage through transparent bodies, with which they do not combine, as the air continues cool even in the focus of the largest burning-glasses, which in a moment vitrifies a particle of opaque matter.

IV. *Of the Organ of Hearing.*

It is generally believed, that the tympanum of the ear vibrates mechanically, when exposed to audible sounds, like the strings of one musical instrument, when the same notes are struck upon another. Nor is this opinion improbable, as the muscles and cartilages of the larynx are employed in producing variety of tones by mechanical vibration: so the muscles and bones of the ear seem adapted to increase or diminish the tension of the tympanum for the purposes of similar mechanical vibrations.

But it appears from dissection, that the tympanum is not the immediate organ of hearing, but that like the humours and cornea of the eye, it is only of use to prepare the object for the immediate organ. For the portio mollis of the auditory nerve is not spread upon the tympanum, but upon the vestibulum, and cochlea, and semicircular canals of the ear; while between the tympanum and the expansion of the auditory nerve the cavity is said by Dr. Cotunnus and Dr. Meckel to be filled with water; as they had frequently observed by freezing the heads of dead animals before they dissected them; and water being a more dense fluid than air is much better adapted to the propagation of vibrations. We may add, that even the external opening of the ear is not absolutely necessary for the perception of sound: for some people, who from these defects would have been completely deaf, have distinguished acute or grave sounds by the tremours of a stick held between their teeth propagated along the bones of the head, (Haller. Phys. T. V. p. 295).

Hence it appears, that the immediate organ of hearing is not affected by the particles of the air themselves, but is stimulated into animal motion by the vibrations of them. And it is probable from the loose bones, which are found in the heads of some fishes, that the vibrations of water are sensible to the inhabitants of that element by a similar organ.

The motions of the atmosphere, which we become acquainted with by the sense of touch, are combined with its solidity, weight, or vis intertiæ; whereas those, that are perceived by this organ, depend alone on its elasticity. But though the vibration of the air is the immediate object of the sense of hearing, yet the ideas, we receive by this sense, like those received from light, are only as a language, which by acquired associations acquaints us with those motions of tangible bodies, which depend on their elasticity; and which we had before learned by our sense of touch.

V. *Of Smell and of Taste.*

The objects of smell are dissolved in the fluid atmosphere, and those of taste in the saliva, or other aqueous fluid, for the better diffusing them on their respective organs, which seem to be stimulated into animal motion perhaps by the chemical affinities of these particles, which constitute the sapidity and odorosity of bodies with the nerves of sense, which perceive them.

Mr. Volta has lately observed a curious circumstance relative to our sense of taste. If a bit of clean lead and a bit of clean silver be separately applied to the tongue and palate no taste is perceived; but by applying them in contact in respect to the parts out of the mouth, and nearly so in respect to the parts, which are immediately applied to the tongue and palate, a saline or acidulous taste is perceived, as of a fluid like a stream of electricity passing from one of them to the other. This new application of the sense of taste deserves further investigation, as it may acquaint us with new properties of matter.

From the experiments above mentioned of Galvani, Volta, Fowler, and others, it appears, that a plate of zinc and a plate of silver have greater effect than lead and silver. If one edge of a plate of silver about the size of half a crown-piece be placed upon the tongue, and one edge of a plate of zinc about the same size beneath the tongue, and if their opposite edges are then brought into contact before the point of the tongue, a taste is perceived at the moment of their coming into contact; secondly, if one of the above plates be put between the upper lip and the gum of the fore-teeth, and the other be placed under the tongue, and their exterior edges be then brought into contact in a darkish room, a flash of light is perceived in the eyes.

These effects I imagine only shew the sensibility of our nerves of sense to very small quantities of the electric fluid, as it passes through them; for I suppose these sensations are occasioned by slight electric shocks produced in the following manner. By the experiments published by Mr. Bennet, with his ingenious doubler of electricity, which is the greatest discovery made in that science since the coated jar, and the eduction of lightning from the skies, it appears that zinc was always found minus, and silver was always found plus, when both of them were in their separate state. Hence, when they are placed in the manner above described, as soon as their exterior edges come nearly into contact, so near as to have an extremely thin plate of air between them, that plate of air becomes charged in the same manner as a plate of coated glass; and is at the same instant discharged through the nerves of taste or of sight, and gives the sensations, as above described, of light or of saporocity; and only shews the great sensibility of these organs of sense to the stimulus of the electric fluid in suddenly passing through them.

VI. Of the Sense of Heat.

There are many experiments in chemical writers, that evince the existence of heat as a fluid element, which covers and pervades all bodies, and is attracted by the solutions of some of them, and is detruded from the combination of others. Thus from the combinations of metals with acids, and from those combinations of animal fluids, which are termed secretions, this fluid matter of heat is given out amongst the neighbouring bodies; and in the solutions of salts in water, or of water in air, it is absorbed from the bodies, that surround them; whilst in its facility in passing through metallic bodies, and its difficulty in pervading resins and glass, it resembles the properties of the electric aura; and is like that excited by friction, and seems like that to gravitate amongst other bodies in its uncombined state, and to find its equilibrium.

There is no circumstance of more consequence in the animal economy than a due proportion of this fluid of heat; for the digestion of our nutriment in the stomach and bowels, and the proper qualities of all our secreted fluids, as they are produced or prepared partly by animal and partly by chemical processes, depend much on the quantity of heat; the excess of which, or its deficiency, alike gives us pain, and induces us to avoid the circumstances that occasion them. And in this the perception of heat essentially differs from the perceptions of the sense of touch, as we receive pain from too great pressure of solid bodies, but none from the absence of it. It is hence probable, that nature has provided us with a set of nerves for the perception of this fluid, which anatomists have not yet attended to.

There may be some difficulty in the proof of this assertion; if we look at a hot fire, we experience no pain of the optic nerve, though the heat along with the light must be concentrated upon it. Nor does warm water or warm oil poured into the ear give pain to the organ of hearing; and hence as these organs of sense do not perceive small excesses or deficiences of heat; and as heat has no greater analogy to the solidity or to the figures of bodies, than it has to their colours or vibrations; there seems no sufficient reason for our ascribing the perception of heat and cold to the sense of touch; to which it has generally been attributed, either because it is diffused beneath the whole skin like the sense of touch, or owing to the inaccuracy of our observations, or the defect of our languages.

There is another circumstance would induce us to believe, that the perceptions of heat and cold do not belong to the organ of touch; since the teeth, which are the least adapted for the perceptions of solidity or figure, are the most sensible to heat or cold; whence we are forewarned from swallowing those materials, whose degree of coldness or of heat would injure our stomachs.

The following is an extract from a letter of Dr. R.W. Darwin, of Shrewsbury, when he was a student at Edinburgh. "I made an experiment yesterday in our hospital, which much favours your opinion, that the sensation of heat and of touch depend on different sets of nerves. A man who had lately recovered from a fever, and was still weak, was seized with violent cramps in his legs and feet; which were removed by opiates, except that one of his feet remained insensible. Mr. Ewart pricked him with a pin in five or six places, and the patient declared he did not feel it in the least, nor was he sensible of a very smart pinch. I then held a red-hot poker at some distance, and brought it gradually nearer till it came within three inches, when he asserted that he felt it quite distinctly. I suppose some violent irritation on the nerves of touch had caused the cramps, and had left them paralytic; while the nerves of heat, having suffered no increased stimulus, retained their irritability."

Add to this, that the lungs, though easily stimulated into inflammation, are not sensible to heat. See Class. III. 1. 1. 10.

VII. Of the Sense of Extension.

The organ of touch is properly the sense of pressure, but the muscular fibres themselves constitute the organ of sense, that feels extension. The sense of pressure is always attended with the ideas of the figure and solidity of the object, neither of which accompany our perception of extension. The whole set of muscles, whether they are hollow ones, as the heart, arteries, and intestines, or longitudinal ones attached to bones, contract themselves, whenever they are stimulated by forcible elongation; and it is observable, that the white muscles, which constitute the arterial system, seem to be excited into contraction from no other kinds of stimulus, according to the experiments of Haller. And hence the violent pain in some inflammations, as in the paronychia, obtains immediate relief by cutting the membrane, that was stretched by the tumour of the subjacent parts.

Hence the whole muscular system may be considered as one organ of sense, and the various attitudes of the body, as ideas belonging to this organ, of many of which we are hourly conscious, while many others, like the irritative ideas of the other senses, are performed without our attention.

When the muscles of the heart cease to act, the refluent blood again distends or elongates them; and thus irritated they contract as before. The same happens to the arterial system, and I suppose to the capillaries, intestines, and various glands of the body.

When the quantity of urine, or of excrement, distends the bladder, or rectum, those parts contract, and exclude their contents, and many other muscles by association act along with them; but if these evacuations are not soon complied with, pain is produced by a little further extension of the muscular fibres: a similar pain is caused in the muscles, when a limb is much extended for the reduction of dislocated bones; and in the punishment of the rack: and in the painful cramps of the calf of the leg, or of other muscles, for a greater degree of contraction of a muscle, than the movement of the two bones, to which its ends are affixed, will admit of, must give similar pain to that, which is produced by extending it beyond its due length. And the pain from punctures or incisions arises from the distention of the fibres, as the knife passes through them; for it nearly ceases as soon as the division is completed.

All these motions of the muscles, that are thus naturally excited by the stimulus of distending bodies, are also liable to be called into strong action by their catenation, with the irritations or sensations produced by the momentum of the progressive particles of blood in the arteries, as in inflammatory fevers, or by acrid substances on other sensible organs, as in the strangury, or tenesmus, or cholera.

We shall conclude this account of the sense of extension by observing, that the want of its object is attended with a disagreeable sensation, as well as the excess of it. In those hollow muscles, which have been accustomed to it, this disagreeable sensation is called faintness, emptiness, and sinking; and, when it arises to a certain degree, is attended with syncope, or a total quiescence of all motions, but the internal irritative ones, as happens from sudden loss of blood, or in the operation of tapping in the dropsy.

VIII. *Of the Appetites of Hunger, Thirst, Heat, Extension, the want of fresh Air, animal Love, and the Suckling of Children.*

Hunger is most probably perceived by those numerous ramifications of nerves that are seen about the upper opening of the stomach; and thirst by the nerves about the fauces, and the top of the gula. The ideas of these senses are few in the generality of mankind, but are more numerous in those, who by disease, or indulgence, desire particular kinds of foods or liquids.

A sense of heat has already been spoken of, which may with propriety be called an appetite, as we painfully desire it, when it is deficient in quantity.

The sense of extension may be ranked amongst these appetites, since the deficiency of its object gives disagreeable sensation; when this happens in the arterial system, it is called faintness, and seems to bear some analogy to hunger and to cold; which like it are attended with emptiness of a part of the vascular system.

The sense of want of fresh air has not been attended to, but is as distinct as the others, and the first perhaps that we experience after our nativity; from the want of the object of this sense many diseases are produced, as the jail-fever, plague, and other epidemic maladies. Animal love is another appetite, which occurs later in life, and the females of lactiferous animals have another natural inlet of pleasure or pain from the suckling their offspring. The want of which either owing to the death of their progeny, or to the fashion of their country, has been fatal to many of the sex. The males have also pectoral glands, which are frequently turgid with a thin milk at their nativity, and are furnished with nipples, which erect on titillation like those of the female; but which seem now to be of no further use, owing perhaps to some change which these animals have undergone in the gradual progression of the formation of the earth, and of all that it inhabit.

These seven last mentioned senses may properly be termed appetites, as they differ from those of touch, sight, hearing, taste, and smell, in this respect; that they are affected with pain as well by the defect of their objects as by the excess of them, which is not so in the latter. Thus cold and hunger give us pain, as well as an excess of heat or satiety; but it is not so with darkness and silence.

IX. Before we conclude this Section on the organs of sense, we must observe, that, as far as we know, there are many more senses, than have been here mentioned, as every gland seems to be influenced to separate from the blood, or to absorb from the cavities of the body, or from the atmosphere, its appropriated fluid, by the stimulus of that fluid on the living gland; and not by mechanical capillary absorption, nor by chemical affinity. Hence it appears, that each of these glands must have a peculiar organ to perceive these irritations, but as these irritations are not succeeded by sensation, they have not acquired the names of senses.

However when these glands are excited into motions stronger than usual, either by the acrimony of their fluids, or by their own irritability being much increased, then the sensation of pain is produced in them as in all the other senses of the body; and these pains are all of different kinds, and hence the glands at this time really become each a different organ of sense, though these different kinds of pain have acquired no names.

Thus a great excess of light does not give the idea of light but of pain, as in forcibly opening the eye when it is much inflamed. The great excess of pressure or distention, as when the point of a pin is pressed upon our skin, produces pain, (and when this pain of the sense of distention is slighter, it is termed itching, or tickling), without any idea of solidity or of figure: an excess of heat produces smarting, of cold another kind of pain; it is probable by this sense of heat the pain produced by caustic bodies is perceived, and of electricity, as all these are fluids, that permeate, distend, or decompose the parts that feel them.

SECT. XV
OF THE CLASSES OF IDEAS

I. 1. *Ideas received in tribes. 2. We combine them further, or abstract from these tribes. 3. Complex ideas. 4. Compounded ideas. 5. Simple ideas, modes, substances, relations, general ideas. 6. Ideas of reflexion. 7. Memory and imagination imperfectly defined. Ideal presence. Memorandum-rings. II. 1. Irritative ideas. Perception. 2. Sensitive ideas, imagination. 3. Voluntary ideas, recollection. 4. Associated ideas, suggestion. III. 1. Definitions of perception, memory. 2. Reasoning, judgment, doubting, distinguishing, comparing. 3. Invention. 4. Consciousness. 5. Identity. 6. Lapse of time. 7. Free-will.*

I. 1. As the constituent elements of the material world are only perceptible to our organs of sense in a state of combination; it follows, that the ideas or sensual motions excited by them, are never received singly, but ever with a greater or less degree of combination. So the colours of bodies or their hardnesses occur with their figures: every smell and taste has its degree of pungency as well as its peculiar flavour: and each note in music is combined with the tone of some instrument. It appears from hence, that we can be sensible of a number of ideas at the same time, such as the whiteness, hardness, and coldness, of a snow-ball, and can experience at the same time many irritative ideas of surrounding bodies, which we do not attend to, as mentioned in Section VII. 3. 2. But those ideas which belong to the same sense, seem to be more easily combined into synchronous tribes, than those which were not received by the same sense, as we can more easily think of the whiteness and figure of a lump of sugar at the same time, than the whiteness and sweetness of it.

2. As these ideas, or sensual motions, are thus excited with greater or less degrees of combination; so we have a power, when we repeat them either by our volition or sensation, to increase or diminish this degree of combination, that is, to form compounded ideas from those, which were more simple; and abstract ones from those, which were more complex, when they were first

excited; that is, we can repeat a part or the whole of those sensual motions, which did constitute our ideas of perception; and the repetition of which now constitutes our ideas of recollection, or of imagination.

3. Those ideas, which we repeat without change of the quantity of that combination, with which we first received them, are called complex ideas, as when you recollect Westminster Abbey, or the planet Saturn: but it must be observed, that these complex ideas, thus re-excited by volition, sensation, or association, are seldom perfect copies of their correspondent perceptions, except in our dreams, where other external objects do not detract our attention.

4. Those ideas, which are more complex than the natural objects that first excited them, have been called compounded ideas, as when we think of a sphinx, or griffin.

5. And those that are less complex than the correspondent natural objects, have been termed abstracted ideas: thus sweetness, and whiteness, and solidity, are received at the same time from a lump of sugar, yet I can recollect any of these qualities without thinking of the others, that were excited along with them.

When ideas are so far abstracted as in the above example, they have been termed simple by the writers of metaphysics, and seem indeed to be more complete repetitions of the ideas or sensual motions, originally excited by external objects.

Other classes of these ideas, where the abstraction has not been so great, have been termed, by Mr. Locke, modes, substances, and relations, but they seem only to differ in their degree of abstraction from the complex ideas that were at first excited; for as these complex or natural ideas are themselves imperfect copies of their correspondent perceptions, so these abstract or general ideas are only still more imperfect copies of the same perceptions. Thus when I have seen an object but once, as a rhinoceros, my abstract idea of this animal is the same as my complex one. I may think more or less distinctly of a rhinoceros, but it is the very rhinoceros that I saw, or some part or property of him, which recurs to my mind.

But when any class of complex objects becomes the subject of conversation, of which I have seen many individuals, as a castle or an army, some property or circumstance belonging to it is peculiarly alluded to; and then I feel in my own mind, that my abstract idea of this complex object is only an idea of that part, property, or attitude of it, that employs the present conversation, and varies with every sentence that is spoken concerning it. So if any one should say, "one may sit upon a horse safer than on a camel," my abstract idea of the two animals includes only an outline of the level back of the one, and the gibbosity on the back of the other. What noise is that in the street?—Some horses trotting over the pavement. Here my idea of the horses includes principally the shape and motion of their legs. So also the abstract ideas of goodness and courage are still more imperfect representations of the objects they were received from; for here we abstract the material parts, and recollect only the qualities.

Thus we abstract so much from some of our complex ideas, that at length it becomes difficult to determine of what perception they partake; and in many instances our idea seems to be no other than of the sound or letters of the word, that stands for the collective tribe, of which we are said to have an abstracted idea, as noun, verb, chimæra, apparition.

6. Ideas have been divided into those of perception and those of reflection, but as whatever is perceived must be external to the organ that perceives it, all our ideas must originally be ideas of perception.

7. Others have divided our ideas into those of memory, and those of imagination; they have said that a recollection of ideas in the order they were received constitutes memory, and without that order imagination; but all the ideas of imagination, excepting the few that are termed simple ideas, are parts of trains or tribes in the order they were received; as if I think of a sphinx, or a griffin, the fair face, bosom, wings, claws, tail, are all complex ideas in the order they were received: and it behoves the writers, who adhere to this definition, to determine, how small the trains must be, that shall be called imagination; and how great those, that shall be called memory.

Others have thought that the ideas of memory have a greater vivacity than those of imagination: but the ideas of a person in sleep, or in a waking reverie, where the trains connected with sensation are uninterrupted, are more vivid and distinct than those of memory, so that they cannot be distinguished by this criterion.

The very ingenious author of the Elements of Criticism has described what he conceives to be a species of memory, and calls it ideal presence; but the instances he produces are the reveries of sensation, and are therefore in truth connections of the imagination, though they are recalled in the order they were received.

The ideas connected by association are in common discourse attributed to memory, as we talk of memorandum-rings, and tie a knot on our handkerchiefs to bring something into our minds at a distance of time. And a school-boy, who can repeat a thousand unmeaning lines in Lilly's Grammar, is said to have a good memory. But these have been already shewn to belong to the class of association; and are termed ideas of suggestion.

II. Lastly, the method already explained of classing ideas into those excited by irritation, sensation, volition, or association, we hope will be found more convenient both for explaining the operations of the mind, and for comparing them with those of the body; and for the illustration and the cure of the diseases of both, and which we shall here recapitulate.

1. Irritative ideas are those, which are preceded by irritation, which is excited by objects external to the organs of sense: as the idea of that tree, which either I attend to, or which I shun in walking near it without attention. In the former case it is termed perception, in the latter it is termed simply an irritative idea.

2. Sensitive ideas are those, which are preceded by the sensation of pleasure or pain; as the ideas, which constitute our dreams or reveries, this is called imagination.

3. Voluntary ideas are those, which are preceded by voluntary exertion, as when I repeat the alphabet backwards: this is called recollection.

4. Associate ideas are those, which are preceded by other ideas or muscular motions, as when we think over or repeat the alphabet by rote in its usual order; or sing a tune we are accustomed to; this is called suggestion.

III. 1. Perceptions signify those ideas, which are preceded by irritation and succeeded by the sensation of pleasure or pain, for whatever excites our attention interests us; that is, it is accompanied with, pleasure or pain; however slight may be the degree or quantity of either of them.

The word memory includes two classes of ideas, either those which, are preceded by voluntary exertion, or those which are suggested by their associations with other ideas.

2. Reasoning is that operation of the sensorium, by which we excite two or many tribes of ideas; and then re-excite the ideas, in which they differ, or correspond. If we determine this difference, it is called judgment; if we in vain endeavour to determine it, it is called doubting.

If we re-excited the ideas, in which they differ, it is called distinguishing. If we re-excite those in which they correspond, it is called comparing.

3. Invention is an operation of the sensorium, by which we voluntarily continue to excite one train of ideas, suppose the design of raising water by a machine; and at the same time attend to all other ideas, which are connected with this by every kind of catenation; and combine or separate them voluntarily for the purpose of obtaining some end.

For we can create nothing new, we can only combine or separate the ideas, which we have already received by our perceptions: thus if I wish to represent a monster, I call to my mind the ideas of every thing disagreeable and horrible, and combine the nastiness and gluttony of a hog, the stupidity and obstinacy of an ass, with the fur and awkwardness of a bear, and call the new combination Caliban. Yet such a monster may exist in nature, as all his attributes are parts of nature. So when I wish to represent every thing, that is excellent, and amiable; when I combine benevolence with cheerfulness, wisdom, knowledge, taste, wit, beauty of person, and elegance of manners, and associate them in one lady as a pattern to the world, it is called invention; yet such a person may exist,—such a person does exist!—It is — — — —, who is as much a monster as Caliban.

4. In respect to consciousness, we are only conscious of our existence, when we think about it; as we only perceive the lapse of time, when we attend to it; when we are busied about other objects, neither the lapse of time nor the consciousness of our own existence can occupy our attention. Hence, when we think of our own existence, we only excite abstracted or reflex ideas (as they are termed), of our principal pleasures or pains, of our desires or aversions, or of the figure, solidity, colour, or other properties of our bodies, and call that act of the sensorium a consciousness of our existence. Some philosopher, I believe it is Des Cartes, has said, "I think, therefore I exist." But this is not right reasoning, because thinking is a mode of existence; and it is thence only saying, "I exist, therefore I exist." For there are three modes of existence, or in the language of grammarians three kinds of verbs. First, simply I am, or exist. Secondly, I am acting, or exist in a state of activity, as I move. Thirdly, I am suffering, or exist in a state of being acted upon, as I am moved. The when, and the where, as applicable to this existence, depends on the successive motions of our own or of other bodies; and on their respective situations, as spoken of Sect. XIV. 2. 5.

5. Our identity is known by our acquired habits or catenated trains of ideas and muscular motions; and perhaps, when we compare infancy with old age, in those alone can our identity be supposed to exist. For what else is there of similitude between the first speck of living entity and the mature man?—every deduction of reasoning, every sentiment or passion, with every fibre of the corporeal part of our system, has been subject almost to annual mutation; while some catenations alone of our ideas and muscular actions have continued in part unchanged.

By the facility, with which we can in our waking hours voluntarily produce certain successive trains of ideas, we know by experience, that we have before reproduced them; that is, we are conscious of a time of our existence previous to the present time; that is, of our identity now and heretofore. It is these habits of action, these catenations of ideas and muscular motions, which begin with life, and only terminate with it; and which we can in some measure deliver to our posterity; as explained in Sect. XXXIX.

6. When the progressive motions of external bodies make a part of our present catenation of ideas, we attend to the lapse of time; which appears the longer, the more frequently we thus attend to it; as when we expect something at a certain hour, which much interests us, whether it be an agreeable or disagreeable event; or when we count the passing seconds on a stop-watch.

When an idea of our own person, or a reflex idea of our pleasures and pains, desires and aversions, makes a part of this catenation, it is termed consciousness; and if this idea of consciousness makes a part of a catenation, which we excite by recollection, and know by the facility with which we excite it, that we have before experienced it, it is called identity, as explained above.

7. In respect to freewill, it is certain, that we cannot will to think of a new train of ideas, without previously thinking of the first link of it; as I cannot will to think of a black swan, without previously thinking of a black swan. But if I now think of a tail, I can voluntarily recollect all animals, which have tails; my will is so far free, that I can pursue the ideas linked to this idea of tail, as far as my knowledge of the subject extends; but to will without motive is to will without desire or aversion; which is as absurd as to feel without pleasure or pain; they are both solecisms in the terms. So far are we governed by the catenations of motions, which affect both the body and the mind of man, and which begin with our irritability, and end with it.

SECT. XVI
OF INSTINCT

Haud equidem credo, quia sit divinitus illis

Ingenium, aut rerum fato prudentia major.—Virg. Georg. L. I. 415.

I. Instinctive actions defined. Of connate passions. II. Of the sensations and motions of the fœtus in the womb. III. Some animals are more perfectly formed than others before nativity. Of learning to walk. IV. Of the swallowing, breathing, sucking, pecking, and lapping of young animals. V. Of the sense of smell, and its uses to animals. Why cats do not eat their kittens. VI. Of the accuracy of sight in mankind, and their sense of beauty. Of the sense of touch in elephants, monkies, beavers, men. VII. Of natural language. VIII. The origin of natural language; 1. the language of fear; 2. of grief; 3. of tender pleasure; 4. of serene pleasure; 5. of anger; 6. of attention. IX. Artificial language of turkies, hens, ducklings, wagtails, cuckoos, rabbits, dogs, and nightingales. X. Of music; of tooth-edge; of a good ear; of architecture. XI. Of acquired knowledge; of foxes, rooks, fieldfares, lapwings, dogs, cats, horses, crows, and pelicans. XII. Of birds of passage, dormice, snakes, bats, swallows, quails, ringdoves, stare, chaffinch, hoopoe, chatterer, hawfinch, crossbill, rails and cranes. XIII. Of birds nests; of the cuckoo; of swallows nests; of the taylor bird. XIV. Of the old soldier; of haddocks, cods, and dog fish; of the remora; of crabs, herrings, and salmon. XV. Of spiders, caterpillars, ants, and the ichneumon. XVI. 1. Of locusts, gnats; 2. bees; 3. dormice, flies, worms, ants, and wasps. XVII. Of the faculty that distinguishes man from the brutes.

I. All those internal motions of animal bodies, which contribute to digest their aliment, produce their secretions, repair their injuries, or increase their growth, are performed without our attention or consciousness. They exist as well in our sleep, as in our waking hours, as well in the fœtus during the time of gestation, as in the infant after nativity, and proceed with equal regularity in the vegetable as in the animal system. These motions have been shewn in a former part of this work to depend on the irritations of peculiar fluids, and as they have never been classed amongst the instinctive actions of animals, are precluded from our present disquisition.

But all those actions of men or animals, that are attended with consciousness, and seem neither to have been directed by their appetites, taught by their experience, nor deduced from observation or tradition, have been referred to the power of instinct. And this power has been explained to be a *divine something*, a kind of inspiration; whilst the poor animal, that possesses it, has been thought little better than *a machine*!

The *irksomeness*, that attends a continued attitude of the body, or the *pains*, that we receive from heat, cold, hunger, or other injurious circumstances, excite us to *general locomotion*: and our senses are so formed and constituted by the hand of nature, that certain objects present us with pleasure, others with pain, and we are induced to approach and embrace these, to avoid and abhor those, as such sensations direct us.

Thus the palates of some animals are gratefully affected by the mastication of fruits, others of grains, and others of flesh; and they are thence instigated to attain, and to consume those materials; and are furnished with powers of muscular motion, and of digestion proper for such purposes.

These *sensations* and *desires* constitute a part of our system, as our *muscles* and *bones* constitute another part: and hence they may alike be termed *natural* or *connate*; but neither of them can properly be termed *instinctive*: as the word instinct in its usual acceptation refers only to the *actions* of animals, as above explained: the origin of these *actions* is the subject of our present enquiry.

The reader is intreated carefully to attend to this definition of *instinctive actions*, lest by using the word instinct without adjoining any accurate idea to it, he may not only include the natural desires of love and hunger, and the natural sensations of pain or pleasure, but the figure and contexture of the body, and the faculty of reason itself under this general term.

II. We experience some sensations, and perform some actions before our nativity; the sensations of cold and warmth, agitation and rest, fulness and inanition, are instances of the former; and the repeated struggles of the limbs of the fœtus, which begin about the middle of gestation, and those motions by which it frequently wraps the umbilical chord around its neck or body, and even sometimes ties it on a knot; are instances of the latter. Smellie's Midwifery, (Vol. I. p. 182.)

By a due attention to these circumstances many of the actions of young animals, which at first sight seemed only referable to an inexplicable instinct, will appear to have been acquired like all other animal actions, that are attended with consciousness, *by the repeated efforts of our muscles under the conduct of our sensations or desires.*

The chick in the shell begins to move its feet and legs on the sixth day of incubation (Mattreican, p. 138); or on the seventh day, (Langley); afterwards they are seen to move themselves gently in the liquid that surrounds them, and to open and shut their mouths, (Harvei, de Generat. p. 62, and 197. Form de Poulet. ii. p. 129). Puppies before the membranes are broken, that involve them, are seen to move themselves, to put out their tongues, and to open and shut their mouths, (Harvey, Gipson, Riolan, Haller). And calves lick themselves and swallow many of their hairs before their nativity: which however puppies do not, (Swammerden, p. 319. Flemyng Phil. Trans. Ann. 1755. 42). And towards the end of gestation, the fœtus of all animals are proved to drink part of the liquid in which they swim, (Haller. Physiol. T. 8. 204). The white of egg is found in the mouth and gizzard of the chick, and is nearly or quite consumed before it is hatched, (Harvie de Generat. 58). And the liquor amnii is found in the mouth and stomach of the human fœtus, and of calves; and how else should that excrement be produced in the intestines of all animals, which is voided in great quantity soon after their birth; (Gipson, Med. Essays, Edinb. V. i. 13. Halleri Physiolog. T. 3. p. 318. and T. 8). In the stomach of a calf the quantity of this liquid amounted to about three pints, and the hairs amongst it were of the same colour with those on its skin, (Blasii Anat. Animal, p.m. 122). These facts are attested by many other writers of credit, besides those above mentioned.

III. It has been deemed a surprising instance of instinct, that calves and chickens should be able to walk by a few efforts almost immediately after their nativity: whilst the human infant in those countries where he is not incumbered with clothes, as in India, is five or six months, and in our climate almost a twelvemonth, before he can safely stand upon his feet.

The struggles of all animals in the womb must resemble their mode of swimming, as by this kind of motion they can best change their attitude in water. But the swimming of the calf and chicken resembles their manner of walking, which they have thus in part acquired before their nativity, and hence accomplish it afterwards with very few efforts, whilst the swimming of the human creature resembles that of the frog, and totally differs from his mode of walking.

There is another circumstance to be attended to in this affair, that not only the growth of those peculiar parts of animals, which are first wanted to secure their subsistence, are in general furthest advanced before their nativity: but some animals come into the world more completely formed throughout their whole system than others: and are thence much forwarder in all their habits of motion. Thus the colt, and the lamb, are much more perfect animals than the blind puppy, and the naked rabbit; and the chick of the pheasant, and the partridge, has more perfect plumage, and more perfect eyes, as well as greater aptitude to locomotion, than the callow nestlings of the dove, and of the wren. The parents of the former only find it necessary to shew them their food, and to teach them to take it up; whilst those of the latter are obliged for many days to obtrude it into their gaping mouths.

IV. From the facts mentioned in No. 2. of this Section, it is evinced that the fœtus learns to swallow before its nativity; for it is seen to open its mouth, and its stomach is found filled with the liquid that surrounds it. It opens its mouth, either instigated by hunger, or by the irksomeness of a continued attitude of the muscles of its face; the liquor amnii, in which it swims, is agreeable to its palate, as it consists of a nourishing material, (Haller Phys. T. 8. p. 204). It is tempted to experience its taste further in the mouth, and by a few efforts learns to swallow, in the same manner as we learn all other animal actions, which are attended with consciousness, *by the repeated efforts of our muscles under the conduct of our sensations or volitions.*

The inspiration of air into the lungs is so totally different from that of swallowing a fluid in which we are immersed, that it cannot be acquired before our nativity. But at this time, when the circulation of the blood is no longer continued through the placenta, that suffocating sensation, which we feel about the precordia, when we are in want of fresh air, disagreeably affects the infant: and all the muscles of the body are excited into action to relieve this oppression; those of the breast, ribs, and diaphragm are found to answer this purpose, and thus respiration is discovered, and is continued throughout our lives, as often as the oppression begins to recur. Many infants, both of the human creature, and of quadrupeds, struggle for a minute after they are born before they begin to breathe, (Haller Phys. T. 8. p. 400. ib pt. 2. p. 1). Mr. Buffon thinks the action of the dry air upon the nerves of smell of new-born animals, by producing an endeavour to sneeze, may contribute to induce this first inspiration, and that the rarefaction of the air by the warmth of the lungs contributes to induce expiration, (Hist. Nat. Tom. 4. p. 174). Which latter it may effect by producing a disagreeable sensation by its delay, and a consequent effort to relieve it. Many children sneeze before they respire, but not all, as far as I have observed, or can learn from others.

At length, by the direction of its sense of smell, or by the officious care of its mother, the young animal approaches the odoriferous rill of its future nourishment, already experienced to swallow. But in the act of swallowing, it is necessary nearly to close the mouth, whether the creature be immersed in the fluid it is about to drink, or not: hence, when the child first attempts to suck, it does not slightly compress the nipple between its lips, and suck as an adult person would do, by absorbing the milk; but it takes the whole nipple into its mouth for this purpose, compresses it between its gums, and thus repeatedly chewing (as it were) the nipple, presses out the milk, exactly in the same manner as it is drawn from the teats of cows by the hands of the milkmaid. The celebrated Harvey observes, that the fœtus in the womb must have sucked in a part of its nourishment, because it knows how to suck the minute it is born, as any one may experience by putting a finger between its lips, and because in a few days it forgets this art of sucking, and cannot without some difficulty again acquire it, (Exercit. de Gener. Anim. 48). The same observation is made by Hippocrates.

A little further experience teaches the young animal to suck by absorption, as well as by compression; that is, to open the chest as in the beginning of respiration, and thus to rarefy the air in the mouth, that the pressure of the denser external atmosphere may contribute to force out the milk.

The chick yet in the shell has learnt to drink by swallowing a part of the white of the egg for its food; but not having experienced how to take up and swallow solid seeds, or grains, is either taught by the felicitous industry of its mother; or by many repeated attempts is enabled at length to distinguish and to swallow this kind of nutriment.

And puppies, though they know how to suck like other animals from their previous experience in swallowing, and in respiration; yet are they long in acquiring the art of lapping with their tongues, which from the flaccidity of their cheeks, and length of their mouths, is afterwards a more convenient way for them to take in water.

V. The senses of smell and taste in many other animals greatly excel those of mankind, for in civilized society, as our victuals are generally prepared by others, and are adulterated with salt, spice, oil, and empyreuma, we do not hesitate about eating whatever is set before us, and neglect to cultivate these senses: whereas other animals try every morsel by the smell, before they take it into their mouths, and by the taste before they swallow it: and are led not only each to his proper nourishment by this organ of sense, but it also at a maturer age directs them in the gratification of their appetite of love. Which may be further understood by considering the sympathies of these parts described in Class IV. 2. 1. 7. While the human animal is directed to the object of his love by his sense of beauty, as mentioned in No. VI. of this Section. Thus Virgil. Georg. III. 250.

> Nonne vides, ut tota tremor pertentat equorum
>
> Corpora, si tantum notas odor attulit auras?
>
> Nonne canis nidum veneris nasutus odore
>
> Quærit, et erranti trahitur sublambere linguâ?
>
> Respuit at gustum cupidus, labiisque retractis
>
> Elevat os, trepidansque novis impellitur æstris
>
> Inserit et vivum felici vomere semen. —
>
> Quam tenui filo cæcos adnectit amores
>
> Docta Venus, vitæque monet renovare favillam! — ANON.

The following curious experiment is related by Galen. "On dissecting a goat great with young I found a brisk embryon, and having detached it from the matrix, and snatching it away before it saw its dam, I brought it into a certain room, where there were many vessels, some filled with wine, others with oil, some with honey, others with milk, or some other liquor; and in others were grains and fruits; we first observed the young animal get upon its feet, and walk; then it shook itself, and afterwards scratched its side with one of its feet: then we saw it smelling to every one of these things, that were set in the room; and when it had smelt to them all, it drank up the milk." L. 6. de locis. cap. 6.

Parturient quadrupeds, as cats, and bitches, and sows, are led by their sense of smell to eat the placenta as other common food; why then do they not devour their whole progeny, as is represented in an antient emblem of TIME? This is said sometimes to happen in the unnatural state in which we confine sows; and indeed nature would seem to have endangered her offspring in this nice circumstance! But at this time the stimulus of the milk in the tumid teats of the mother excites her to look out for, and to desire some unknown circumstance to relieve her. At the same time the smell of the milk attracts the exertions of the young animals towards its source, and thus the delighted mother discovers a new appetite, as mentioned in Sect. XIV. 8. and her little progeny are led to receive and to communicate pleasure by this most beautiful contrivance.

VI. But though the human species in some of their sensations are much inferior to other animals, yet the accuracy of the sense of touch, which they possess in so eminent a degree, gives them a great superiority of understanding; as is well observed by the ingenious Mr. Buffon. The extremities of other animals terminate in horns, and hoofs, and claws, very unfit for the sensation of touch; whilst the human hand is finely adapted to encompass its object with this organ of sense.

The elephant is indeed endued with a fine sense of feeling at the extremity of his proboscis, and hence has acquired much more accurate ideas of touch and of sight than most other creatures. The two following instances of the sagacity of these animals may entertain the reader, as they were told me by some gentlemen of distinct observation, and undoubted veracity, who had been much conversant with our eastern settlements. First, the elephants that are used to carry the baggage of our armies, are put each under the care of one of the natives of Indostan, and whilst himself and his wife go into the woods to collect leaves and branches of trees for his food, they fix him to the ground by a length of chain, and frequently leave a child yet unable to walk, under his protection: and the intelligent animal not only defends it, but as it creeps about, when it arrives near the extremity of his chain, he wraps his trunk gently round its body, and brings it again into the centre of his circle. Secondly, the traitor elephants are taught to walk on a narrow path between two pit-falls, which are covered with turf, and then to go into the woods, and to seduce the wild elephants to come that way, who fall into these wells, whilst he passes safe between them: and it is universally observed, that those wild elephants that escape the snare, pursue the traitor with the utmost vehemence, and if they can overtake him, which sometimes happens, they always beat him to death.

The monkey has a hand well enough adapted for the sense of touch, which contributes to his great facility of imitation; but in taking objects with his hands, as a stick or an apple, he puts his thumb on the same side of them with his fingers, instead of counteracting the pressure of his fingers with it: from this neglect he is much slower in acquiring the figures of objects, as he is less able to determine the distances or diameters of their parts, or to distinguish their vis inertiæ from their hardness. Helvetius adds, that the shortness of his life, his being fugitive before mankind, and his not inhabiting all climates, combine to prevent his improvement. (De l'Esprit. T. 1. p.) There is however at this time an old monkey shewn in Exeter Change, London, who having lost his teeth, when nuts are given him, takes a stone into his hand, and cracks them with it one by one; thus using tools to effect his purpose like mankind.

The beaver is another animal that makes much use of his hands, and if we may credit the reports of travellers, is possessed of amazing ingenuity. This however, M. Buffon affirms, is only where they exist in large numbers, and in countries thinly peopled with men; while in France in their solitary state they shew no uncommon ingenuity.

Indeed all the quadrupeds, that have collar-bones, (claviculæ) use their fore-limbs in some measure as we use our hands, as the cat, squirrel, tyger, bear and lion; and as they exercise the sense of touch more universally than other animals, so are they more sagacious in watching and surprising their prey. All those birds, that use their claws for hands, as the hawk, parrot, and cuckoo, appear to be more docile and intelligent; though the gregarious tribes of birds have more acquired knowledge.

Now as the images, that are painted on the retina of the eye, are no other than signs, which recall to our imaginations the objects we had before examined by the organ of touch, as is fully demonstrated by Dr. Berkley in his treatise on vision; it follows that the human creature has greatly more accurate and distinct sense of vision than that of any other animal. Whence as he advances to maturity he gradually acquires a sense of female beauty, which at this time directs him to the object of his new passion.

Sentimental love, as distinguished from the animal passion of that name, with which it is frequently accompanied, consists in the desire or sensation of beholding, embracing, and saluting a beautiful object.

The characteristic of beauty therefore is that it is the object of love; and though many other objects are in common language called beautiful, yet they are only called so metaphorically, and ought to be termed agreeable. A Grecian temple may give us the pleasurable idea of sublimity, a Gothic temple may give us the pleasurable idea of variety, and a modern house the pleasurable idea of utility; music and poetry may inspire our love by association of ideas; but none of these, except metaphorically, can be termed beautiful; as we have no wish to embrace or salute them.

Our perception of beauty consists in our recognition by the sense of vision of those objects, first, which have before inspired our love by the pleasure, which they have afforded to many of our senses: as to our sense of warmth, of touch, of smell, of taste, hunger and thirst; and, secondly, which bear any analogy of form to such objects.

When the babe, soon after it is born into this cold world, is applied to its mother's bosom; its sense of perceiving warmth is first agreeably affected; next its sense of smell is delighted with the odour of her milk; then its taste is gratified by the flavour of it: afterwards the appetites of hunger and of thirst afford pleasure by the possession of their objects, and by the subsequent digestion of the aliment; and, lastly, the sense of touch is delighted by the softness and smoothness of the milky fountain, the source of such variety of happiness.

All these various kinds of pleasure at length become associated with the form of the mother's breast; which the infant embraces with its hands, presses with its lips, and watches with its eyes; and thus acquires more accurate ideas of the form of its mother's bosom, than of the odour and flavour or warmth, which it perceives by its other senses. And hence at our maturer years, when any object of vision is presented to us, which by its waving or spiral lines bears any similitude to the form of the female bosom, whether it be found in a landscape with soft gradations of rising and descending surface, or in the forms of some antique vases, or in other works of the pencil or the chissel, we feel a general glow of delight, which seems to influence all our senses; and, if the object be not too large, we experience an attraction to embrace it with our arms, and to salute it with our lips, as we did in our early infancy the bosom of our mother. And thus we find, according to the ingenious idea of Hogarth, that the waving lines of beauty were originally taken from the temple of Venus.

This animal attraction is love; which is a sensation, when the object is present; and a desire, when it is absent. Which constitutes the purest source of human felicity, the cordial drop in the otherwise vapid cup of life, and which overpays mankind for the care and labour, which are attached to the pre-eminence of his situation above other animals.

It should have been observed, that colour as well as form sometimes enters into our idea of a beautiful object, as a good complexion for instance, because a fine or fair colour is in general a sign of health, and conveys to us an idea of the warmth of the object; and a pale countenance on the contrary gives an idea of its being cold to the touch.

It was before remarked, that young animals use their lips to distinguish the forms of things, as well as their fingers, and hence we learn the origin of our inclination to salute beautiful objects with our lips. For a definition of Grace, see Class III. 1. 2. 4.

VII. There are two ways by which we become acquainted with the passions of others: first, by having observed the effects of them, as of fear or anger, on our own bodies, we know at sight when others are under the influence of these affections. So when two cocks are preparing to fight, each feels the feathers rise round his own neck, and knows from the same sign the disposition of his adversary: and children long before they can speak, or understand the language of their parents, may be frightened by an angry countenance, or soothed by smiles and blandishments.

Secondly, when we put ourselves into the attitude that any passion naturally occasions, we soon in some degree acquire that passion; hence when those that scold indulge themselves in loud oaths, and violent actions of the arms, they increase their anger by the mode of expressing themselves: and on the contrary the counterfeited smile of pleasure in disagreeable company soon brings along with it a portion of the reality, as is well illustrated by Mr. Burke. (Essay on the Sublime and Beautiful.)

This latter method of entering into the passions of others is rendered of very extensive use by the pleasure we take in imitation, which is every day presented before our eyes, in the actions of children, and indeed in all the customs and fashions of the world. From this our aptitude to imitation, arises what is generally understood by the word sympathy so well explained by Dr. Smith of Glasgow. Thus the appearance of a cheerful countenance gives us pleasure, and of a melancholy one makes us sorrowful. Yawning and sometimes vomiting are thus propagated by sympathy, and some people of delicate fibres, at the presence of a spectacle of misery, have felt pain in the same parts of their own bodies, that were diseased or mangled in the other. Amongst the writers of antiquity Aristotle thought this aptitude to imitation an essential property of the human species, and calls man an imitative animal. Το ζωον μιμωμενον.

These then are the natural signs by which we understand each other, and on this slender basis is built all human language. For without some natural signs, no artificial ones could have been invented or understood, as is very ingeniously observed by Dr. Reid. (Inquiry into the Human Mind.)

VIII. The origin of this universal language is a subject of the highest curiosity, the knowledge of which has always been thought utterly inaccessible. A part of which we shall however here attempt.

Light, sound, and odours, are unknown to the fœtus in the womb, which, except the few sensations and motions already mentioned, sleeps away its time insensible of the busy world. But the moment he arrives into day, he begins to experience many vivid pains and pleasures; these are at the same time attended with certain muscular motions, and from this their early, and individual association, they acquire habits of occurring together, that are afterwards indissoluble.

1. Of Fear.

As soon as the young animal is born, the first important sensations, that occur to him, are occasioned by the oppression about his precordia for want of respiration, and by his sudden transition from ninety-eight degrees of heat into so cold a climate.—He trembles, that is, he exerts alternately all the muscles of his body, to enfranchise himself from the oppression about his bosom, and begins to breathe with frequent and short respirations; at the same time the cold contracts his red skin, gradually turning it pale; the contents of the bladder and of the bowels are evacuated: and from the experience of these first disagreeable sensations the passion of fear is excited, which is no other than the expectation of disagreeable sensations. This early association of motions and sensations persists throughout life; the passion of fear produces a cold and pale skin, with tremblings, quick respiration, and an evacuation of the bladder and bowels, and thus constitutes the natural or universal language of this passion.

On observing a Canary bird this morning, January 28, 1772, at the house of Mr. Harvey, near Tutbury, in Derbyshire, I was told it always fainted away, when its cage was cleaned, and desired to see the experiment. The cage being taken from the ceiling, and its bottom drawn out, the bird began to tremble, and turned quite white about the root of his bill: he then opened his mouth as if for breath, and respired quick, stood straighter up on his perch, hung his wings, spread his tail, closed his eyes, and appeared quite stiff and cataleptic for near half an hour, and at length with much trembling and deep respirations came gradually to himself.

2. *Of Grief.*

That the internal membrane of the nostrils may be kept always moist, for the better perception of odours, there are two canals, that conduct the tears after they have done their office in moistening and cleaning the ball of the eye into a sack, which is called the lacrymal sack; and from which there is a duct, that opens into the nostrils: the aperture of this duct is formed of exquisite sensibility, and when it is stimulated by odorous particles, or by the dryness or coldness of the air, the sack contracts itself, and pours more of its contained moisture on the organ of smell. By this contrivance the organ is rendered more fit for perceiving such odours, and is preserved from being injured by those that are more strong or corrosive. Many other receptacles of peculiar fluids disgorge their contents, when the ends of their ducts are stimulated; as the gall bladder, when the contents of the duodenum stimulate the extremity of the common bile duct: and the salivary glands, when the termination of their ducts in the mouth are excited by the stimulus of the food we masticate. Atque vesiculæ seminales suum exprimunt fluidum glande penis fricatâ.

The coldness and dryness of the atmosphere, compared with the warmth and moisture, which the new-born infant had just before experienced, disagreeably affects the aperture of this lacrymal sack: the tears, that are contained in this sack, are poured into the nostrils, and a further supply is secreted by the lacrymal glands, and diffused upon the eye-balls; as is very visible in the eyes and nostrils of children soon after their nativity. The same happens to us at our maturer age, for in severe frosty weather, snivelling and tears are produced by the coldness and dryness of the air.

But the lacrymal glands, which separate the tears from the blood, are situated on the upper external part of the globes of each eye; and, when a greater quantity of tears are wanted, we contract the forehead, and bring down the eye-brows, and use many other distortions of the face, to compress these glands.

Now as the suffocating sensation, that produces respiration, is removed almost as soon as perceived, and does not recur again: this disagreeable irritation of the lacrymal ducts, as it must frequently recur, till the tender organ becomes used to variety of odours, is one of the first pains that is repeatedly attended to: and hence throughout our infancy, and in many people throughout their lives, all disagreeable sensations are attended with snivelling at the nose, a profusion of tears, and some peculiar distortions of countenance: according to the laws of early association before mentioned, which constitutes the natural or universal language of grief.

You may assure yourself of the truth of this observation, if you will attend to what passes, when you read a distressful tale alone; before the tears overflow your eyes, you will invariably feel a titillation at that extremity of the lacrymal duct, which terminates in the nostril, then the compression of the eyes succeeds, and the profusion of tears.

Linnæus asserts, that the female bear sheds tears in grief; the same has been said of the hind, and some other animals.

3. *Of Tender Pleasure.*

The first most lively impression of pleasure, that the infant enjoys after its nativity, is excited by the odour of its mother's milk. The organ of smell is irritated by this perfume, and the lacrymal sack empties itself into the nostrils, as before explained, and an increase of tears is poured into the eyes. Any one may observe this, when very young infants are about to suck; for at those early periods of life, the sensation affects the organ of smell, much more powerfully, than after the repeated habits of smelling has inured it to odours of common strength: and in our adult years, the stronger smells, though they are at the same time agreeable to us, as of volatile spirits, continue to produce an increased secretion of tears.

This pleasing sensation of smell is followed by the early affection of the infant to the mother that suckles it, and hence the tender feelings of gratitude and love, as well as of hopeless grief, are ever after joined with the titillation of the extremity of the lacrymal ducts, and a profusion of tears.

Nor is it singular, that the lacrymal sack should be influenced by pleasing ideas, as the sight of agreeable food produces the same effect on the salivary glands. Ac dum vidimus insomniis lascivæ puellæ simulacrum tenditur penis.

Lambs shake or wriggle their tails, at the time when they first suck, to get free of the hard excrement, which had been long lodged in their bowels. Hence this becomes afterwards a mark of pleasure in them, and in dogs, and other tailed animals. But cats gently extend and contract their paws when they are pleased, and purr by drawing in their breath, both which resemble their manner of sucking, and thus become their language of pleasure, for these animals having collar-bones use their paws like hands when they suck, which dogs and sheep do not.

4. *Of Serene Pleasure.*

In the action of sucking, the lips of the infant are closed around the nipple of its mother, till he has filled his stomach, and the pleasure occasioned by the stimulus of this grateful food succeeds. Then the sphincter of the mouth, fatigued by the continued action of sucking, is relaxed; and the antagonist muscles of the face gently acting, produce the smile of pleasure: as cannot but be seen by all who are conversant with children.

Hence this smile during our lives is associated with gentle pleasure; it is visible in kittens, and puppies, when they are played with, and tickled; but more particularly marks the human features. For in children this expression of pleasure is much encouraged, by their imitation of their parents, or friends; who generally address them with a smiling countenance: and hence some nations are more remarkable for the gaiety, and others for the gravity of their looks.

5. *Of Anger.*

The actions that constitute the mode of fighting, are the immediate language of anger in all animals; and a preparation for these actions is the natural language of threatening. Hence the human creature clenches his fist, and sternly surveys his adversary, as if meditating where to make the attack; the ram, and the bull, draws himself some steps backwards, and levels his horns; and the horse, as he most frequently fights by striking with his hinder feet, turns his heels to his foe, and bends back his ears, to listen out the place of his adversary, that the threatened blow may not be ineffectual.

6. *Of Attention.*

The eye takes in at once but half our horizon, and that only in the day, and our smell informs us of no very distant objects, hence we confide principally in the organ of hearing to apprize us of danger: when we hear any the smallest sound, that we cannot immediately account for, our fears are alarmed, we suspend our steps, hold every muscle still, open our mouths a little, erect our ears, and listen to gain further information: and this by habit becomes the general language of attention to objects of sight, as well as of hearing; and even to the successive trains of our ideas.

The natural language of violent pain, which is expressed by writhing the body, grinning, and screaming; and that of tumultuous pleasure, expressed in loud laughter; belong to Section XXXIV. on Diseases from Volition.

IX. It must have already appeared to the reader, that all other animals, as well as man, are possessed of this natural language of the passions, expressed in signs or tones; and we shall endeavour to evince, that those animals, which have preserved themselves from being enslaved by mankind, and are associated in flocks, are also possessed of some artificial language, and of some traditional knowledge.

The mother-turkey, when she eyes a kite hovering high in air, has either seen her own parents thrown into fear at his presence, or has by observation been acquainted with his dangerous designs upon her young. She becomes agitated with fear, and uses the natural language of that passion, her young ones catch the fear by imitation, and in an instant conceal themselves in the grass.

At the same time that she shews her fears by her gesture and deportment, she uses a certain exclamation, Koe-ut, Koe-ut, and the young ones afterwards know, when they hear this note, though they do not see their dam, that the presence of their adversary is denounced, and hide themselves as before.

The wild tribes of birds have very frequent opportunities of knowing their enemies, by observing the destruction they make among their progeny, of which every year but a small part escapes to maturity: but to our domestic birds these opportunities so rarely occur, that their knowledge of their distant enemies must frequently be delivered by tradition in the manner above explained, through many generations.

This note of danger, as well as the other notes of the mother-turkey, when she calls her flock to their food, or to sleep under her wings, appears to be an artificial language, both as expressed by the mother, and as understood by the progeny. For a hen teaches this language with equal ease to the ducklings, she has hatched from supposititious eggs, and educates as her own offspring: and the wagtails, or hedge-sparrows, learn it from the young cuckoo their softer nursling, and supply him with food long after he can fly about, whenever they hear his cuckooing, which Linnæus tells us, is his call of hunger, (Syst. Nat.) And all our domestic animals are readily taught to come to us for food, when we use one tone of voice, and to fly from our anger, when we use another.

Rabbits, as they cannot easily articulate sounds, and are formed into societies, that live under ground, have a very different method of giving alarm. When danger is threatened, they thump on the ground with one of their hinder feet, and produce a sound, that can be heard a great way by animals near the surface of the earth, which would seem to be an artificial sign both from its singularity and its aptness to the situation of the animal.

The rabbits on the island of Sor, near Senegal, have white flesh, and are well tasted, but do not burrow in the earth, so that we may suspect their digging themselves houses in this cold climate is an acquired art, as well as their note of alarm, (Adanson's Voyage to Senegal).

The barking of dogs is another curious note of alarm, and would seem to be an acquired language, rather than a natural sign: for "in the island of Juan Fernandes, the dogs did not attempt to bark, till some European dogs were put among them, and then they gradually begun to imitate them, but in a strange manner at first, as if they were learning a thing that was not natural to them," (Voyage to South America by Don G. Juan, and Don Ant. de Ulloa. B. 2. c. 4).

Linnæus also observes, that the dogs of South America do not bark at strangers, (Syst. Nat.) And the European dogs, that have been carried to Guinea, are said in three or four generations to cease to bark, and only howl, like the dogs that are natives of that coast, (World Displayed, Vol. XVII. p. 26.)

A circumstance not dissimilar to this, and equally curious, is mentioned by Kircherus, de Musurgia, in his Chapter de Lusciniis, "That the young nightingales, that are hatched under other birds, never sing till they are instructed by the company of other nightingales." And Jonston affirms, that the nightingales that visit Scotland, have not the same harmony as those of Italy, (Pennant's Zoology, octavo, p. 255); which would lead us to suspect that the singing of birds, like human music, is an artificial language rather than a natural expression of passion.

X. Our music like our language, is perhaps entirely constituted of artificial tones, which by habit suggest certain agreeable passions. For the same combination of notes and tones do not excite devotion, love, or poetic melancholy in a native of Indostan and of Europe. And "the Highlander has the same warlike ideas annexed to the sound of a bagpipe (an instrument which an Englishman derides), as the Englishman has to that of a trumpet or fife," (Dr. Brown's Union of Poetry and Music, p. 58.) So "the music of the Turks is very different from the Italian, and the people of Fez and Morocco have again a different kind, which to us appears very rough and horrid, but is highly pleasing to them," (L'Arte Armoniaca a Giorgio Antoniotto). Hence we see why the Italian opera does not delight an untutored Englishman; and why those, who are unaccustomed to music, are more pleased with a tune, the second or third time they hear it, than the first. For then the same melodious train of sounds excites the melancholy, they had learned from the song; or the same vivid combination of them recalls all the mirthful ideas of the dance and company.

Even the sounds, that were once disagreeable to us, may by habit be associated with other ideas, so as to become agreeable. Father Lasitau, in his account of the Iroquois, says "the music and dance of those Americans, have something in them extremely barbarous, which at first disgusts. We grow reconciled to them by degrees, and in the end partake of them with pleasure, the savages themselves are fond of them to distraction," (Mœurs des Savages, Tom. ii.)

There are indeed a few sounds, that we very generally associate with agreeable ideas, as the whistling of birds, or purring of animals, that are delighted; and some others, that we as generally associate with disagreeable ideas, as the cries of animals in pain, the hiss of some of them in anger, and the midnight howl of beasts of prey. Yet we receive no terrible or sublime ideas from the lowing of a cow, or the braying of an ass. Which evinces, that these emotions are owing to previous associations. So if the rumbling of a carriage in the street be for a moment mistaken for thunder, we receive a sublime sensation, which ceases as soon as we know it is the noise of a coach and six.

There are other disagreeable sounds, that are said to set the teeth on edge; which, as they have always been thought a necessary effect of certain discordant notes, become a proper subject of our enquiry. Every one in his childhood has repeatedly bit a part of the glass or earthen vessel, in which his food has been given him, and has thence had a very disagreeable sensation in the teeth, which sensation was designed by nature to prevent us from exerting them on objects harder than themselves. The jarring sound produced between the cup and the teeth is always attendant on this disagreeable sensation: and ever after when such a sound is accidentally produced by the conflict of two hard bodies, we feel by association of ideas the concomitant disagreeable sensation in our teeth.

Others have in their infancy frequently held the corner of a silk handkerchief in their mouth, or the end of the velvet cape of their coat, whilst their companions in play have plucked it from them, and have given another disagreeable sensation to their teeth, which has afterwards recurred on touching those materials. And the sight of a knife drawn along a china plate, though no sound is excited by it, and even the imagination of such a knife and plate so scraped together, I know by repeated experience will produce the same disagreeable sensation of the teeth.

These circumstances indisputably prove, that this sensation of the tooth-edge is owing to associated ideas; as it is equally excitable by sight, touch, hearing, or imagination.

In respect to the artificial proportions of sound excited by musical instruments, those, who have early in life associated them with agreeable ideas, and have nicely attended to distinguish them from each other, are said to have a good ear, in that country where such proportions are in fashion: and not from any superior perfection in the organ of hearing, or any intuitive sympathy between certain sounds and passions.

I have observed a child to be exquisitely delighted with music, and who could with great facility learn to sing any tune that he heard distinctly, and yet whole organ of hearing was so imperfect, that it was necessary to speak louder to him in common conversation than to others.

Our music, like our architecture, seems to have no foundation in nature, they are both arts purely of human creation, as they imitate nothing. And the professors of them have only classed those circumstances, that are most agreeable to the accidental taste of their age, or country; and have called it Proportion. But this proportion must always fluctuate, as it rests on the caprices, that are introduced into our minds by our various modes of education. And these fluctuations of taste must become more frequent in the present age, where mankind have enfranchised themselves from the blind obedience to the rules of antiquity in perhaps every science, but that of architecture. See Sect. XII. 7. 3.

XI. There are many articles of knowledge, which the animals in cultivated countries seem to learn very early in their lives, either from each other, or from experience, or observation: one of the most general of these is to avoid mankind. There is so great a resemblance in the natural language of the passions of all animals, that we generally know, when they are in a pacific, or in a malevolent humour, they have the same knowledge of us; and hence we can scold them from us by some tones and gestures, and could possibly attract them to us by others, if they were not already apprized of our general malevolence towards them. Mr. Gmelin, Professor at Petersburg, assures us, that in his journey into Siberia, undertaken by order of the Empress of Russia, he saw foxes, that expressed no fear of himself or companions, but permitted him to come quite near them, having never seen the human creature before. And Mr. Bongainville relates, that at his arrival at the Malouine, or Falkland's Islands, which were not inhabited by men, all the animals came about himself and his people; the fowls settling upon their heads and shoulders, and the quadrupeds running about their feet. From the difficulty of acquiring the confidence of old animals, and the ease of taming young ones, it appears that the fear, they all conceive at the sight of mankind, is an acquired article of knowledge.

This knowledge is more nicely understood by rooks, who are formed into societies, and build, as it were, cities over our heads; they evidently distinguish, that the danger is greater when a man is armed with a gun. Every one has seen this, who in the spring of the year has walked under a rookery with a gun in his hand: the inhabitants of the trees rise on their wings, and scream to the unfledged young to shrink into their nests from the sight of the enemy. The vulgar observing this circumstance so uniformly to occur, assert that rooks can smell gun-powder.

The fieldfares, (turdus pilarus) which breed in Norway, and come hither in the cold season for our winter berries; as they are associated in flocks, and are in a foreign country, have evident marks of keeping a kind of watch, to remark and announce the appearance of danger. On approaching a tree, that is covered with them, they continue fearless till one at the extremity of the bush rising on his wings gives a loud and peculiar note of alarm, when they all immediately fly, except one other, who continues till you approach still nearer, to certify as it were the reality of the danger, and then he also flies off repeating the note of alarm.

And in the woods about Senegal there is a bird called uett-uett by the negroes, and squallers by the French, which, as soon as they see a man, set up a loud scream, and keep flying round him, as if their intent was to warn other birds, which upon hearing the cry immediately take wing. These birds are the bane of sportsmen, and frequently put me into a passion, and obliged me to shoot them, (Adanson's Voyage to Senegal, 78). For the same intent the lesser birds of our climate seem to fly after a hawk, cuckoo, or owl, and scream to prevent their companions from being surprised by the general enemies of themselves, or of their eggs and progeny.

But the lapwing, (charadrius pluvialis Lin.) when her unfledged offspring run about the marshes, where they were hatched, not only gives the note of alarm at the approach of men or dogs, that her young may conceal themselves; but flying and screaming near the adversary, she appears more felicitous and impatient, as he recedes from her family, and thus endeavours to mislead him, and frequently succeeds in her design. These last instances are so apposite to the situation, rather than to the natures of the creatures, that use them; and are so similar to the actions of men in the same circumstances, that we cannot but believe, that they proceed from a similar principle.

Miss M.E. Jacson acquainted me, that she witnessed this autumn an agreeable instance of sagacity in a little bird, which seemed to use the means to obtain an end; the bird repeatedly hopped upon a poppy-stem, and shook the head with its bill, till many seeds were scattered, then it settled on the ground, and eat the seeds, and again repeated the same management. Sept. 1, 1794.

On the northern coast of Ireland a friend of mine saw above a hundred crows at once preying upon muscles; each crow took a muscle up into the air twenty or forty yards high, and let it fall on the stones, and thus by breaking the shell, got possession of the animal.—A certain philosopher (I think it was Anaxagoras) walking along the sea-shore to gather shells, one of these unlucky birds mistaking his bald head for a stone, dropped a shell-fish upon it, and killed at once a philosopher and an oyster.

Our domestic animals, that have some liberty, are also possessed of some peculiar traditional knowledge: dogs and cats have been forced into each other's society, though naturally animals of a very different kind, and have hence learned from each other to eat dog's grass (agrostis canina) when they are sick, to promote vomiting. I have seen a cat mistake the blade of barley for this grass, which evinces it is an acquired knowledge. They have also learnt of each other to cover their excrement and urine;—about a spoonful of water was spilt upon my hearth from the tea-kettle, and I observed a kitten cover it with ashes. Hence this must also be an acquired art, as the creature mistook the application of it.

To preserve their fur clean, and especially their whiskers, cats wash their faces, and generally quite behind their ears, every time they eat. As they cannot lick those places with their tongues, they first wet the inside of the leg with saliva, and then repeatedly wash their faces with it, which must originally be an effect of reasoning, because a means is used to produce an effect; and seems afterwards to be taught or acquired by imitation, like the greatest part of human arts.

These animals seem to possess something like an additional sense by means of their whiskers; which have perhaps some analogy to the antennæ of moths and butterflies. The whiskers of cats consist not only of the long hairs on their upper lips, but they have also four or five long hairs standing up from each eyebrow, and also two or three on each cheek; all which, when the animal erects them, make with their points so many parts of the periphery of a circle, of an extent at least equal to the circumference of any part of their own bodies. With this instrument, I conceive, by a little experience, they can at once determine, whether any aperture amongst hedges or shrubs, in which animals of this genus live in their wild state, is large enough to admit their bodies; which to them is a matter of the greatest consequence, whether pursuing or pursued. They have likewise a power of erecting and bringing forward the whiskers on their lips; which probably is for the purpose of feeling, whether a dark hole be further permeable.

The antennæ, or horns, of butterflies and moths, who have awkward wings, the minute feathers of which are very liable to injury, serve, I suppose, a similar purpose of measuring, as they fly or creep amongst the leaves of plants and trees, whither their wings can pass without touching them.

Mr. Leonard, a very intelligent friend of mine, saw a cat catch a trout by darting upon it in a deep clear water at the mill at Weaford, near Lichfield. The cat belonged to Mr. Stanley, who had often seen her catch fish in the same manner in summer, when the mill-pool was drawn so low, that the fish could be seen. I have heard of other cats taking fish in shallow water, as they stood on the bank. This seems a natural art of taking their prey in cats, which their acquired delicacy by domestication has in general prevented them from using, though their desire of eating fish continues in its original strength.

Mr. White, in his ingenious History of Selbourn, was witness to a cat's suckling a young hare, which followed her about the garden, and came jumping to her call of affection. At Elford, near Lichfield, the Rev. Mr. Sawley had taken the young ones out of a hare, which was shot; they were alive, and the cat, who had just lost her own kittens, carried them away, as it was supposed, to eat them; but it presently appeared, that it was affection not hunger which incited her, as she suckled them, and brought them up as their mother.

Other instances of the mistaken application of what has been termed instinct may be observed in flies in the night, who mistaking a candle for day-light, approach and perish in the flame. So the putrid smell of the stapelia, or carrion-flower, allures the large flesh-fly to deposit its young worms on its beautiful petals, which perish there for want of nourishment. This therefore cannot be a necessary instinct, because the creature mistakes the application of it.

Though in this country horses shew little vestiges of policy, yet in the deserts of Tartary, and Siberia, when hunted by the Tartars they are seen to form a kind of community, set watches to prevent their being surprised, and have commanders, who direct, and hasten their flight, Origin of Language, Vol. I. p. 212. In this country, where four or five horses travel in a line, the first always points his ears forward, and the last points his backward, while the intermediate ones seem quite careless in this respect; which seems a part of policy to prevent surprise. As all animals depend most on the ear to apprize them of the approach of danger, the eye taking in only half the horizon at once, and horses possess a great nicety of this sense; as appears from their mode of fighting mentioned No. 8. 5. of this Section, as well as by common observation.

There are some parts of a horse, which he cannot conveniently rub, when they itch, as about the shoulder, which he can neither bite with his teeth, nor scratch with his hind foot; when this part itches, he goes to another horse, and gently bites him in the part which he wishes to be bitten, which is immediately done by his intelligent friend. I once observed a young foal thus bite its large mother, who did not choose to drop the grass she had in her mouth, and rubbed her nose against the foal's neck instead of biting it; which evinces that she knew the design of her progeny, and was not governed by a necessary instinct to bite where she was bitten.

Many of our shrubs, which would otherwise afford an agreeable food to horses, are armed with thorns or prickles, which secure them from those animals; as the holly, hawthorn, gooseberry, gorse. In the extensive moorlands of Staffordshire, the horses have learnt to stamp upon a gorse-bush with one of their fore-feet for a minute together, and when the points are broken, they eat it without injury. The horses in the new forest in Hampshire are affirmed to do the same by Mr. Gilpin. Forest Scenery, II. 251, and 112. Which is an art other horses in the fertile parts of the country do not possess, and prick their mouths till they bleed, if they are induced by hunger or caprice to attempt eating gorse.

Swine have a sense of touch as well as of smell at the end of their nose, which they use as a hand, both to root up the soil, and to turn over and examine objects of food, somewhat like the proboscis of an elephant. As they require shelter from the cold in this climate, they have learnt to collect straw in their mouths to make their nest, when the wind blows cold; and to call their companions by repeated cries to assist in the work, and add to their warmth by their numerous bedfellows. Hence these animals, which are esteemed so unclean, have also learned never to befoul their dens, where they have liberty, with their own excrement; an art, which cows and horses, which have open hovels to run into, have never acquired. I have observed great sagacity in swine; but the short lives we allow them, and their general confinement, prevents their improvement, which might probably be otherwise greater than that of dogs.

Instances of the sagacity and knowledge of animals are very numerous to every observer, and their docility in learning various arts from mankind, evinces that they may learn similar arts from their own species, and thus be possessed of much acquired and traditional knowledge.

A dog whose natural prey is sheep, is taught by mankind, not only to leave them unmolested, but to guard them; and to hunt, to set, or to destroy other kinds of animals, as birds, or vermin; and in some countries to catch fish, in others to find truffles, and to practise a great variety of tricks; is it more surprising that the crows should teach each other, that the hawk can catch less birds, by the superior swiftness of his wing, and if two of them follow him, till he succeeds in his design, that they can by force share a part of the capture? This I have formerly observed with attention and astonishment.

There is one kind of pelican mentioned by Mr. Osbeck, one of Linnæus's travelling pupils (the pelicanus aquilus), whose food is fish; and which it takes from other birds, because it is not formed to catch them itself; hence it is called by the English a Man-of-war-bird, Voyage to China, p. 88. There are many other interesting anecdotes of the pelican and cormorant, collected from authors of the best authority, in a well-managed Natural History for Children, published by Mr. Galton. Johnson. London.

And the following narration from the very accurate Mons. Adanson, in his Voyage to Senegal, may gain credit with the reader: as his employment in this country was solely to make observations in natural history. On the river Niger, in his road to the island Griel, he saw a great number of pelicans, or wide throats. "They moved with great state like swans upon the water, and are the largest bird next to the ostrich; the bill of the one I killed was upwards of a foot and half long, and the bag fastened underneath it held two and twenty pints of water. They swim in flocks, and form a large circle, which they contract afterwards, driving the fish before them with their legs: when they see the fish in sufficient number confined in this space, they plunge their bill wide open into the water, and shut it again with great quickness. They thus get fish into their throat-bag, which they eat afterwards on shore at their leisure." P. 247.

XII. The knowledge and language of those birds, that frequently change their climate with the seasons, is still more extensive: as they perform these migrations in large societies, and are less subject to the power of man, than the resident tribes of birds. They are said to follow a leader during the day, who is occasionally changed, and to keep a continual cry during the night to keep themselves together. It is probable that these emigrations were at first undertaken as accident directed, by the more adventurous of their species, and learned from one another like the discoveries of mankind in navigation. The following circumstances strongly support this opinion.

1. Nature has provided these animals, in the climates where they are produced, with another resource: when the season becomes too cold for their constitutions, or the food they were supported with ceases to be supplied, I mean that of sleeping. Dormice, snakes, and bats, have not the means of changing their country; the two former from the want of wings, and the latter from his being not able to bear the light of the day. Hence these animals are obliged to make use of this resource, and sleep during the winter. And those swallows that have been hatched too late in the year to acquire their full strength of pinion, or that have been maimed by accident or disease, have been frequently found in the hollows of rocks on the sea coasts, and even under water in this torpid state, from which they have been revived by the warmth of a fire. This torpid state of swallows is testified by innumerable evidences both of antient and modern names. Aristotle speaking of the swallows says, "They pass into warmer climates in winter, if such places are at no great distance; if they are, they bury themselves in the climates where they dwell," (8. Hist. c. 16. See also Derham's Phys. Theol. v. ii. p. 177.)

Hence their emigrations cannot depend on a *necessary* instinct, as the emigrations themselves are not *necessary*.

2. When the weather becomes cold, the swallows in the neighbourhood assemble in large flocks; that is, the unexperienced attend those that have before experienced the journey they are about to undertake: they are then seen some time to hover on the coast, till there is calm whether, or a wind, that suits the direction of their flight. Other birds of passage have been drowned by thousands in the sea, or have settled on ships quite exhausted with fatigue. And others, either by mistaking their course, or by distress of weather, have arrived in countries where they were never seen before: and thus are evidently subject to the same hazards that the human species undergo, in the execution of their artificial purposes.

3. The same birds are emigrant from some countries and not so from others: the swallows were seen at Goree in January by an ingenious philosopher of my acquaintance, and he was told that they continued there all the year; as the warmth of the climate was at all seasons sufficient for their own constitutions, and for the production of the flies that supply them with nourishment. Herodotus says, that in Libya, about the springs of the Nile, the swallows continue all the year. (L. 2.)

Quails (tetrao corturnix, Lin.) are birds of passage from the coast of Barbary to Italy, and have frequently settled in large shoals on ships fatigued with their flight. (Ray, Wisdom of God, p. 129. Derham. Physic. Theol. v. ii. p. 178,) Dr. Ruffel, in his History of Aleppo, observes that the swallows visit that country about the end of February, and having hatched their young disappear about the end of July; and returning again about the beginning of October, continue about a fortnight, and then again disappear. (P. 70.)

When my late friend Dr. Chambres, of Derby, was on the island of Caprea in the bay of Naples, he was informed that great flights of quails annually settle on that island about the beginning of May, in their passage trom Africa to Europe. And that they always come when the south-east wind blows, are fatigued when they rest on this island, and are taken in such amazing quantities and sold to the Continent, that the inhabitants pay the bishop his stipend out of the profits arising from the sale of them.

The flights of these birds across the Mediterranean are recorded near three thousand years ago. "There went forth a wind from the Lord and brought quails from the sea, and let them fall upon the camp, a day's journey round about it, and they were two cubits above the earth," (Numbers, chap. ii. ver. 31.)

In our country, Mr. Pennant informs us, that some quails migrate, and others only remove from the internal parts of the island to the coasts, (Zoology, octavo, 210.) Some of the ringdoves and stares breed here, others migrate, (ibid. 510, ii.) And the slender billed small birds do not all quit these kingdoms in the winter, though the difficulty of procuring the worms and insects, that they feed on, supplies the same reason for migration to them all, (ibid. 511.)

Linnæus has observed, that in Sweden the female chaffinches quit that country in September, migrating into Holland, and leave their mates behind till their return in spring. Hence he has called them Fringilla cælebs, (Amæn. Acad. ii. 42. iv. 595.) Now in our climate both sexes of them are perennial birds. And Mr. Pennant observes that the hoopoe, chatterer, hawfinch, and crossbill, migrate into England so rarely, and at such uncertain times, as not to deserve to be ranked among our birds of passage, (ibid. 511.)

The water fowl, as geese and ducks, are better adapted for long migrations, than the other tribes of birds, as, when the weather is calm, they can not only rest themselves, or sleep upon the ocean, but possibly procure some kind of food from it.

Hence in Siberia, as soon as the lakes are frozen, the water fowl, which are very numerous, all disappear, and are supposed to fly to warmer climates, except the rail, which, from its inability for long flights, probably sleeps, like our bat, in their winter. The following account from the Journey of Professor Gmelin, may entertain the reader. "In the neighbourhood of Krasnoiark, amongst many other emigrant water fowls, we observed a great number of rails, which when pursued never took flight, but endeavoured to escape by running. We enquired how these birds, that could not fly, could retire into other countries in the winter, and were told, both by the Tartars and Assanians, that they well knew those birds could not alone pass into other countries: but when the cranes (les grues) retire in autumn, each one takes a rail (un rale) upon his back, and carries him to a warmer climate."

Recapitulation.

1. All birds of passage can exist in the climates, where they are produced.

2. They are subject in their migrations to the same accidents and difficulties, that mankind are subject to in navigation.

3. The same species of birds migrate from some countries, and are resident in others.

From all these circumstances it appears that the migrations of birds are not produced by a necessary instinct, but are accidental improvements, like the arts among mankind, taught by their cotemporaries, or delivered by tradition from one generation of them to another.

XIII. In that season of the year which supplies the nourishment proper for the expected brood, the birds enter into a contract of marriage, and with joint labour construct a bed for the reception of their offspring. Their choice of the proper season, their contracts of marriage, and the regularity with which they construct their nests, have in all ages excited the admiration of naturalists; and have always been attributed to the power of instinct, which, like the occult qualities of the antient philosophers, prevented all further enquiry. We shall consider them in their order.

Their Choice of the Season.

Our domestic birds, that are plentifully supplied throughout the year with their adapted food, and are covered with houses from the inclemency of the weather, lay their eggs at any season: which evinces that the spring of the year is not pointed out to them by a necessary instinct.

Whilst the wild tribes of birds choose this time of the year from their acquired knowledge, that the mild temperature of the air is more convenient for hatching their eggs, and is soon likely to supply that kind of nourishment, that is wanted for their young.

If the genial warmth of the spring produced the passion of love, as it expands the foliage of trees, all other animals should feel its influence as well as birds: but, the viviparous creatures, as they suckle their young, that is, as they previously digest the natural food, that it may better suit the tender stomachs of their offspring, experience the influence of this passion at all seasons of the year, as cats and bitches. The graminivorous animals indeed generally produce their young about the time when grass is supplied in the greatest plenty, but this is without any degree of exactness, as appears from our cows, sheep, and hares, and may be a part of the traditional knowledge, which they learn from the example of their parents.

Their Contracts of Marriage.

Their mutual passion, and the acquired knowledge, that their joint labour is necessary to procure sustenance for their numerous family, induces the wild birds to enter into a contract of marriage, which does not however take place among the ducks, geese, and fowls, that are provided with their daily food from our barns.

An ingenious philosopher has lately denied, that animals can enter into contracts, and thinks this an essential difference between them and the human creature:—but does not daily observation convince us, that they form contracts of friendship with each other, and with mankind? When puppies and kittens play together, is there not a tacit contract, that they will not hurt each other? And does not your favorite dog expect you should give him his daily food, for his services and attention to you? And thus barters his love for your protection? In the same manner that all contracts are made amongst men, that do not understand each others arbitrary language.

Construction of their Nests.

1. They seem to be instructed how to build their nests from their observation of that, in which they were educated, and from their knowledge of those things, that are most agreeable to their touch in respect: to warmth, cleanliness, and stability. They choose their situations from their ideas of safety from their enemies, and of shelter from the weather. Nor is the colour of their nests a circumstance unthought of; the finches, that build in green hedges, cover their habitations with green moss; the swallow or martin, that builds against rocks and houses, covers her's with clay, whilst the lark chooses vegetable straw nearly of the colour of the ground she inhabits: by this contrivance, they are all less liable to be discovered by their adversaries.

2. Nor are the nests of the same species of birds constructed always of the same materials, nor in the same form; which is another circumstance that ascertains, that they are led by observation.

In the trees before Mr. Levet's house in Lichfield, there are annually nests built by sparrows, a bird which usually builds under the tiles of houses, or the thatch of barns. Not finding such convenient situations for their nests, they build a covered nest bigger than a man's head, with an opening like a mouth at the side, resembling that of a magpie, except that it is built with straw and hay, and lined with feathers, and so nicely managed as to be a defence against both wind and rain.

The following extract from a Letter of the Rev. Mr. J. Darwin, of Carleton Scroop in Lincolnshire, authenticates a curious fact of this kind. "When I mentioned to you the circumstance of crows or rooks building in the spire of Welbourn church, you expressed a desire of being well informed of the certainty of the fact. Welbourn is situated in the road from Grantham to Lincoln on the Cliff row; I yesterday took a ride thither, and enquired of the rector, Mr. Ridgehill, whether the report was true, that rooks built in the spire of his church. He assured me it was true, and that they had done so time immemorial, as his parishioners affirmed. There was a common tradition, he said, that formerly a rookery in some high trees adjoined the church yard, which being cut down (probably in the spring, the building season), the rooks removed to the church, and built their nests on the outside of the spire on the tops of windows, which by their projection a little from the spire made them convenient room, but that they built also on the inside. I saw two nests made with sticks on the outside, and in the spires, and Mr. Ridgehill said there were always a great many.

"I spent the day with Mr. Wright, a clergyman, at Fulbeck, near Welbourn, and in the afternoon Dr. Ellis of Headenham, about two miles from Welbourn, drank tea at Mr. Wright's, who said he remembered, when Mr. Welby lived at Welbourn, that he received a letter from an acquaintance in the west of England, desiring an answer, whether the report of rooks building in Welbourn church was true, as a wager was depending on that subject; to which he returned an answer ascertaining the fact, and decided the wager." Aug. 30, 1794.

So the jackdaw (corvus monedula) generally builds in church-steeples, or under the roofs of high houses; but at Selbourn, in Southamptonshire, where towers and steeples are not sufficiently numerous, these birds build in forsaken rabbit burrows. See a curious account of these subterranean nests in White's History of Selbourn, p. 59. Can the skilful change of architecture in these birds and the sparrows above mentioned be governed by instinct? Then they must have two instincts, one for common, and the other for extraordinary occasions.

I have seen green worsted in a nest, which no where exists in nature: and the down of thistles in those nests, that were by some accident constructed later in the summer, which material could not be procured for the earlier nests: in many different climates they cannot procure the same materials, that they use in ours. And it is well known, that the canary birds, that are propagated in this country, and the finches, that are kept tame, will build their nests of any flexile materials, that are given them. Plutarch, in his Book on Rivers, speaking of the Nile, says, "that the swallows collect a material, when the waters recede, with which they form nests, that are impervious to water." And in India there is a swallow that collects a glutinous substance for this purpose, whose nest is esculent, and esteemed a principal rarity amongst epicures, (Lin. Syst. Nat.) Both these must be constructed of very different materials from those used by the swallows of our country.

In India the birds exert more artifice in building their nests on account of the monkeys and snakes: some form their pensile nests in the shape of a purse, deep and open at top; others with a hole in the side; and others, still more cautious, with an entrance at the very bottom, forming their lodge near the summit. But the taylor-bird will not ever trust its nest to the extremity of a tender twig, but makes one more advance to safety by fixing it to the leaf itself. It picks up a dead leaf, and sews it to the side of a living one, its slender bill being its needle, and its thread some fine fibres; the lining consists of feathers, gossamer, and down; its eggs are white, the colour of the bird light yellow, its length three inches, its weight three sixteenths of an ounce; so that the materials of the nest, and the weight of the bird, are not likely to draw down an habitation so slightly suspended. A nest of this bird is preserved in the British Museum, (Pennant's Indian Zoology). This calls to one's mind the Mosaic account of the origin of mankind, the first dawning of art there ascribed to them, is that of sewing leaves together. For many other curious kinds of nests see Natural History for Children, by Mr. Galton. Johnson. London. Part I. p. 47. Gen. Oriolus.

3. Those birds that are brought up by our care, and have had little communication with others of their own species, are very defective in this acquired knowledge; they are not only very awkward in the construction of their nests, but generally scatter their eggs in various parts of the room or cage, where they are confined, and seldom produce young ones, till, by failing in their first attempt, they have learnt something from their own observation.

4. During the time of incubation birds are said in general to turn their eggs every day; some cover them, when they leave the nest, as ducks and geese; in some the male is said to bring food to the female, that she may have less occasion of absence, in others he is said to take her place, when she goes

in quest of food; and all of them are said to leave their eggs a shorter time in cold weather than in warm. In Senegal the ostrich sits on her eggs only during the night, leaving them in the day to the heat of the sun; but at the Cape of Good Hope, where the heat is less, she sits on them day and night.

If it should be asked, what induces a bird to sit weeks on its first eggs unconscious that a brood of young ones will be the product? The answer must be, that it is the same passion that induces the human mother to hold her offspring whole nights and days in her fond arms, and press it to her bosom, unconscious of its future growth to sense and manhood, till observation or tradition have informed her.

5. And as many ladies are too refined to nurse their own children, and deliver them to the care and provision of others; so is there one instance of this vice in the feathered world. The cuckoo in some parts of England, as I am well informed by a very distinct and ingenious gentleman, hatches and educates her own young; whilst in other parts she builds no nest, but uses that of some lesser bird, generally either of the wagtail, or hedge sparrow, and depositing one egg in it, takes no further care of her progeny.

As the Rev. Mr. Stafford was walking in Glosop Dale, in the Peak of Derbyshire, he saw a cuckoo rise from its nest. The nest was on the stump of a tree, that had been some time felled, among some chips that were in part turned grey, so as much to resemble the colour of the bird, in this nest were two young cuckoos: tying a string about the leg of one of them, he pegged the other end of it to the ground, and very frequently for many days beheld the old cuckoo feed these her young, as he stood very near them.

The following extract of a Letter from the Rev. Mr. Wilmot, of Morley, near Derby, strengthens the truth of the fact above mentioned, of the cuckoo sometimes making a nest, and hatching her own young.

"In the beginning of July 1792, I was attending some labourers on my farm, when one of them said to me, "There is a bird's nest upon one of the Coal-slack Hills; the bird is now sitting, and is exactly like a cuckoo. They say that cuckoo's never hatch their own eggs, otherwise I should have sworn it was one." He took me to the spot, it was in an open fallow ground; the bird was upon the nest, I stood and observed her some time, and was perfectly satisfied it was a cuckoo; I then put my hand towards her, and she almost let me touch her before she rose from the nest, which she appeared to quit with great uneasiness, skimming over the ground in the manner that a hen partridge does when disturbed from a new hatched brood, and went only to a thicket about forty or fifty yards from the nest; and continued there as long as I staid to observe her, which was not many minutes. In the nest, which was barely a hole scratched out of the coal-slack in the manner of a plover's nest, I observed three eggs, but did not touch them. As I had labourers constantly at work in that field, I went thither every day, and always looked to see if the bird was there, but did not disturb her for seven or eight days, when I was tempted to drive her from the nest, and found *two* young ones, that appeared to have been hatched some days, but there was no appearance of the third egg. I then mentioned this extraordinary circumstance (for such I thought it) to Mr. and Mrs. Holyoak of Bidford Grange, Warwickshire, and to Miss M. Willes, who were on a visit at my house, and who all went to see it. Very lately I reminded Mr. Holyoak of it, who told me he had a perfect recollection of the whole, and that, considering it a curiosity, he walked to look at it several times, was perfectly satisfied as to its being a cuckoo, and thought her more attentive to her young, than any other bird he ever observed, having always found her brooding her young. In about a week after I first saw the young ones, one of them was missing, and I rather suspected my plough-boys having taken it; though it might possibly have been taken by a hawk, some time when the old one was seeking food. I never found her off her nest but once, and that was the last time I saw the remaining young one, when it was almost full feathered. I then went from home for two or three days, and, when I returned, the young one was gone, which I take for granted had flown. Though during this time I frequently saw cuckoos in the thicket I mention, I never observed any one, that I supposed to be the cock-bird, paired with this hen."

Nor is this a new observation, though it is entirely overlooked by the modern naturalists, for Aristotle speaking of the cuckoo, asserts that she sometimes builds her nest among broken rocks, and on high mountains, (L. 6. H. c. 1.) but adds in another place that she generally possesses the nest of another bird, (L. 6. H. c. 7.) And Niphus says that cuckoos rarely build for themselves, most frequently laying their eggs in the nests of other birds, (Gesner, L. 3. de Cuculo.)

The Philosopher who is acquainted with these facts concerning the cuckoo, would seem to have very little *reason* himself, if he could imagine this neglect of her young to be a necessary *instinct*!

XIV. The deep recesses of the ocean are inaccessible to mankind, which prevents us from having much knowledge of the arts and government of its inhabitants.

1. One of the baits used by the fisherman is an animal called an Old Soldier, his size and form are somewhat like the craw-fish, with this difference, that his tail is covered with a tough membrane instead of a shell; and to obviate this defect, he seeks out the uninhabited shell of some dead

fish, that is large enough to receive his tail, and carries it about with him as part of his clothing or armour.

2. On the coasts about Scarborough, where the haddocks, cods, and dog-fish, are in great abundance, the fishermen universally believe that the dog-fish make a line, or semicircle, to encompass a shoal of haddocks and cod, confining them within certain limits near the shore, and eating them as occasion requires. For the haddocks and cod are always found near the shore without any dog-fish among them, and the dog-fish further off without any haddocks or cod; and yet the former are known to prey upon the latter, and in some years devour such immense quantities as to render this fishery more expensive than profitable.

3. The remora, when he wishes to remove his situation, as he is a very slow swimmer, is content to take an outside place on whatever conveyance is going his way; nor can the cunning animal be tempted to quit his hold of a ship when she is sailing, not even for the lucre of a piece of pork, lest it should endanger the loss of his passage: at other times he is easily caught with the hook.

4. The crab-fish, like many other testaceous animals, annually changes its shell; it is then in a soft state, covered only with a mucous membrane, and conceals itself in holes in the sand or under weeds; at this place a hard shelled crab always stands centinel, to prevent the sea insects from injuring the other in its defenceless state; and the fishermen from his appearance know where to find the soft ones, which they use for baits in catching other fish.

And though the hard shelled crab, when he is on this duty, advances boldly to meet the foe, and will with difficulty quit the field; yet at other times he shews great timidity, and has a wonderful speed in attempting his escape; and, if often interrupted, will pretend death like the spider, and watch an opportunity to sink himself into the sand, keeping only his eyes above. My ingenious friend Mr. Burdett, who favoured me with these accounts at the time he was surveying the coasts, thinks the commerce between the sexes takes place at this time, and inspires the courage of the creature.

5. The shoals of herrings, cods, haddocks, and other fish, which approach our shores at certain seasons, and quit them at other seasons without leaving one behind; and the salmon, that periodically frequent our rivers, evince, that there are vagrant tribes of fish, that perform as regular migrations as the birds of passage already mentioned.

6. There is a cataract on the river Liffey in Ireland about nineteen feet high: here in the salmon season many of the inhabitants amuse themselves

in observing these fish leap up the torrent. They dart themselves quite out of the water as they ascend, and frequently fall back many times before they surmount it, and baskets made of twigs are placed near the edge of the stream to catch them in their fall.

I have observed, as I have sat by a spout of water, which descends from a stone trough about two feet into a stream below, at particular seasons of the year, a great number of little fish called minums, or pinks, throw themselves about twenty times their own length out of the water, expecting to get into the trough above.

This evinces that the storgee, or attention of the dam to provide for the offspring, is strongly exerted amongst the nations of fish, where it would seem to be the most neglected; as these salmon cannot be supposed to attempt so difficult and dangerous a task without being conscious of the purpose or end of their endeavours.

It is further remarkable, that most of the old salmon return to the sea before it is proper for the young shoals to attend them, yet that a few old ones continue in the rivers so late, that they become perfectly emaciated by the inconvenience of their situation, and this apparently to guide or to protect the unexperienced brood.

Of the smaller water animals we have still less knowledge, who nevertheless probably possess many superior arts; some of these are mentioned in Botanic Garden, P. I. Add. Note XXVII. and XXVIII. The nympha of the water-moths of our rivers, which cover themselves with cases of straw, gravel, and shell, contrive to make their habitations, nearly in equilibrium with the water; when too heavy, they add a bit of wood or straw; when too light, a bit of gravel. Edinb. Trans.

All these circumstances bear a near resemblance to the deliberate actions of human reason.

XV. We have a very imperfect acquaintance with the various tribes of insects: their occupations, manner of life, and even the number of their senses, differ from our own, and from each other; but there is reason to imagine, that those which possess the sense of touch in the most exquisite degree, and whole occupations require the most constant exertion of their powers, are induced with a greater proportion or knowledge and ingenuity.

The spiders of this country manufacture nets of various forms, adapted to various situations, to arrest the flies that are their food; and some of them have a house or lodging-place in the middle of the net, well contrived for warmth, security, or concealment. There is a large spider in South America, who constructs nets of so strong a texture as to entangle small birds, particularly the humming bird. And in Jamaica there is another spider, who digs a hole in the earth obliquely downwards, about three inches in length, and one inch in diameter, this cavity she lines with a tough thick web, which when taken out resembles a leathern purse: but what is most curious, this house has a door with hinges, like the operculum of some sea shells; and herself and family, who tenant this nest, open and shut the door, whenever they pass or repass. This history was told me, and the nest with its operculum shewn me by the late Dr. Butt of Bath, who was some years physician in Jamaica.

The production of these nets is indeed a part of the nature or conformation of the animal, and their natural use is to supply the place of wings, when she wishes to remove to another situation. But when she employs them to entangle her prey, there are marks of evident design, for she adapts the form of each net to its situation, and strengthens those lines, that require it, by joining others to the middle of them, and attaching those others to distant objects, with the same individual art, that is used by mankind in supporting the masts and extending the sails of ships. This work is executed with more mathematical exactness and ingenuity by the field spiders, than by those in our houses, as their constructions are more subjected to the injuries of dews and tempests.

Besides the ingenuity shewn by these little creatures in taking their prey, the circumstance of their counterfeiting death, when they are put into terror, is truly wonderful; and as soon as the object of terror is removed, they recover and run away. Some beetles are also said to possess this piece of hypocrisy.

The curious webs, or chords, constructed by some young caterpillars to defend themselves from cold, or from insects of prey; and by silk-worms and some other caterpillars, when they transmigrate into aureliæ or larvæ, have deservedly excited the admiration of the inquisitive. But our ignorance of their manner of life, and even of the number of their senses, totally precludes us from understanding the means by which they acquire this knowledge.

The care of the salmon in choosing a proper situation for her spawn, the structure of the nests of birds, their patient incubation, and the art of the cuckoo in depositing her egg in her neighbour's nursery, are instances of great sagacity in those creatures: and yet they are much inferior to the arts exerted by many of the insect tribes on similar occasions. The hairy excrescences on briars, the oak apples, the blasted leaves of trees, and the lumps on the backs of cows, are situations that are rather produced than chosen by the mother insect for the convenience of her offspring. The cells of bees, wasps, spiders, and of the various coralline insects, equally astonish us, whether we attend to the materials or to the architecture.

But the conduct of the ant, and of some species of the ichneumon fly in the incubation of their eggs, is equal to any exertion of human science. The ants many times in a day move their eggs nearer the surface of their habitation, or deeper below it, as the heat of the weather varies; and in colder days lie upon them in heaps for the purpose of incubation: if their mansion is too dry, they carry them to places where there is moisture, and you may distinctly see the little worms move and suck up the water. When too much moisture approaches their nest, they convey their eggs deeper in the earth, or to some other place of safety. (Swammerd. Epil. ad Hist. Insects, p. 153. Phil. Trans. No. 23. Lowthrop. V. 2. p. 7.)

There is one species of ichneumon-fly, that digs a hole in the earth, and carrying into it two or three living caterpillars, deposits her eggs, and nicely closing up the nest leaves them there; partly doubtless to assist the incubation, and partly to supply food to her future young, (Derham. B. 4, c. 13. Aristotle Hist. Animal, L. 5. c. 20.)

A friend of mine put about fifty large caterpillars collected from cabbages on some bran and a few leaves into a box, and covered it with gauze to prevent their escape. After a few days we saw, from more than three fourths of them, about eight or ten little caterpillars of the ichneumon-fly come out of their backs, and spin each a small cocoon of silk, and in a few days the large caterpillars died. This small fly it seems lays its egg in the back of the cabbage caterpillar, which when hatched preys upon the material, which is produced there for the purpose of making silk for the future nest of the cabbage caterpillar; of which being deprived, the creature wanders about till it dies, and thus our gardens are preserved by the ingenuity of this cruel fly. This curious property of producing a silk thread, which is common to some sea animals, see Botanic Garden, Part I. Note XXVII. and is designed for the purpose of their transformation as in the silk-worm, is used for conveying themselves from higher branches to lower ones of trees by some caterpillars, and to make themselves temporary nests or tents, and by the spider for entangling his prey. Nor is it strange that so much knowledge should be acquired by such small animals; since there is reason to imagine, that these insects have the sense of touch, either in their proboscis, or their antennæ, to a great degree of perfection; and thence may possess, as far as their sphere extends, as accurate knowledge, and as subtle invention, as the discoverers of human arts.

XVI. 1. If we were better acquainted with the histories of those insects that are formed into societies, as the bees, wasps, and ants, I make no doubt but we should find, that their arts and improvements are not so similar and uniform as they now appear to us, but that they arose in the same manner from experience and tradition, as the arts of our own species; though their reasoning is from fewer ideas, is busied about fewer objects, and is exerted with less energy.

There are some kinds of insects that migrate like the birds before mentioned. The locust of warmer climates has sometimes come over to England; it is shaped like a grasshopper, with very large wings, and a body above an inch in length. It is mentioned as coming into Egypt with an east wind, "The lord brought an east wind upon the land all that day and night, and in the morning the east wind brought the locusts, and covered the face of the earth, so that the land was dark," Exod. x. 13. The migrations of these insects are mentioned in another part of the scripture, "The locusts have no king, yet go they forth all of them in bands," Prov. xxx. 27.

The accurate Mr. Adanson, near the river Gambia in Africa, was witness to the migration of these insects. "About eight in the morning, in the month of February, there suddenly arose over our heads a thick cloud, which darkened the air, and deprived us of the rays of the sun. We found it was a cloud of locusts raised about twenty or thirty fathoms from the ground, and covering an extent of several leagues; at length a shower of these insects descended, and after devouring every green herb, while they rested, again resumed their flight. This cloud was brought by a strong east-wind, and was all the morning in passing over the adjacent country." (Voyage to Senegal, 158.)

In this country the gnats are sometimes seen to migrate in clouds, like the musketoes of warmer climates, and our swarms of bees frequently travel many miles, and are said in North America always to fly towards the south. The prophet Isaiah has a beautiful allusion to these migrations, "The Lord shall call the fly from the rivers of Egypt, and shall hiss for the bee that is in the land of Assyria," Isa. vii. 18. which has been lately explained by Mr. Bruce, in his travels to discover the source of the Nile.

2. I am well informed that the bees that were carried into Barbadoes, and other western islands, ceased to lay up any honey after the first year, as they found it not useful to them: and are now become very troublesome to the inhabitants of those islands by infesting their sugar houses; but those in Jamaica continue to make honey, as the cold north winds, or rainy seasons of that island, confine them at home for several weeks together. And the bees of Senegal, which differ from those of Europe only in size, make their honey not only superior to ours in delicacy of flavour, but it has this singularity, that it never concretes, but remains liquid as syrup, (Adanson). From some

observations of Mr. Wildman, and of other people of veracity, it appears, that during the severe part of the winter season for weeks together the bees are quite benumbed and torpid from the cold, and do not consume any of their provision. This state of sleep, like that of swallows and bats, seems to be the natural resource of those creatures in cold climates, and the making of honey to be an artificial improvement.

As the death of our hives of bees appears to be owning to their being kept so warm, as to require food when their stock is exhausted; a very observing gentleman at my request put two hives for many weeks into a dry cellar, and observed, during all that time, they did not consume any of their provision, for their weight did not decrease as it had done when they were kept in the open air. The same observation is made in the Annual Register for 1768, p. 113. And the Rev. Mr. White, in his Method of preserving Bees, adds, that those on the north side of his house consumed less honey in the winter than those on the south side.

There is another observation on bees well ascertained, that they at various times, when the season begins to be cold, by a general motion of their legs as they hang in clusters produce a degree of warmth, which is easily perceptible by the hand. Hence by this ingenious exertion, they for a long time prevent the torpid state they would naturally fall into.

According to the late observations of Mr. Hunter, it appears that the bee's-wax is not made from the dust of the anthers of flowers, which they bring home on their thighs, but that this makes what is termed bee-bread, and is used for the purpose of feeding the bee-maggots; in the same manner butterflies live on honey, but the previous caterpillar lives on vegetable leaves, while the maggots of large flies require flesh for their food, and those of the ichneumon fly require insects for their food. What induces the bee who lives on honey to lay up vegetable powder for its young? What induces the butterfly to lay its eggs on leaves, when itself feeds on honey? What induces the other flies to seek a food for their progeny different from what they consume themselves? If these are not deductions from their own previous experience or observation, all the actions of mankind must be resolved into instinct.

3. The dormouse consumes but little of its food during the rigour of the season, for they roll themselves up, or sleep, or lie torpid the greatest part of the time; but on warm sunny days experience a short revival, and take a little food, and then relapse into their former state." (Pennant Zoolog. p. 67.) Other animals, that sleep in winter without laying up any provender, are observed to go into their winter beds fat and strong, but return to day-light in the spring season very lean and feeble. The common flies sleep

during the winter without any provision for their nourishment, and are daily revived by the warmth of the sun, or of our fires. These whenever they see light endeavour to approach it, having observed, that by its greater vicinity they get free from the degree of torpor, that the cold produces; and are hence induced perpetually to burn themselves in our candles: deceived, like mankind, by the misapplication of their knowledge. Whilst many of the subterraneous insects, as the common worms, seem to retreat so deep into the earth as not to be enlivened or awakened by the difference of our winter days; and stop up their holes with leaves or straws, to prevent the frosts from injuring them, or the centipes from devouring them. The habits of peace, or the stratagems of war, of these subterranean nations are covered from our view; but a friend of mine prevailed on a distressed worm to enter the hole of another worm on a bowling-green, and he presently returned much wounded about his head. And I once saw a worm rise hastily out of the earth into the sunshine, and observed a centipes hanging at its tail: the centipes nimbly quitted the tail, and seizing the worm about its middle cut it in half with its forceps, and preyed upon one part, while the other escaped. Which evinces they have design in stopping the mouths of their habitations.

4. The wasp of this country fixes his habitation under ground, that he may not be affected with the various changes of our climate; but in Jamaica he hangs it on the bough of a tree, where the seasons are less severe. He weaves a very curious paper of vegetable fibres to cover his nest, which is constructed on the same principle with that of the bee, but with a different material; but as his prey consists of flesh, fruits, and insects, which are perishable commodities, he can lay up no provender for the winter.

M. de la Loubiere, in his relation of Siam, says, "That in a part of that kingdom, which lies open to great inundations, all the ants make their settlements upon trees; no ants' nests are to be seen any where else." Whereas in our country the ground is their only situation. From the scriptual account of these insects, one might be led to suspect, that in some climates they lay up a provision for the winter. Origen affirms the same, (Cont. Cels. L. 4.) But it is generally believed that in this country they do not, (Prov. vi. 6. xxx. 25.) The white ants of the coast of Africa make themselves pyramids eight or ten feet high, on a base of about the same width, with a smooth surface of rich clay, excessively hard and well built, which appear at a distance like an assemblage of the huts of the negroes, (Adanson). The history of these has been lately well described in the Philosoph. Transactions, under the name of termes, or termites. These differ very much from the nest of our large ant; but the real history of this creature, as well as of the wasp, is yet very imperfectly known.

Wasps are said to catch large spiders, and to cut off their legs, and carry their mutilated bodies to their young, Dict. Raison. Tom. I. p. 152.

One circumstance I shall relate which fell under my own eye, and shewed the power or reason in a wasp, as it is exercised among men. A wasp, on a gravel walk, had caught a fly nearly as large as himself; kneeling on the ground I observed him separate the tail and the head from the body part, to which the wings were attached. He then took the body part in his paws, and rose about two feet from the ground with it; but a gentle breeze wafting the wings of the fly turned him round in the air, and he settled again with his prey upon the gravel. I then distinctly observed him cut off with his mouth, first one of the wings, and then the other, after which he flew away with it unmolested by the wind.

Go, thou sluggard, learn arts and industry from the bee, and from the ant!

Go, proud reasoner, and call the worm thy sister!

XVII. *Conclusion.*

It was before observed how much the superior accuracy of our sense of touch contributes to increase our knowledge; but it is the greater energy and activity of the power of volition (as explained in the former Sections of this work) that marks mankind, and has given him the empire of the world.

There is a criterion by which we may distinguish our voluntary acts or thoughts from those that are excited by our sensations: "The former are always employed about the *means* to acquire pleasureable objects, or to avoid painful ones: while the latter are employed about the *possession* of those that are already in our power."

If we turn our eyes upon the fabric of our fellow animals, we find they are supported with bones, covered with skins, moved by muscles; that they possess the same senses, acknowledge the same appetites, and are nourished by the same aliment with ourselves; and we should hence conclude from the strongest analogy, that their internal faculties were also in some measure similar to our own.

Mr. Locke indeed published an opinion, that other animals possessed no abstract or general ideas, and thought this circumstance was the barrier between the brute and the human world. But these abstracted ideas have been since demonstrated by Bishop Berkley, and allowed by Mr. Hume, to have no existence in nature, not even in the mind of their inventor, and we are hence necessitated to look for some other mark of distinction.

The ideas and actions of brutes, like those of children, are almost perpetually produced by their present pleasures, or their present pains; and, except in the few instances that have been mentioned in this Section, they seldom busy themselves about the *means* of procuring future bliss, or of avoiding future misery.

Whilst the acquiring of languages, the making of tools, and the labouring for money; which are all only the *means* of procuring pleasure; and the praying to the Deity, as another *means* to procure happiness, are characteristic of human nature.

SECT. XVII
THE CATENATION OF MOTIONS

I. 1. Catenations of animal motion. 2. Are produced by irritations, by sensations, by volitions. 3. They continue some time after they have been excited. Cause of catenation. 4. We can then exert our attention on other objects. 5. Many catenations of motions go on together. 6. Some links of the catenations of motions may be left out without disuniting the chain. 7. Interrupted circles of motion continue confusedly till they come to the part of the circle, where they were disturbed. 8. Weaker catenations are dissevered by stronger. 9. Then new catenations take place. 10. Much effort prevents their reuniting. Impediment of speech. 11. Trains more easily dissevered than circles. 12. Sleep destroys volition and external stimulus. II. Instances of various catenations in a young lady playing on the harpsichord. III. 1. What catenations are the strongest. 2. Irritations joined with associations from strongest connexions. Vital motions. 3. New links with increased force, cold fits of fever produced. 4. New links with decreased force. Cold bath. 5. Irritation joined with sensation. Inflammatory fever. Why children cannot tickle themselves. 6 . Volition joined with sensation. Irritative ideas of sound become sensible. 7. Ideas of imagination, dissevered by irritations, by volition, production of surprise.

I. 1. To investigate with precision the catenations of animal motions, it would be well to attend to the manner of their production; but we cannot begin this disquisition early enough for this purpose, as the catenations of motion seem to begin with life, and are only extinguishable with it; We have spoken of the power of irritation, of sensation, of volition, and of association, as preceding the fibrous motions; we now step forwards, and consider, that conversely they are in their turn preceded by those motions; and that all the successive trains or circles of our actions are composed of this twofold concatenation. Those we shall call trains of action, which continue to proceed without any stated repetitions; and those circles of action, when the parts of them return at certain periods, though the trains, of which they consist, are not exactly similar. The reading an epic poem is a train of actions; the reading a song with a chorus at equal distances in the measure constitutes so many circles of action.

2. Some catenations of animal motion are produced by reiterated successive irritations, as when we learn to repeat the alphabet in its order by frequently reading the letters of it. Thus the vermicular motions of the bowels were originally produced by the successive irritations of the passing aliment; and the succession of actions of the auricles and ventricles of the heart was originally formed by successive stimulus of the blood, these afterwards become part of the diurnal circles of animal actions, as appears by the periodical returns of hunger, and the quickened pulse of weak people in the evening.

Other catenations of animal motion are gradually acquired by successive agreeable sensations, as in learning a favourite song or dance; others by disagreeable sensations, as in coughing or nictitation; these become associated by frequent repetition, and afterwards compose parts of greater circles of action like those above mentioned.

Other catenations of motions are gradually acquired by frequent voluntary repetitions; as when we deliberately learn to march, read, fence, or any mechanic art, the motions of many of our muscles become gradually linked together in trains, tribes, or circles of action. Thus when any one at first begins to use the tools in turning wood or metals in a lathe, he wills the motions of his hand or fingers, till at length these actions become so connected with the effect, that he seems only to will the point of the chisel. These are caused by volition, connected by association like those above described, and afterwards become parts of our diurnal trains or circles of action.

3. All these catenations of animal motions, are liable to proceed some time after they are excited, unless they are disturbed or impeded by other irritations, sensations, or volitions; and in many instances in spite of our endeavours to stop them; and this property of animal motions is probably the cause of their catenation. Thus when a child revolves some minute on one foot, the spectra of the ambient objects appear to circulate round him some time after he falls upon the ground. Thus the palpitation of the heart continues some time after the object of fear, which occasioned it, is removed. The blush of shame, which is an excess of sensation, and the glow of anger, which is an excess of volition, continue some time, though the affected person finds, that those emotions were caused by mistaken facts, and endeavours to extinguish their appearance. See Sect. XII. 1. 5.

4. When a circle of motions becomes connected, by frequent repetitions as above, we can exert our attention strongly on other objects, and the concatenated circle of motions will nevertheless proceed in due order; as whilst you are thinking on this subject, you use variety of muscles in walking about your parlour, or in sitting at your writing-table.

5. Innumerable catenations of motions may proceed at the same time, without incommoding each other. Of these are the motions of the heart and arteries; those of digestion and glandular secretion; of the ideas, or sensual motions; those of progression, and of speaking; the great annual circle of actions so apparent in birds in their times of breeding and moulting; the monthly circles of many female animals; and the diurnal circles of sleeping and waking, of fulness and inanition.

6. Some links of successive trains or of synchronous tribes of action may be left out without disjoining the whole. Such are our usual trains of recollection; after having travelled through an entertaining country, and viewed many delightful lawns, rolling rivers, and echoing rocks; in the recollection of our journey we leave out the many districts, that we crossed, which were marked with no peculiar pleasure. Such also are our complex ideas, they are catenated tribes of ideas, which do not perfectly resemble their correspondent perceptions, because some of the parts are omitted.

7. If an interrupted circle of actions is not entirely dissevered, it will continue to proceed confusedly, till it comes to the part of the circle, where it was interrupted.

The vital motions in a fever from drunkenness, and in other periodical diseases, are instances of this circumstance. The accidental inebriate does not recover himself perfectly till about the same hour on the succeeding day. The accustomed drunkard is disordered, if he has not his usual potation of fermented liquor. So if a considerable part of a connected tribe of action be disturbed, that whole tribe goes on with confusion, till the part of the tribe affected regains its accustomed catenations. So vertigo produces vomiting, and a great secretion of bile, as in sea-sickness, all these being parts of the tribe of irritative catenations.

8. Weaker catenated trains may be dissevered by the sudden exertion of the stronger. When a child first attempts to walk across a room, call to him, and he instantly falls upon the ground. So while I am thinking over the virtues of my friends, if the tea-kettle spurt out some hot water on my stocking; the sudden pain breaks the weaker chain of ideas, and introduces a new group of figures of its own. This circumstance is extended to some unnatural trains of action, which have not been confirmed by long habit; as the hiccough, or an ague-fit, which are frequently curable by surprise.

A young lady about eleven years old had for five days had a contraction of one muscle in her fore arm, and another in her arm, which occurred four or five times every minute; the muscles were seen to leap, but without bending the arm. To counteract this new morbid habit, an issue was placed over the convulsed muscle of her arm, and an adhesive plaster wrapped tight like a bandage over the whole fore arm, by which the new motions were immediately destroyed, but the means were continued some weeks to prevent a return.

9. If any circle of actions is dissevered, either by omission of some of the links, as in sleep, or by insertion of other links, as in surprise, new catenations take place in a greater or less degree. The last link of the broken chain of actions becomes connected with the new motion which has broken it, or with that which was nearest the link omitted; and these new catenations proceed instead of the old ones. Hence the periodic returns of ague-fits, and the chimeras of our dreams.

10. If a train of actions is dissevered, much effort of volition or sensation will prevent its being restored. Thus in the common impediment of speech, when the association of the motions of the muscles of enunciation with the idea of the word to be spoken is disordered, the great voluntary efforts, which distort the countenance, prevent the rejoining of the broken associations. See No. II. 10. of this Section. It is thus likewise observable in some inflammations of the bowels, the too strong efforts made by the muscles to carry forwards the offending material fixes it more firmly in its place, and prevents the cure. So in endeavouring to recal to our memory some particular word of a sentence, if we exert ourselves too strongly about it, we are less likely to regain it.

11. Catenated trains or tribes of action are easier dissevered than catenated circles of action. Hence in epileptic fits the synchronous connected tribes of action, which keep the body erect, are dissevered, but the circle of vital motions continues undisturbed.

12. Sleep destroys the power of volition, and precludes the stimuli of external objects, and thence dissevers the trains, of which these are a part; which confirms the other catenations, as those of the vital motions, secretions, and absorptions; and produces the new trains of ideas, which constitute our dreams.

II. 1. All the preceding circumstances of the catenations of animal motions will be more clearly understood by the following example of a person learning music; and when we recollect the variety of mechanic arts, which are performed by associated trains of muscular actions catenated with the effects they produce, as in knitting, netting, weaving; and the greater variety of associated trains of ideas caused or catenated by volitions or sensations, as in our hourly modes of reasoning, or imagining, or recollecting, we shall gain some idea of the innumerable catenated trains and circles of action, which form the tenor of our lives, and which began, and will only cease entirely with them.

2. When a young lady begins to learn music, she voluntarily applies herself to the characters of her music-book, and by many repetitions endeavours to catenate them with the proportions of sound, of which they are symbols. The ideas excited by the musical characters are slowly connected with the keys of the harpsichord, and much effort is necessary to produce every note with the proper finger, and in its due place and time; till at length a train of voluntary exertions becomes catenated with certain irritations. As the various notes by frequent repetitions become connected in the order, in which they are produced, a new catenation of sensitive exertions becomes mixed with the voluntary ones above described; and not only the musical symbols of crotchets and quavers, but the auditory notes and tones at the same time, become so many successive or synchronous links in this circle of catenated actions.

At length the motions of her fingers become catenated with the musical characters; and these no sooner strike the eye, than the finger presses down the key without any voluntary attention between them; the activity of the hand being connected with the irritation of the figure or place of the musical symbol on the retina; till at length by frequent repetitions of the same tune the movements of her fingers in playing, and the muscles of the larynx in singing, become associated with each other, and form part of those intricate trains and circles of catenated motions, according with the second article of the preceding propositions in No. 1. of this Section.

3. Besides the facility, which by habit attends the execution of this musical performance, a curious circumstance occurs, which is, that when our young musician has began a tune, she finds herself inclined to continue it; and that even when she is carelessly singing alone without attending to her own song; according with the third preceding article.

4. At the same time that our young performer continues to play with great exactness this accustomed tune, she can bend her mind, and that intensely, on some other object, according with the fourth article of the preceding proportions.

The manuscript copy of this work was lent to many of my friends at different times for the purpose of gaining their opinions and criticisms on many parts of it, and I found the following anecdote written with a pencil opposite to this page, but am not certain by whom. "I remember seeing the pretty young actress, who succeeded Mrs. Arne in the performance of the celebrated Padlock, rehearse the musical parts at her harpsichord under the eye of her master with great taste and accuracy; though I observed her countenance full of emotion, which I could not account for; at last she suddenly burst into tears; for she had all this time been eyeing a beloved canary bird, suffering great agonies, which at that instant fell dead from its perch."

5. At the same time many other catenated circles of action are going on in the person of our fair musician, as well as the motions of her fingers, such as the vital motions, respiration, the movements of her eyes and eyelids, and of the intricate muscles of vocality, according with the fifth preceding article.

6. If by any strong impression on the mind of our fair musician she should be interrupted for a very inconsiderable time, she can still continue her performance, according to the sixth article.

7. If however this interruption be greater, though the chain of actions be not dissevered, it proceeds confusedly, and our young performer continues indeed to play, but in a hurry without accuracy and elegance, till she begins the tune again, according to the seventh of the preceding articles.

8. But if this interruption be still greater, the circle of actions becomes entirely dissevered, and she finds herself immediately under the necessity to begin over again to recover the lost catenation, according to the eighth preceding article.

9. Or in trying to recover it she will sing some dissonant notes, or strike some improper keys, according to the ninth preceding article.

10. A very remarkable thing attends this breach of catenation, if the performer has forgotten some word of her song, the more energy of mind she uses about it, the more distant is she from regaining it; and artfully employs her mind in part on some other object, or endeavours to dull its perceptions, continuing to repeat, as it were inconsciously, the former part of the song, that she remembers, in hopes to regain the lost connexion.

For if the activity of the mind itself be more energetic, or takes its attention more, than the connecting word, which is wanted; it will not perceive the slighter link of this lost word; as who listens to a feeble sound, must be very silent and motionless; so that in this case the very vigour of the mind itself seems to prevent it from regaining the lost catenation, as well as the too great exertion in endeavouring to regain it, according to the tenth preceding article.

We frequently experience, when we are doubtful about the spelling of a word, that the greater voluntary exertion we use, that is the more intensely we think about it, the further are we from regaining the lost association between the letters of it, but which readily recurs when we have become careless about it. In the same manner, after having for an hour laboured to recollect the name of some absent person, it shall seem, particularly after sleep, to come into the mind as it were spontaneously; that is the word we are in search of, was joined to the preceding one by association; this association being dissevered, we endeavour to recover it by volition; this very action of the mind strikes our attention more, than the faint link of association, and we find it impossible by this means to retrieve the lost word. After sleep, when volition is entirely suspended, the mind becomes capable of perceiving the fainter link of association, and the word is regained.

On this circumstance depends the impediment of speech before mentioned; the first syllable of a word is causable by volition, but the remainder of it is in common conversation introduced by its associations with this first syllable acquired by long habit. Hence when the mind of the stammerer is vehemently employed on some idea of ambition of shining, or fear of not succeeding, the associations of the motions of the muscles of articulation with each other become dissevered by this greater exertion, and he endeavours in vain by voluntary efforts to rejoin the broken association. For this purpose he continues to repeat the first syllable, which is causable by volition, and strives in vain, by various distortions of countenance, to produce the next links, which are subject to association. See Class IV. 3. 1. 1.

11. After our accomplished musician has acquired great variety of tunes and songs, so that some of them begin to cease to be easily recollected, she finds progressive trains of musical notes more frequently forgotten, than those which are composed of reiterated circles, according with the eleventh preceding article.

12. To finish our example with the preceding articles we must at length suppose, that our fair performer falls asleep over her harpsichord; and thus by the suspension of volition, and the exclusion of external stimuli, she dissevers the trains and circles of her musical exertions.

III. 1. Many of these circumstances of catenations of motions receive an easy explanation from the four following consequences to the seventh law of animal causation in Sect. IV. These are, first, that those successions or combinations of animal motions, whether they were united by causation, association, or catenation, which have been most frequently repeated, acquire the strongest connection. Secondly, that of these, those, which have been less frequently mixed with other trains or tribes of motion, have the strongest connection. Thirdly, that of these, those, which were first formed, have the strongest connection. Fourthly, that if an animal motion be excited by more than one causation, association, or catenation, at the same time, it will be performed with greater energy.

2. Hence also we understand, why the catenations of irritative motions are more strongly connected than those of the other classes, where the quantity of unmixed repetition has been equal; because they were first formed. Such are those of the secerning and absorbent systems of vessels, where the action of the gland produces a fluid, which stimulates the mouths of its correspondent absorbents. The associated motions seem to be the next most strongly united, from their frequent repetition; and where both these circumstances unite, as in the vital motions, their catenations are indissoluble but by the destruction of the animal.

3. Where a new link has been introduced into a circle of actions by some accidental defect of stimulus; if that defect of stimulus be repeated at the same part of the circle a second or a third time, the defective motions thus produced, both by the repeated defect of stimulus and by their catenation with the parts of the circle of actions, will be performed with less and less energy. Thus if any person is exposed to cold at a certain hour to-day, so long as to render some part of the system for a time torpid; and is again exposed to it at the same hour to-morrow, and the next day; he will be more and more affected by it, till at length a cold fit of fever is completely formed, as happens at the beginning of many of those fevers, which are called nervous or low fevers. Where the patient has slight periodical shiverings and paleness for many days before the febrile paroxysm is completely formed.

4. On the contrary, if the exposure to cold be for so short a time, as not to induce any considerable degree of torpor or quiescence, and is repeated daily as above mentioned, it loses its effect more and more at every repetition, till the constitution can bear it without inconvenience, or indeed without being conscious of it. As in walking into the cold air in frosty weather. The same rule is applicable to increased stimulus, as of heat, or of vinous spirit, within certain limits, as is applied in the two last paragraphs to Deficient Stimulus; as is further explained in Sect. XXXVI. on the Periods of Diseases.

5. Where irritation coincides with sensation to produce the same catenations of motion, as in inflammatory fevers, they are excited with still greater energy than by the irritation alone. So when children expect to be tickled in play, by a feather lightly passed over the lips, or by gently vellicating the soles of their feet, laughter is most vehemently excited; though they can stimulate these parts with their own fingers unmoved. Here the pleasureable idea of playfulness coincides with the vellication; and there is no voluntary exertion used to diminish the sensation, as there would be, if a child should endeavour to tickle himself. See Sect. XXXIV. 1. 4.

6. And lastly, the motions excited by the junction of voluntary exertion with irritation are performed with more energy, than those by irritation singly; as when we listen to small noises, as to the ticking of a watch in the night, we perceive the most weak sounds, that are at other times unheeded. So when we attend to the irritative ideas of sound in our ears, which are generally not attended to, we can hear them; and can see the spectra of objects, which remain in the eye, whenever we please to exert our voluntary power in aid of those weak actions of the retina, or of the auditory nerve.

7. The temporary catenations of ideas, which are caused by the sensations of pleasure or pain, are easily dissevered either by irritations, as when a sudden noise disturbs a day-dream; or by the power of volition, as when we awake from sleep. Hence in our waking hours, whenever an idea occurs, which is incongruous to our former experience, we instantly dissever the train of imagination by the power of volition, and compare the incongruous idea with our previous knowledge of nature, and reject it. This operation of the mind has not yet acquired a specific name, though it is exerted every minute of our waking hours; unless it may be termed INTUITIVE ANALOGY. It is an act of reasoning of which we are unconscious except from its effects in preserving the congruity of our ideas, and bears the same relation to the sensorial power of volition, that irritative ideas, of which we are inconscious except by their effects, do to the sensorial power of irritation; as the former is produced by volition without our attention to it, and the latter by irritation without our attention to them.

If on the other hand a train of imagination or of voluntary ideas are excited with great energy, and passing on with great vivacity, and become dissevered by some violent stimulus, as the discharge of a pistol near one's ear, another circumstance takes place, which is termed SURPRISE; which by exciting violent irritation, and violent sensation, employs for a time the whole sensorial energy, and thus dissevers the passing trains of ideas, before the power of volition has time to compare them with the usual phenomena of nature. In this case fear is generally the companion of surprise, and adds to our embarrassment, as every one experiences in some degree when he hears a noise in the dark, which he cannot instantly account for. This catenation of fear with surprise is owing to our perpetual experience of injuries from external bodies in motion, unless we are upon our guard against them. See Sect. XVIII. 17. XIX. 2.

Many other examples of the catenations of animal motions are explained in Sect. XXXVI. on the Periods of Diseases.

SECT. XVIII
OF SLEEP

1. There are four situations of our system, which in their moderate degrees are not usually termed diseases, and yet abound with many very curious and instructive phenomena; these are sleep, reverie, vertigo, drunkenness. These we shall previously consider, before we step forwards

to develop the causes and cures of diseases with the modes of the operation of medicines.

As all those trains and tribes of animal motion, which are subjected to volition, were the last that were caused, their connection is weaker than that of the other classes; and there is a peculiar circumstance attending this causation, which is, that it is entirely suspended during sleep; whilst the other classes of motion, which are more immediately necessary to life, as those caused by internal stimuli, for instance the pulsations of the heart and arteries, or those catenated with pleasurable sensation, as the powers of digestion, continue to strengthen their habits without interruption. Thus though man in his sleeping state is a much less perfect animal, than in his waking hours; and though he consumes more than one third of his life in this his irrational situation; yet is the wisdom of the Author of nature manifest even in this seeming imperfection of his work!

The truth of this assertion with respect to the large muscles of the body, which are concerned in locomotion, is evident; as no one in perfect sanity walks about in his sleep, or performs any domestic offices: and in respect to the mind, we never exercise our reason or recollection in dreams; we may sometimes seem distracted between contending passions, but we never compare their objects, or deliberate about the acquisition of those objects, if our sleep is perfect. And though many synchronous tribes or successive trains of ideas may represent the houses or walks, which have real existence, yet are they here introduced by their connection with our sensations, and are in truth ideas of imagination, not of recollection.

2. For our sensations of pleasure and pain are experienced with great vivacity in our dreams; and hence all that motley group of ideas, which are caused by them, called the ideas of imagination, with their various associated trains, are in a very vivid manner acted over in the sensorium; and these sometimes call into action the larger muscles, which have been much associated with them; as appears from the muttering sentences, which some people utter in their dreams, and from the obscure barking of sleeping dogs, and the motions of their feet and nostrils.

This perpetual flow of the trains of ideas, which constitute our dreams, and which are caused by painful or pleasurable sensation, might at first view be conceived to be an useless expenditure of sensorial power. But it has been shewn, that those motions, which are perpetually excited, as those of the arterial system by the stimulus of the blood, are attended by a great accumulation of sensorial power, after they have been for a time suspended; as the hot-fit of fever is the consequence of the cold one. Now as these trains of ideas caused by sensation are perpetually excited during our waking hours, if they were to be suspended in sleep like the voluntary motions, (which are exerted only by intervals during our waking hours,) an accumulation of sensorial power would follow; and on our awaking a delirium would supervene, since these ideas caused by sensation would be produced with such energy, that we should mistake the trains of imagination for ideas excited by irritation; as perpetually happens to people debilitated by fevers on their first awaking; for in these fevers with debility the general quantity of irritation being diminished, that of sensation is increased. In like manner if the actions of the stomach, intestines, and various glands, which are perhaps in part at least caused by or catenated with agreeable sensation, and which perpetually exist during our waking hours, were like the voluntary motions suspended in our sleep; the great accumulation of sensorial power, which would necessarily follow, would be liable to excite inflammation in them.

3. When by our continued posture in sleep, some uneasy sensations are produced, we either gradually awake by the exertion of volition, or the muscles connected by habit with such sensations alter the position of the body; but where the sleep is uncommonly profound, and those uneasy sensations great, the disease called the incubus, or nightmare, is produced. Here the desire of moving the body is painfully exerted, by the power of moving it, or volition, is incapable of action, till we awake. Many less disagreeable struggles in our dreams, as when we wish in vain to fly from terrifying objects, constitute a slighter degree of this disease. In awaking from the nightmare I have more than once observed, that there was no disorder in my pulse; nor do I believe the respiration is laborious, as some have affirmed. It occurs to people whose sleep is too profound, and some disagreeable sensation exists, which at other times would have awakened them, and have thence prevented the disease of nightmare; as after great fatigue or hunger with too large a supper and wine, which occasion our sleep to be uncommonly profound. See No. 14, of this Section.

4. As the larger muscles of the body are much more frequently excited by volition than by sensation, they are but seldom brought into action in our sleep: but the ideas of the mind are by habit much more frequently connected with sensation than with volition; and hence the ceaseless flow of our ideas in dreams. Every one's experience will teach him this truth, for we all daily exert much voluntary muscular motion: but few of mankind can bear the fatigue of much voluntary thinking.

5. A very curious circumstance attending these our sleeping imaginations is, that we seem to receive them by the senses. The muscles, which are subservient to the external organs of sense, are connected with volition, and cease to act in sleep; hence the eyelids are closed, and the

tympanum of the ear relaxed; and it is probable a similarity of voluntary exertion may be necessary for the perceptions of the other nerves of sense; for it is observed that the papillæ of the tongue can be seen to become erected, when we attempt to taste any thing extremely grateful. Hewson Exper. Enquir. V. 2. 186. Albini Annot. Acad. L. i. c. 15. Add to this, that the immediate organs of sense have no objects to excite them in the darkness and silence of the night, but their nerves of sense nevertheless continue to possess their perfect activity subservient to all their numerous sensitive connections. This vivacity of our nerves of sense during the time of sleep is evinced by a circumstance, which almost every one must at some time or other have experienced; that is, if we sleep in the daylight, and endeavour to see some object in our dream, the light is exceedingly painful to our eyes; and after repeated struggles we lament in our sleep, that we cannot see it. In this case I apprehend the eyelid is in some degree opened by the vehemence of our sensations; and, the iris being dilated, the optic nerve shews as great or greater sensibility than in our waking hours. See No. 15. of this Section.

When we are forcibly waked at midnight from profound sleep, our eyes are much dazzled with the light of the candle for a minute or two, after there has been sufficient time allowed for the contraction of the iris; which is owing to the accumulation of sensorial power in the organ of vision during its state of less activity. But when we have dreamt much of visible objects, this accumulation of sensorial power in the organ of vision is lessened or prevented, and we awake in the morning without being dazzled with the light, after the iris has had time to contract itself. This is a matter of great curiosity, and may be thus tried by any one in the day-light. Close your eyes, and cover them with your hat; think for a minute on a tune, which you are accustomed to, and endeavour to sing it with as little activity of mind as possible. Suddenly uncover and open your eyes, and in one second of time the iris will contract itself, but you will perceive the day more luminous for several seconds, owing to the accumulation of sensorial power in the optic nerve.

Then again close and cover your eyes, and think intensely on a cube of ivory two inches diameter, attending first to the north and south sides of it, and then to the other four sides of it; then get a clear image in your mind's eye of all the sides of the same cube coloured red; and then of it coloured green; and then of it coloured blue; lastly, open your eyes as in the former experiment, and after the first second of time allowed for the contraction of the iris, you will not perceive any increase of the light of the day, or dazzling; because now there is no accumulation of sensorial power in the optic nerve; that having been expended by its action in thinking over visible objects.

This experiment is not easy to be made at first, but by a few patient trials the fact appears very certain; and shews clearly, that our ideas of imagination are repetitions of the motions of the nerve, which were originally occasioned by the stimulus of external bodies; because they equally expend the sensorial power in the organ of sense. See Sect. III. 4. which is analogous to our being as much fatigued by thinking as by labour.

6. Nor is it in our dreams alone, but even in our waking reveries, and in great efforts of invention, so great is the vivacity of our ideas, that we do not for a time distinguish them from the real presence of substantial objects; though the external organs of sense are open, and surrounded with their usual stimuli. Thus whilst I am thinking over the beautiful valley, through which I yesterday travelled, I do not perceive the furniture of my room: and there are some, whose waking imaginations are so apt to run into perfect reverie, that in their common attention to a favourite idea they do not hear the voice of the companion, who accosts them, unless it is repeated with unusual energy.

This perpetual mistake in dreams and reveries, where our ideas of imagination are attended with a belief of the presence of external objects, evinces beyond a doubt, that all our ideas are repetitions of the motions of the nerves of sense, by which they were acquired; and that this belief is not, as some late philosophers contend, an instinct necessarily connected only with our perceptions.

7. A curious question demands our attention in this place; as we do not distinguish in our dreams and reveries between our perceptions of external objects, and our ideas of them in their absence, how do we distinguish them at any time? In a dream, if the sweetness of sugar occurs to my imagination, the whiteness and hardness of it, which were ideas usually connected with the sweetness, immediately follow in the train; and I believe a material lump of sugar present before my senses: but in my waking hours, if the sweetness occurs to my imagination, the stimulus of the table to my hand, or of the window to my eye, prevents the other ideas of the hardness and whiteness of the sugar from succeeding; and hence I perceive the fallacy, and disbelieve the existence of objects correspondent to those ideas, whose tribes or trains are broken by the stimulus of other objects. And further in our waking hours, we frequently exert our volition in comparing present appearances with such, as we have usually observed; and thus correct the errors of one sense by our general knowledge of nature by intuitive analogy. See Sect. XVII. 3. 7. Whereas in dreams the power of volition is suspended, we can recollect and compare our present ideas with none of our acquired knowledge, and are hence incapable of observing any absurdities in them.

By this criterion we distinguish our waking from our sleeping hours, we can voluntarily recollect our sleeping ideas, when we are awake, and compare them with our waking ones; but we cannot in our sleep *voluntarily* recollect our waking ideas at all.

8. The vast variety of scenery, novelty of combination, and distinctness of imagery, are other curious circumstances of our sleeping imaginations. The variety of scenery seems to arise from the superior activity and excellence of our sense of vision; which in an instant unfolds to the mind extensive fields of pleasurable ideas; while the other senses collect their objects slowly, and with little combination; add to this, that the ideas, which this organ presents us with, are more frequently connected with our sensation than those of any other.

9. The great novelty of combination is owing to another circumstance; the trains of ideas, which are carried on in our waking thoughts, are in our dreams dissevered in a thousand places by the suspension of volition, and the absence of irritative ideas, and are hence perpetually falling into new catenations. As explained in Sect. XVII. 1. 9. For the power of volition is perpetually exerted during our waking hours in comparing our passing trains of ideas with our acquired knowledge of nature, and thus forms many intermediate links in their catenation. And the irritative ideas excited by the stimulus of the objects, with which we are surrounded, are every moment intruded upon us, and form other links of our unceasing catenations of ideas.

10. The absence of the stimuli of external bodies, and of volition, in our dreams renders the organs of sense liable to be more strongly affected by the powers of sensation, and of association. For our desires or aversions, or the obtrusions of surrounding bodies, dissever the sensitive and associate tribes of ideas in our waking hours by introducing those of irritation and volition amongst them. Hence proceeds the superior distinctness of pleasurable or painful imagery in our sleep; for we recal the figure and the features of a long lost friend, whom we loved, in our dreams with much more accuracy and vivacity than in our waking thoughts. This circumstance contributes to prove, that our ideas of imagination are reiterations of those motions of our organs of sense, which were excited by external objects; because while we are exposed to the stimuli of present objects, our ideas of absent objects cannot be so distinctly formed.

11. The rapidity of the succession of transactions in our dreams is almost inconceivable; insomuch that, when we are accidentally awakened by the jarring of a door, which is opened into our bed-chamber, we sometimes dream a whole history of thieves or fire in the very instant of awaking.

During the suspension of volition we cannot compare our other ideas with those of the parts of time in which they exist; that is, we cannot compare the imaginary scene, which is before us, with those changes of it, which precede or follow it: because this act of comparing requires recollection or voluntary exertion. Whereas in our waking hours, we are perpetually making this comparison, and by that means our waking ideas are kept confident with each other by intuitive analogy; but this companion retards the succession of them, by occasioning their repetition. Add to this, that the transactions of our dreams consist chiefly of visible ideas, and that a whole history of thieves and fire may be *beheld* in an instant of time like the figures in a picture.

12. From this incapacity of attending to the parts of time in our dreams, arises our ignorance of the length of the night; which, but from our constant experience to the contrary, we should conclude was but a few minutes, when our sleep is perfect. The same happens in our reveries; thus when we are possessed with vehement joy, grief, or anger, time appears short, for we exert no volition to compare the present scenery with the past or future; but when we are compelled to perform those exercises of mind or body, which, are unmixed with passion, as in travelling over a dreary country, time appears long; for our desire to finish our journey occasions us more frequently to compare our present situation with the parts of time or place, which are before and behind us.

So when we are enveloped in deep contemplation of any kind, or in reverie, as in reading a very interesting play or romance, we measure time very inaccurately; and hence, if a play greatly affects our passions, the absurdities of passing over many days or years, and or perpetual changes of place, arc not perceived by the audience; as is experienced by every one, who reads or sees some plays of the immortal Shakespear; but it is necessary for inferior authors to observe those rules of the πιθανον and πρεπον inculcated by Aristotle, because their works do not interest the passions sufficiently to produce complete reverie.

Those works, however, whether a romance or a sermon, which do not interest us so much as to induce reverie, may nevertheless incline us to sleep. For those pleasurable ideas, which are presented to us, and are too gentle to excite laughter, (which is attended with interrupted voluntary exertions, as explained Sect. XXXIV. 1. 4.) and which are not accompanied with any other emotion, which usually excites some voluntary exertion, as anger, or fear, are liable to produce sleep; which consists in a suspension of all voluntary power. But if the ideas thus presented to us, and interest our attention, are accompanied with so much pleasurable or painful sensation as to excite our voluntary exertion at the same time, reverie is the consequence. Hence an interesting play produces reverie, a tedious one produces sleep: in the latter we become exhausted by attention, and are not excited to any voluntary exertion, and therefore sleep; in the former we are excited by some emotion, which prevents by its pain the suspension of volition, and in as much as it interests us, induces reverie, as explained in the next Section.

But when our sleep is imperfect, as when we have determined to rise in half an hour, time appears longer to us than in most other situations. Here our solicitude not to oversleep the determined time induces us in this imperfect sleep to compare the quick changes of imagined scenery with the parts of time or place, they would have taken up, had they real exigence; and that more frequently than in our waking hours; and hence the time appears longer to us: and I make no doubt, but the permitted time appears long to a man going to the gallows, as the fear of its quick lapse will make him think frequently about it.

13. As we gain our knowledge of time by comparing the present scenery with the past and future, and of place by comparing the situations of objects with each other; so we gain our idea of consciousness by comparing ourselves with the scenery around us; and of identity by comparing our present consciousness with our past consciousness: as we never think of time or place, but when we make the companions above mentioned, so we never think of consciousness, but when we compare our own existence with that of other objects; nor of identity, but when we compare our present and our past consciousness. Hence the consciousness of our own existence, and of our identity, is owing to a voluntary exertion of our minds: and on that account in our complete dreams we neither measure time, are surprised at the sudden changes of place, nor attend to our own existence, or identity; because our power of volition is suspended. But all these circumstances are more or less observable in our incomplete ones; for then we attend a little to the lapse of time, and the changes of place, and to our own existence; and even to our identity of person; for a lady seldom dreams, that she is a soldier; nor a man, that he is brought to bed.

14. As long as our sensations only excite their sensual motions, or ideas, our sleep continues sound; but as soon as they excite desires or aversions, our sleep becomes imperfect; and when that desire or aversion is so strong, as to produce voluntary motions, we begin to awake; the larger muscles of the body are brought into action to remove that irritation or sensation, which a continued posture has caused; we stretch our limbs, and yawn, and our sleep is thus broken by the accumulation of voluntary power.

Sometimes it happens, that the act of waking is suddenly produced, and this soon after the commencement of sleep; which is occasioned by some sensation so disagreeable, as instantaneously to excite the power of volition; and a temporary action of all the voluntary motions suddenly succeeds, and we start awake. This is sometimes accompanied with loud noise in the ears, and with some degree of fear; and when it is in great excess, so as to produce continued convulsive motions of those muscles, which are generally subservient to volition, it becomes epilepsy: the fits of which in some patients generally commence during sleep. This differs from the nightmare described in No. 3. of this Section, because in that the disagreeable sensation is not so great as to excite the power of volition into action; for as soon as that happens, the disease ceases.

Another circumstance, which sometimes awakes people soon after the commencement of their sleep, is where the voluntary power is already so great in quantity as almost to prevent them from falling asleep, and then a little accumulation of it soon again awakens them; this happens in cases of insanity, or where the mind has been lately much agitated by fear or anger. There is another circumstance in which sleep is likewise of short duration, which arises from great debility, as after great over-fatigue, and in some fevers, where the strength of the patient is greatly diminished, as in these cases the pulse intermits or flutters, and the respiration is previously affected, it seems to originate from the want of some voluntary efforts to facilitate respiration, as when we are awake. And is further treated of in Vol. II. Class I. 2. 1. 2. on the Diseases of the Voluntary Power. Art. Somnus interruptus.

15. We come now to those motions which depend on irritation. The motions of the arterial and glandular systems continue in our sleep, proceeding slower indeed, but stronger and more uniformly, than in our waking hours, when they are incommoded by external stimuli, or by the movements of volition; the motions of the muscles subservient to respiration continue to be stimulated into action, and the other internal senses of hunger, thirst, and lust, are not only occasionally excited in our sleep, but their irritative motions are succeeded by their usual sensations, and make a part of the farrago of our dreams. These sensations of the want of air, of hunger, thirst, and lust, in our dreams, contribute to prove, that the nerves of the external senses are also alive and excitable in our sleep; but as the stimuli of external objects are either excluded from them by the darkness and silence of the night, or their access to them is prevented by the suspension of volition, these nerves of sense fall more readily into their connexions with sensation and with association; because much sensorial power, which during the day was expended in moving the external organs of sense in consequence of irritation from external stimuli, or in consequence of volition, becomes now in some degree accumulated, and renders the internal or immediate organs

of sense more easily excitable by the other sensorial powers. Thus in respect to the eye, the irritation from external stimuli, and the power of volition during our waking hours, elevate the eye-lids, adapt the aperture of the iris to the quantity of light, the focus of the crystalline humour, and the angle of the optic axises to the distance of the object, all which perpetual activity during the day expends much sensorial power, which is saved during our sleep.

Hence it appears, that not only those parts of the system, which are always excited by internal stimuli, as the stomach, intestinal canal, bile-ducts, and the various glands, but the organs of sense also may be more violently excited into action by the irritation from internal stimuli, or by sensation, during our sleep than in our waking hours; because during the suspension of volition, there is a greater quantity of the spirit of animation to be expended by the other sensorial powers. On this account our irritability to internal stimuli, and our sensibility to pain or pleasure, is not only greater in sleep, but increases as our sleep is prolonged. Whence digestion and secretion are performed better in sleep, than in our waking hours, and our dreams in the morning have greater variety and vivacity, as our sensibility increases, than at night when we first lie down. And hence epileptic fits, which are always occasioned by some disagreeable sensation, so frequently attack those, who are subject to them, in their sleep; because at this time the system is more excitable by painful sensation in consequence of internal stimuli; and the power of volition is then suddenly exerted to relieve this pain, as explained Sect. XXXIV. 1. 4.

There is a disease, which frequently affects children in the cradle, which is termed ecstasy, and seems to consist in certain exertions to relieve painful sensation, in which the voluntary power is not so far excited as totally to awaken them, and yet is sufficient to remove the disagreeable sensation, which excites it; in this case changing the posture of the child frequently relieves it.

I have at this time under my care an elegant young man about twenty-two years of age, who seldom sleeps more than an hour without experiencing a convulsion fit; which ceases in about half a minute without any subsequent stupor. Large doses of opium only prevented the paroxysms, so long as they prevented him from sleeping by the intoxication, which they induced. Other medicines had no effect on him. He was gently awakened every half hour for one night, but without good effect, as he soon slept again, and the fit returned at about the same periods of time, for the accumulated sensorial power, which occasioned the increased sensibility to pain, was not thus exhausted. This case evinces, that the sensibility of the system to internal excitation increases, as our sleep is prolonged; till the pain thus occasioned produces voluntary exertion; which, when it is in its usual degree, only awakens us; but when it is more violent, it occasions convulsions.

The cramp in the calf of the leg is another kind of convulsion, which generally commences in sleep, occasioned by the continual increase of irritability from internal stimuli, or of sensibility, during that state of our existence. The cramp is a violent exertion to relieve pain, generally either of the skin from cold, or of the bowels, as in some diarrhœas, or from the muscles having been previously overstretched, as in walking up or down steep hills. But in these convulsions of the muscles, which form the calf of the leg, the contraction is so violent as to occasion another pain in consequence of their own too violent contraction; as soon as the original pain, which caused the contraction, is removed. And hence the cramp, or spasm, of these muscles is continued without intermission by this new pain, unlike the alternate convulsions and remissions in epileptic fits. The reason, that the contraction of these muscles of the calf of the leg is more violent during their convulsion than that of others, depends on the weakness of their antagonist muscles; for after these have been contracted in their usual action, as at every step in walking, they are again extended, not, as most other muscles are, by their antagonists, but by the weight of the whole body on the balls of the toes; and that weight applied to great mechanical advantage on the heel, that is, on the other end of the bone of the foot, which thus acts as a lever.

Another disease, the periods of which generally commence during our sleep, is the asthma. Whatever may be the remote cause of paroxysms of asthma, the immediate cause of the convulsive respiration, whether in the common asthma, or in what is termed the convulsive asthma, which are perhaps only different degrees of the same disease, must be owing to violent voluntary exertions to relieve pain, as in other convulsions; and the increase of irritability to internal stimuli, or of sensibility, during sleep must occasion them to commence at this time.

Debilitated people, who have been unfortunately accustomed to great ingurgitation of spirituous potation, frequently part with a great quantity of water during the night, but with not more than usual in the day-time. This is owing to a beginning torpor of the absorbent system, and precedes anasarca, which commences in the day, but is cured in the night by the increase of the irritability of the absorbent system during sleep, which thus imbibes from the cellular membrane the fluids, which had been accumulated there during the day; though it is possible the horizontal position of the body may contribute something to this purpose, and also the greater irritability of some branches of the absorbent vessels, which open their mouths in the cells of the cellular membrane, than that of other branches.

As soon as a person begins to sleep, the irritability and sensibility of the system begins to increase, owing to the suspension of volition and the exclusion of external stimuli. Hence the actions of the vessels in obedience to internal stimulation become stronger and more energetic, though less frequent in respect to number. And as many of the secretions are increased, so the heat of the system is gradually increased, and the extremities of feeble people, which had been cold during the day, become warm. Till towards morning many people become so warm, as to find it necessary to throw off some of their bed-clothes, as soon as they awake; and in others sweats are so liable to occur towards morning during their sleep.

Thus those, who are not accustomed to sleep in the open air, are very liable to take cold, if they happen to fall asleep on a garden bench, or in a carriage with the window open. For as the system is warmer during sleep, as above explained, if a current of cold air affects any part of the body, a torpor of that part is more effectually produced, as when a cold blast of air through a key-hole or casement falls upon a person in a warm room. In those cases the affected part possesses less irritability in respect to heat, from its having previously been exposed to a greater stimulus of heat, as in the warm room, or during sleep; and hence, when the stimulus of heat is diminished, a torpor is liable to ensue; that is, we take cold. Hence people who sleep in the open air, generally feel chilly both at the approach of sleep, and on their awaking; and hence many people are perpetually subject to catarrhs if they sleep in a less warm head-dress, than that which they wear in the day.

16. Not only the sensorial powers of irritation and of sensation, but that of association also appear to act with greater vigour during the suspension of volition in sleep. It will be shewn in another place, that the gout generally first attacks the liver, and that afterwards an inflammation of the ball of the great toe commences by association, and that of the liver ceases. Now as this change or metastasis of the activity of the system generally commences in sleep, it follows, that these associations of motion exist with greater energy at that time; that is, that the sensorial faculty of association, like those of irritation and of sensation, becomes in some measure accumulated during the suspension of volition.

Other associate tribes and trains of motions, as well as the irritative and sensitive ones, appear to be increased in their activity during the suspension of volition in sleep. As those which contribute to circulate the blood, and to perform the various secretions; as well as the associate tribes and trains of ideas, which contribute to furnish the perpetual dreams of our dreaming imaginations.

In sleep the secretions have generally been supposed to be diminished, as the expectorated mucus in coughs, the fluids discharged in diarrhœas, and in salivation, except indeed the secretion of sweat, which is often visibly increased. This error seems to have arisen from attention to the excretions rather than to the secretions. For the secretions, except that of sweat, are generally received into reservoirs, as the urine into the bladder, and the mucus of the intestines and lungs into their respective cavities; but these reservoirs do not exclude these fluids immediately by their stimulus, but require at the same time some voluntary efforts, and therefore permit them to remain during sleep. And as they thus continue longer in those receptacles in our sleeping hours, a greater part is absorbed from them, and the remainder becomes thicker, and sometimes in less quantity, though at the time it was secreted the fluid was in greater quantity than in our waking hours. Thus the urine is higher coloured after long sleep; which shews that a greater quantity has been secreted, and that more of the aqueous and saline part has been reabsorbed, and the earthy part left in the bladder; hence thick urine in fevers shews only a greater action of the vessels which secrete it in the kidneys, and of those which absorb it from the bladder.

The same happens to the mucus expectorated in coughs, which is thus thickened by absorption of its aqueous and saline parts; and the same of the feces of the intestines. From hence it appears, and from what has been said in No. 15. of this Section concerning the increase of irritability and of sensibility during sleep, that the secretions are in general rather increased than diminished during these hours of our existence; and it is probable that nutrition is almost entirely performed in sleep; and that young animals grow more at this time than in their waking hours, as young plants have long since been observed to grow more in the night, which is their time of sleep.

17. Two other remarkable circumstances of our dreaming ideas are their inconsistency, and the total absence of surprise. Thus we seem to be present at more extraordinary metamorphoses of animals or trees, than are to be met with in the fables of antiquity; and appear to be transported from place to place, which seas divide, as quickly as the changes of scenery are performed in a play-house; and yet are not sensible of their inconsistency, nor in the least degree affected with surprise.

We must consider this circumstance more minutely. In our waking trains of ideas, those that are inconsistent with the usual order of nature, so rarely have occurred to us, that their connexion is the slightest of all others: hence, when a consistent train of ideas is exhausted, we attend to the external stimuli, that usually surround us, rather than to any inconsistent idea, which might otherwise present itself; and if an inconsistent idea should intrude itself, we immediately compare it with the preceding one, and voluntarily reject the train it would introduce; this appears further in the Section on Reverie, in which state of the mind external stimuli are not attended to, and yet the streams of ideas are kept consistent by the efforts of volition. But as our faculty of volition is suspended, and all external stimuli are excluded in sleep, this slighter connexion of ideas takes place; and the train is said to be inconsistent; that is, dissimilar to the usual order of nature.

But, when any consistent train of sensitive or voluntary ideas is flowing along, if any external stimulus affects us so violently, as to intrude irritative ideas forcibly into the mind, it disunites the former train of ideas, and we are affected with surprise. These stimuli of unusual energy or novelty not only disunite our common trains of ideas, but the trains of muscular motions also, which have not been long established by habit, and disturb those that have. Some people become motionless by great surprise, the fits of hiccup and or ague have been often removed by it, and it even affects the movements of the heart, and arteries; but in our sleep, all external stimuli are excluded, and in consequence no surprise can exist. See Section XVII. 3. 7.

18. We frequently awake with pleasure from a dream, which has delighted us, without being able to recollect the transactions of it; unless perhaps at a distance of time, some analogous idea may introduce afresh this forgotten train: and in our waking reveries we sometimes in a moment lose the train of thought, but continue to feel the glow of pleasure, or the depression of spirits, it occasioned: whilst at other times we can retrace with ease these histories of our reveries and dreams.

The above explanation of surprise throws light upon this subject. When we are suddenly awaked by any violent stimulus, the surprise totally disunites the trains of our sleeping ideas from these of our waking ones; but if we gradually awake, this does not happen; and we readily unravel the preceding trains of imagination.

19. There are various degrees of surprise; the more intent we are upon the train of ideas, which we are employed about, the more violent must be the stimulus that interrupts them, and the greater is the degree of surprise. I have observed dogs, who have slept by the fire, and by their obscure barking and struggling have appeared very intent on their prey, that shewed great surprise for a few seconds after their awaking by looking eagerly around them; which they did not do at other times of waking. And an intelligent

friend of mine has remarked, that his lady, who frequently speaks much and articulately in her sleep, could never recollect her dreams in the morning, when this happened to her: but that when she did not speak in her sleep, she could always recollect them.

Hence, when our sensations act so strongly in sleep as to influence the larger muscles, as in those, who talk or struggle in their dreams; or in those, who are affected with complete reverie (as described in the next Section), great surprise is produced, when they awake; and these as well as those, who are completely drunk or delirious, totally forget afterwards their imaginations at those times.

20. As the immediate cause of sleep consists in the suspension of volition, it follows, that whatever diminishes the general quantity of sensorial power, or derives it from the faculty of volition, will constitute a remote cause of sleep; such as fatigue from muscular or mental exertion, which diminishes the general quantity of sensorial power; or an increase of the sensitive motions, as by attending to soft music, which diverts the sensorial power from the faculty of volition; or lastly, by increase of the irritative motions, as by wine, or food; or warmth; which not only by their expenditure of sensorial power diminish the quantity of volition; but also by their producing pleasureable sensations (which occasion other muscular or sensual motions in consequence), doubly decrease the voluntary power, and thus more forcibly produce sleep. See Sect. XXXIV. 1. 4.

Another method of inducing sleep is delivered in a very ingenious work lately published by Dr. Beddoes. Who, after lamenting that opium frequently occasions restlessness, thinks, "that in most cases it would be better to induce sleep by the abstraction of stimuli, than by exhausting the excitability;" and adds, "upon this principle we could not have a better soporific than an atmosphere with a diminished proportion of oxygene air, and that common air might be admitted after the patient was asleep." (Observ. on Calculus, &c. by Dr. Beddoes, Murray.) If it should be found to be true, that the excitability of the system depends on the quantity of oxygene absorbed by the lungs in respiration according to the theory of Dr. Beddoes, and of M. Girtanner, this idea of sleeping in an atmosphere with less oxygene in its composition might be of great service in epileptic cases, and in cramp, and even in fits of the asthma, where their periods commence from the increase of irritability during sleep.

Sleep is likewise said to be induced by mechanic pressure on the brain in the cases of spina bifida. Where there has been a defect of one of the vertebræ of the back, a tumour is protruded in consequence; and, whenever this tumour has been compressed by the hand, sleep is said to be induced, because the whole of the brain both within the head and spine becomes compressed by the retrocession of the fluid within the tumour. But by what means a compression of the brain induces sleep has not been explained, but probably by diminishing the secretion of sensorial power, and then the voluntary motions become suspended previously to the irritative ones, as occurs in most dying persons.

Another way of procuring sleep mechanically was related to me by Mr. Brindley, the famous canal engineer, who was brought up to the business of a mill-wright; he told me, that he had more than once seen the experiment of a man extending himself across the large stone of a corn-mill, and that by gradually letting the stone whirl, the man fell asleep, before the stone had gained its full velocity, and he supposed would have died without pain by the continuance or increase of the motion. In this case the centrifugal motion of the head and feet must accumulate the blood in both those extremities of the body, and thus compress the brain.

Lastly, we should mention the application of cold; which, when in a less degree, produces watchfulness by the pain it occasions, and the tremulous convulsions of the subcutaneous muscles; but when it is applied in great degree, is said to produce sleep. To explain this effect it has been said, that as the vessels of the skin and extremities become first torpid by the want of the stimulus of heat, and as thence less blood is circulated through them, as appears from their paleness, a greater quantity of blood poured upon the brain produces sleep by its compression of that organ. But I should rather imagine, that the sensorial power becomes exhausted by the convulsive actions in consequence of the pain of cold, and of the voluntary exercise previously used to prevent it, and that the sleep is only the beginning to die, as the suspension of voluntary power in lingering deaths precedes for many hours the extinction of the irritative motions.

21. The following are the characteristic circumstances attending perfect sleep.

1. The power of volition is totally suspended.

2. The trains of ideas caused by sensation proceed with greater facility and vivacity; but become inconsistent with the usual order of nature. The muscular motions caused by sensation continue; as those concerned in our evacuations during infancy, and afterwards in digestion, and in priapismus.

3. The irritative muscular motions continue, as those concerned in the circulation, in secretion, in respiration. But the irritative sensual motions, or ideas, are not excited; as the immediate organs of sense are not stimulated into action by external objects, which are excluded by the external organs of sense; which are not in sleep adapted to their reception by the power of volition, as in our waking hours.

4. The associate motions continue; but their first link is not excited into action by volition, or by external stimuli. In all respects, except those above mentioned, the three last sensorial powers are somewhat increased in energy during the suspension of volition, owing to the consequent accumulation of the spirit of animation.

SECT. XIX
OF REVERIE

1. Various degrees of reverie. 2. Sleep-walkers. Case of a young lady. Great surprise at awaking. And total forgetfulness of what passed in reverie. 3. No suspension of volition in reverie. 4. Sensitive motions continue, and are consistent. 5. Irritative motions continue, but are not succeeded by sensation. 6. Volition necessary for the perception of feeble impressions. 7. Associated motions continue. 8. Nerves of sense are irritable in sleep, but not in reverie. 9. Somnambuli are not asleep. Contagion received but once. 10. Definition of reverie.

1. When we are employed with great sensation of pleasure, or with great efforts of volition, in the pursuit of some interesting train of ideas, we cease to be conscious of our existence, are inattentive to time and place, and do not distinguish this train of sensitive and voluntary ideas from the irritative ones excited by the presence of external objects, though our organs of sense are furnished with their accustomed stimuli, till at length this interesting train of ideas becomes exhausted, or the appulses of external objects are applied with unusual violence, and we return with surprise, or with regret, into the common track of life. This is termed reverie or studium.

In some constitutions these reveries continue a considerable time, and are not to be removed without greater difficulty, but are experienced in a less degree by us all; when we attend earnestly to the ideas excited by volition or sensation, with their associated connexions, but are at the same time conscious at intervals of the stimuli of surrounding bodies. Thus in being present at a play, or in reading a romance, some persons are so totally absorbed as to forget their usual time of sleep, and to neglect their meals; while others are said to have been so involved in voluntary study as not to have heard the discharge of artillery; and there is a story of an Italian politician, who could think so intensely on other subjects, as to be insensible to the torture of the rack.

From hence it appears, that these catenations of ideas and muscular motions, which form the trains of reverie, are composed both of voluntary and sensitive associations of them; and that these ideas differ from those of delirium or of sleep, as they are kept consistent by the power of volition; and they differ also from the trains of ideas belonging to insanity, as they are as frequently excited by sensation as by volition. But lastly, that the whole sensorial power is so employed on these trains of complete reverie, that like the violent efforts of volition, as in convulsions or insanity; or like the great activity of the irritative motions in drunkenness; or of the sensitive motions in delirium; they preclude all sensation consequent to external stimulus.

2. Those persons, who are said to walk in their sleep, are affected with reverie to so great a degree, that it becomes a formidable disease; the essence of which consists in the inaptitude of the mind to attend to external stimuli. Many histories of this disease have been published by medical writers; of which there is a very curious one in the Lausanne Transactions. I shall here subjoin an account of such a case, with its cure, for the better illustration of this subject.

A very ingenious and elegant young lady, with light eyes and hair, about the age of seventeen, in other respects well, was suddenly seized soon after her usual menstruation with this very wonderful malady. The disease began with vehement convulsions of almost every muscle of her body, with great but vain efforts to vomit, and the most violent hiccoughs, that can be conceived: these were succeeded in about an hour with a fixed spasm; in which one hand was applied to her head, and the other to support it: in about half an hour these ceased, and the reverie began suddenly, and was at first manifest by the look of her eyes and countenance, which seemed to express attention. Then she conversed aloud with imaginary persons with her eyes open, and could not for about an hour be brought to attend to the stimulus of external objects by any kind of violence, which it was proper to use; these symptoms returned in this order every day for five or six weeks.

These conversations were quite consistent, and we could understand, what she supposed her imaginary companions to answer, by the continuation of her part of the discourse. Sometimes she was angry, at other times shewed much wit and vivacity, but was most frequently inclined to melancholy. In these reveries she sometimes sung over some music with accuracy, and repeated whole pages from the English poets. In repeating some lines from Mr. Pope's works she had forgot one word, and began again, endeavouring to recollect it; when she came to the forgotten word, it was shouted aloud in her ear, and this repeatedly, to no purpose; but by many trials she at length regained it herself.

These paroxysms were terminated with the appearance of inexpressible surprise, and great fear, from which she was some minutes in recovering herself, calling on her sister with great agitation, and very frequently underwent a repetition of convulsions, apparently from the pain of fear. See Sect. XVII. 3. 7.

After having thus returned for about an hour every day for two or three weeks, the reveries seemed to become less complete, and some of their circumstances varied; so that she could walk about the room in them without running against any of the furniture; though these motions were at first very unsteady and tottering. And afterwards she once drank a dish of tea, when the whole apparatus of the tea-table was set before her; and expressed some suspicion, that a medicine was put into it, and once seemed to smell of a tuberose, which was in flower in her chamber, and deliberated aloud about breaking it from the stem, saying, "it would make her sister so charmingly angry." At another time in her melancholy moments she heard the sound of a passing bell, "I wish I was dead," she cried, listening to the bell, and then taking off one of her shoes, as she sat upon the bed, "I love the colour black," says she, "a little wider, and a little longer, even this might make me a coffin!" — Yet it is evident, she was not sensible at this time, any more than formerly, of seeing or hearing any person about her; indeed when great light was thrown upon her by opening the shutters of the window, her trains of ideas seemed less melancholy; and when I have forcibly held her hands, or covered her eyes, she appeared to grow impatient, and would say, she could not tell what to do, for she could neither see nor move. In all these circumstances her pulse continued unaffected as in health. And when the paroxysm was over, she could never recollect a single idea of what had passed in it.

This astonishing disease, after the use of many other medicines and applications in vain, was cured by very large doses of opium given about an hour before the expected returns of the paroxysms; and after a few relapses, at the intervals of three or four months, entirely disappeared. But she continued at times to have other symptoms of epilepsy.

3. We shall only here consider, what happened during the time of her reveries, as that is our present subject; the fits of convulsion belong to another part of this treatise. Sect. XXXIV. 1. 4.

There seems to have been no suspension of volition during the fits of reverie, because she endeavoured to regain the lost idea in repeating the lines of poetry, and deliberated about breaking the tuberose, and suspected the tea to have been medicated.

4. The ideas and muscular movements depending on sensation were exerted with their usual vivacity, and were kept from being inconsistent by

the power of volition, as appeared from her whole conversation, and was explained in Sect. XVII. 3. 7. and XVIII. 16.

5. The ideas and motions dependant on irritation during the first weeks of her disease, whilst the reverie was complete, were never succeeded by the sensation of pleasure or pain; as she neither saw, heard, nor felt any of the surrounding objects. Nor was it certain that any irritative motions succeeded the stimulus of external objects, till the reverie became less complete, and then she could walk about the room without running against the furniture of it. Afterwards, when the reverie became still less complete from the use of opium, some few irritations were at times succeeded by her attention to them. As when she smelt at a tuberose, and drank a dish of tea, but this only when she seemed voluntarily to attend to them.

6. In common life when we listen to distant sounds, or wish to distinguish objects in the night, we are obliged strongly to exert our volition to dispose the organs of sense to perceive them, and to suppress the other trains of ideas, which might interrupt these feeble sensations. Hence in the present history the strongest stimuli were not perceived, except when the faculty of volition was exerted on the organ of sense; and then even common stimuli were sometimes perceived: for her mind was so strenuously employed in pursuing its own trains of voluntary or sensitive ideas, that no common stimuli could so far excite her attention as to disunite them; that is, the quantity of volition or of sensation already existing was greater than any, which could be produced in consequence of common degrees of stimulation. But the few stimuli of the tuberose, and of the tea, which she did perceive, were such, as accidentally coincided with the trains of thought, which were passing in her mind; and hence did not disunite those trains, and create surprise. And their being perceived at all was owing to the power of volition preceding or coinciding with that of irritation.

This explication is countenanced by a fact mentioned concerning a somnambulist in the Lausanne Transactions, who sometimes opened his eyes for a short time to examine, where he was, or where his ink-pot stood, and then shut them again, dipping his pen into the pot every now and then, and writing on, but never opening his eyes afterwards, although he wrote on from line to line regularly, and corrected some errors of the pen, or in spelling: so much easier was it to him to refer to his ideas of the positions of things, than to his perceptions of them.

7. The associated motions persisted in their usual channel, as appeared by the combinations of her ideas, and the use of her muscles, and the equality of her pulse; for the natural motions of the arterial system, though originally excited like other motions by stimulus, seem in part to continue

by their association with each other. As the heart of a viper pulsates long after it is cut out of the body, and removed from the stimulus of the blood.

8. In the section on sleep, it was observed that the nerves of sense are equally alive and susceptible to irritation in that state, as when we are awake; but that they are secluded from stimulating objects, or rendered unfit to receive them: but in complete reverie the reverse happens, the immediate organs of sense are exposed to their usual stimuli; but are either not excited into action at all, or not into so great action, as to produce attention or sensation.

The total forgetfulness of what passes in reveries; and the surprise on recovering from them, are explained in Section XVIII. 19. and in Section XVII. 3. 7.

9. It appears from hence, that reverie is a disease of the epileptic or cataleptic kind, since the paroxysms of this young lady always began and frequently terminated with convulsions; and though in its greatest degree it has been called somnambulation, or sleep-walking, it is totally different from sleep; because the essential character of sleep consists in the total suspension of volition, which in reverie is not affected; and the essential character of reverie consists not in the absence of those irritative motions of our senses, which are occasioned by the stimulus of external objects, but in their never being productive of sensation. So that during a fit of reverie that strange event happens to the whole system of nerves, which occurs only to some particular branches of them in those, who are a second time exposed to the action of contagious matter. If the matter of the small-pox be inserted into the arm of one, who has previously had that disease, it will stimulate the wound, but the general sensation or inflammation of the system does not follow, which constitutes the disease. See Sect. XII. 3. 6. XXXIII. 2. 8.

10. The following is the definition or character of complete reverie. 1. The irritative motions occasioned by internal stimuli continue, those from the stimuli of external objects are either not produced at all, or are never succeeded by sensation or attention, unless they are at the same time excited by volition. 2. The sensitive motions continue, and are kept consistent by the power of volition. 3. The voluntary motions continue undisturbed. 4. The associate motions continue undisturbed.

Two other cases of reverie are related in Section XXXIV. 3. which further evince, that reverie is an effort of the mind to relieve some painful sensation, and is hence allied to convulsion, and to insanity. Another case is related in Class III. 1. 2. 2.

SECT. XX
OF VERTIGO

1. We determine our perpendicularity by the apparent motions of objects. A person hood-winked cannot walk in a straight line. Dizziness in looking from a tower, in a room stained with uniform lozenges, on riding over snow. 2. Dizziness from moving objects. A whirling-wheel. Fluctuations of a river. Experiment with a child. 3. Dizziness from our own motions and those of other objects. 4. Riding over a broad stream. Sea-sickness. 5. Of turning round on one foot. Dervises in Turkey. Attention of the mind prevents slight sea-sickness. After a voyage ideas of vibratory motions are still perceived on shore. 6. Ideas continue some time after they are excited. Circumstances of turning on one foot, standing on a tower, and walking in the dark, explained. 7. Irritative ideas of apparent motions. Irritative ideas of sounds. Battèment of the sound of bells and organ-pipes. Vertiginous noise in the head. Irritative motions of the stomach, intestines, and glands. 8. Symptoms that accompany vertigo. Why vomiting comes on in strokes of the palsy. By the motion of a ship. By injuries on the head. Why motion makes sick people vomit. 9. Why drunken people are vertiginous. Why a stone in the ureter, or bile-duct, produces vomiting. 10. Why after a voyage ideas of vibratory motions are perceived on shore. 11. Kinds of vertigo and their cure. 12. Definition of vertigo.

1. In learning to walk we judge of the distances of the objects, which we approach, by the eye; and by observing their perpendicularity determine our own. This circumstance not having been attended to by the writers on vision, the disease called vertigo or dizziness has been little understood.

When any person loses the power of muscular action, whether he is erect or in a sitting posture, he sinks down upon the ground; as is seen in fainting fits, and other instances of great debility. Hence it follows, that some exertion of muscular power is necessary to preserve our perpendicular attitude. This is performed by proportionally exerting the antagonist muscles of the trunk, neck, and limbs; and if at any time in our locomotions we find ourselves inclining to one side, we either restore our equilibrium by the efforts of the muscles on the other side, or by moving one of our feet extend the base, which we rest upon, to the new center of gravity.

But the most easy and habitual manner of determining our want of perpendicularity, is by attending to the apparent motion of the objects within the sphere of distinct vision; for this apparent motion of objects, when we incline from our perpendicularity, or begin to fall, is as much greater than the real motion of the eye, as the diameter of the sphere of distinct vision is to our perpendicular height.

Hence no one, who is hood-winked, can walk in a straight line for a hundred steps together; for he inclines so greatly, before he is warned of his want of perpendicularity by the sense of touch, not having the apparent motions of ambient objects to measure this inclination by, that he is necessitated to move one of his feet outwards, to the right or to the left, to support the new centre of gravity, and thus errs from the line he endeavours to proceed in.

For the same reason many people become dizzy, when they look from the summit of a tower, which is raised much above all other objects, as these objects are out of the sphere of distinct vision, and they are obliged to balance their bodies by the less accurate feelings of their muscles.

There is another curious phenomenon belonging to this place, if the circumjacent visible objects are so small, that we do not distinguish their minute parts; or so similar, that we do not know them from each other; we cannot determine our perpendicularity by them. Thus in a room hung with a paper, which is coloured over with similar small black lozenges or rhomboids, many people become dizzy; for when they begin to fall, the next and the next lozenge succeeds upon the eye; which they mistake for the first, and are not aware, that they have any apparent motion. But if you fix a sheet of paper, or draw any other figure, in the midst of these lozenges, the charm ceases, and no dizziness is perceptible.—The same occurs, when we ride over a plain covered with snow without trees or other eminent objects.

2. But after having compared visible objects at rest with the sense of touch, and learnt to distinguish their shapes and shades, and to measure our want of perpendicularity by their apparent motions, we come to consider them in real motion. Here a new difficulty occurs, and we require some experience to learn the peculiar mode of motion of any moving objects, before we can make use of them for the purposes of determining our perpendicularity. Thus some people become dizzy at the sight of a whirling wheel, or by gazing on the fluctuations of a river, if no steady objects are at the same time within the sphere of their distinct vision; and when a child first can stand erect upon his legs, if you gain his attention to a white handkerchief steadily extended like a sail, and afterwards make it undulate, he instantly loses his perpendicularity, and tumbles on the ground.

3. A second difficulty we have to encounter is to distinguish our own real movements from the apparent motions of objects. Our daily practice of walking and riding on horseback soon instructs us with accuracy to discern these modes of motion, and to ascribe the apparent motions of the ambient objects to ourselves; but those, which we have not acquired by repeated habit, continue to confound us. So as we ride on horseback the trees and cottages, which occur to us, appear at rest; we can measure their distances with our eye, and regulate our attitude by them; yet if we carelessly attend to distant hills or woods through a thin hedge, which is near us, we observe the jumping and progressive motions of them; as this is increased by the paralax of these objects; which we have not habituated ourselves to attend to. When first an European mounts an elephant sixteen feet high, and whose mode of motion he is not accustomed to, the objects seem to undulate, as he passes, and he frequently becomes sick and vertiginous, as I am well informed. Any other unusual movement of our bodies has the same effect, as riding backwards in a coach, swinging on a rope, turning round swiftly on one leg, scating on the ice, and a thousand others. So after a patient has been long confined to his bed, when he first attempts to walk, he finds himself vertiginous, and is obliged by practice to learn again the particular modes of the apparent motions of objects, as he walks by them.

4. A third difficulty, which occurs to us in learning to balance ourselves by the eye, is, when both ourselves and the circumjacent objects are in real motion. Here it is necessary, that we should be habituated to both these modes of motion in order to preserve our perpendicularity. Thus on horseback we accurately observe another person, whom we meet, trotting towards us, without confounding his jumping and progressive motion with our own, because we have been accustomed to them both; that is, to undergo the one, and to see the other at the same time. But in riding over a broad and fluctuating stream, though we are well experienced in the motions of our horse, we are liable to become dizzy from our inexperience in that of the water. And when first we go on ship-board, where the movements of ourselves, and the movements of the large waves are both new to us, the vertigo is almost unavoidable with the terrible sickness, which attends it. And this I have been assured has happened to several from being removed from a large ship into a small one; and again from a small one into a man of war.

5. From the foregoing examples it is evident, that, when we are surrounded with unusual motions, we lose our perpendicularity: but there are some peculiar circumstances attending this effect of moving objects, which we come now to mention, and shall hope from the recital of them to gain some insight into the manner of their production.

When a child moves round quick upon one foot, the circumjacent objects become quite indistinct, as their distance increases their apparent motions; and this great velocity confounds both their forms, and their colours, as is seen in whirling round a many coloured wheel; he then loses his usual method of balancing himself by vision, and begins to stagger, and attempts to recover himself by his muscular feelings. This staggering adds to the instability of the visible objects by giving a vibratory motion besides their rotatory one. The child then drops upon the ground, and the neighbouring objects seem to continue for some seconds of time to circulate around him, and the earth under him appears to librate like a balance. In some seconds of time these sensations of a continuation of the motion of objects vanish; but if he continues turning round somewhat longer, before he falls, sickness and vomiting are very liable to succeed. But none of these circumstances affect those who have habituated themselves to this kind of motion, as the dervises in Turkey, amongst whom these swift gyrations are a ceremony of religion.

In an open boat passing from Leith to Kinghorn in Scotland, a sudden change of the wind shook the undistended sail, and stopt our boat; from this unusual movement the passengers all vomited except myself. I observed, that the undulation of the ship, and the instability of all visible objects, inclined me strongly to be sick; and this continued or increased, when I closed my eyes, but as often as I bent my attention with energy on the management and mechanism of the ropes and sails, the sickness ceased; and recurred again, as often as I relaxed this attention; and I am assured by a gentleman of observation and veracity, that he has more than once observed, when the vessel has been in immediate danger, that the sea-sickness of the passengers has instantaneously ceased, and recurred again, when the danger was over.

Those, who have been upon the water in a boat or ship so long, that they have acquired the necessary habits of motion upon that unstable element, at their return on land frequently think in their reveries, or between sleeping and waking, that they observe the room, they sit in, or some of its furniture, to librate like the motion of the vessel. This I have experienced myself, and have been told, that after long voyages, it is some time before these ideas entirely vanish. The same is observable in a less degree after having travelled some days in a stage coach, and particularly when we lie down in bed, and compose ourselves to sleep; in this case it is observable, that the rattling noise of the coach, as well as the undulatory motion, haunts us. The drunken vertigo, and the vulgar custom of rocking children, will be considered in the next Section.

6. The motions, which are produced by the power of volition, may be immediately stopped by the exertion of the same power on the antagonist

muscles; otherwise these with all the other classes of motion continue to go on, some time after they are excited, as the palpitation of the heart continues after the object of fear, which occasioned it, is removed. But this circumstance is in no class of motions more remarkable than in those dependent on irritation; thus if any one looks at the sun, and then covers his eyes with his hand, he will for many seconds of time, perceive the image of the sun marked on his retina: a similar image of all other visible objects would remain some time formed on the retina, but is extinguished by the perpetual change of the motions of this nerve in our attention to other objects. To this must be added, that the longer time any movements have continued to be excited without fatigue to the organ, the longer will they continue spontaneously, after the excitement is withdrawn: as the taste of tobacco in the mouth after a person has been smoking it.

This taste remains so strong, that if a person continues to draw air through a tobacco pipe in the dark, after having been smoking some time, he cannot distinguish whether his pipe be lighted or not.

From these two considerations it appears, that the dizziness felt in the head, after seeing objects in unusual motion, is no other than a continuation of the motions of the optic nerve excited by those objects and which engage our attention. Thus on turning round on one foot, the vertigo continues for some seconds of time after the person is fallen on the ground; and the longer he has continued to revolve, the longer will continue these successive motions of the parts of the optic nerve.

Additional Observations on VERTIGO.

> After revolving with your eyes open till you become vertiginous, as soon as you cease to revolve, not only the circum-ambient objects appear to circulate round you in a direction contrary to that, in which you have been turning, but you are liable to roll your eyes forwards and backwards; as is well observed, and ingeniously demonstrated by Dr. Wells in a late publication on vision. The same occurs, if you revolve with your eyes closed, and open them immediately at the time of your ceasing to turn; and even during the whole time of revolving, as may be felt by your hand pressed lightly on your closed eyelids. To these movements of the eyes, of which he supposes the observer to be inconscious, Dr. Wells ascribes the apparent circumgyration of objects on ceasing to revolve.

> The cause of thus turning our eyes forwards, and then back again, after our body is at rest, depends, I imagine, on the

same circumstance, which induces us to follow the indistinct spectra, which are formed on one side of the center of the retina, when we observe them apparently on clouds, as described in Sect. XL. 2. 2.; and then not being able to gain a more distinct vision of them, we turn our eyes back, and again and again pursue the flying shade.

But this rolling of the eyes, after revolving till we become vertiginous, cannot cause the apparent circumgyration of objects, in a direction contrary to that in which we have been revolving, for the following reasons. 1. Because in pursuing a spectrum in the sky, or on the ground, as above mentioned, we perceive no retrograde motions of objects. 2. Because the apparent retrograde motions of objects, when we have revolved till we are vertiginous, continues much longer than the rolling of the eyes above described.

3. When we have revolved from right to left, the apparent motion of objects, when we stop, is from left to right; and when we have revolved from left to right, the apparent circulation of objects is from right to left; yet in both these cases the eyes of the revolver are seen equally to roll forwards and backwards.

4. Because this rolling of the eyes backwards and forwards takes place during our revolving, as may be perceived by the hand lightly pressed on the closed eyelids, and therefore exists before the effect ascribed to it.

And fifthly, I now come to relate an experiment, in which the rolling of the eyes does not take place at all after revolving, and yet the vertigo is more distressing than in the situations above mentioned. If any one looks steadily at a spot in the ceiling over his head, or indeed at his own finger held up high over his head, and in that situation turns round till he becomes giddy; and then stops, and looks horizontally; he now finds, that the apparent rotation of objects is from above downwards, or from below upwards; that is, that the apparent circulation of objects is now vertical instead of horizontal, making part of a circle round the axis of his *eye*; and this without any rolling of his eyeballs. The reason of there being no rolling of the eyeballs, perceived after this experiment, is, because the images of objects are formed in rotation round the axis of the eye, and not from one side to the other of the axis of it; so that, as the eyeball has not power to turn in its socket round its own axis, it cannot follow the apparent motions of these evanescent spectra, either before or after the body is at rest. From all which arguments it is manifest, that these apparent retrograde gyrations of objects are not caused by the rolling of the eyeballs; first, because no apparent retrogression of objects is observed in other rollings of the eyes: secondly, because the apparent retrogression of objects continues many seconds after the rolling of the eyeballs ceases. Thirdly, because the apparent retrogression of objects is sometimes one way, and sometimes another, yet the rolling of the eyeballs is the same. Fourthly, because the rolling of the eyeballs exists before the apparent retrograde motions of objects is observed; that is, before the revolving person stops. And fifthly, because the apparent retrograde gyration of objects is produced, when there is no rolling of the eyeballs at all.

Doctor Wells imagines, that no spectra can be gained in the eye, if a person revolves with his eyelids closed, and thinks this a sufficient argument against the opinion, that the apparent progression of the spectra of light or colours in the eye can cause the apparent retrogression of objects in the vertigo above described; but it is certain, when any person revolves in a light room with his eyes closed, that he nevertheless perceives differences of light both in quantity and colour through his eyelids, as he turns round; and readily gains spectra of those differences. And these spectra are not very different except in vivacity from those, which he acquires, when he revolves with unclosed eyes, since if he then revolves very rapidly the colours and forms of surrounding objects are as it were mixed together in his eye;. as when, the prismatic colours are painted on a wheel, they appear white as they revolve. The truth of this is evinced by the staggering or vertigo of men perfectly blind, when they turn round; which is not attended with apparent circulation of objects, but is a vertiginous disorder of the sense of touch. Blind men balance themselves by their sense of touch; which, being less adapted for perceiving small deviations from their perpendicular, occasions them to carry themselves more erect in walking. This method of balancing themselves by the direction of their pressure against the floor, becomes disordered by the unusual mode of action in turning round, and they begin to lose their perpendicularity, that is, they become vertiginous; but without any apparent circular motions of visible objects.

It will appear from the following experiments, that the apparent progression of the ocular spectra of light or colours is the cause of the apparent retrogression of objects, after a person has revolved, till he is vertiginous.

First, when a person turns round in a light room with his eyes open, but closes them before he stops, he will seem to be carried forwards in the direction he was turning for a short time after he stops. But if he opens his eyes again, the objects before him instantly appear to move in a retrograde direction, and he loses the sensation of being carried forwards. The same occurs if a person revolves in a light room with his eyes closed; when he stops, he seems to be for a time carried forwards, if his eyes are still closed; but the instant he opens them, the surrounding objects appear to move in retrograde gyration. From hence it may be concluded, that it is the sensation or imagination of our continuing to go forwards in the direction in which we were turning, that causes the apparent retrograde circulation of objects.

Secondly, though there is an audible vertigo, as is known by the battement, or undulations of sound in the ears, which many vertiginous people experience; and though there is also a tangible vertigo, as when a blind person turns round, as mentioned above; yet as this circumgyration of objects is an hallucination or deception of the sense of sight, we are to look for the cause of our appearing to move forward, when we stop with our eyes closed after gyration, to some affection of this sense. Now, thirdly, if the spectra formed in the eye during our rotation, continue to change, when we stand still, like the spectra described in Sect. III. 3. 6. such changes must suggest to us the idea or sensation of our still continuing to turn round; as is the case, when we revolve in a light room, and close our eyes before we stop. And lastly, on opening our eyes in the situation above described, the objects we chance to view amid these changing spectra in the eye, must seem to move in a contrary direction; as the moon sometimes appears to move retrograde, when swift-gliding clouds are passing forwards so much nearer the eye of the beholder.

To make observations on faint ocular spectra requires some degree of habit, and composure of mind, and even patience; some of those described in Sect. XL. were found difficult to see, by many, who tried them; now it happens, that the mind, during the confusion of vertigo, when all the other irritative tribes of motion, as well as those of vision, are in some degree disturbed, together with the fear of falling, is in a very unfit state for the contemplation of such weak sensations, as are occasioned by faint ocular spectra. Yet after frequently revolving, both with my eyes closed, and with them open, and attending to the spectra remaining in them, by shading the light from my eyelids more or less with my hand, I at length ceased to have the idea of going forward, after I stopped with my eyes closed; and saw changing spectra in my eyes, which seemed to move, as it were, over the field of vision; till at length, by repeated trials on sunny days, I persuaded myself, on opening my eyes, after revolving some time, on a shelf of gilded books in my library, that I could perceive the spectra in my eyes move forwards over one or two of the books, like the vapours in the air of a summer's day; and could so far undeceive myself, as to perceive the books to stand still. After more trials I sometimes brought myself to believe, that I saw changing spectra of lights and shades moving in my eyes, after turning round for some time, but did not imagine either the spectra or the objects to be in a state of gyration. I speak, however, with diffidence of these facts, as I could not always make the experiments succeed, when there was not a strong light in my room, or when my eyes were not in the most proper state for such observations.

The ingenious and learned M. Sauvage has mentioned other theories to account for the apparent circumgyration of objects in vertiginous people. As the retrograde motions of the particles of blood in the optic arteries, by spasm, or by fear, as is seen in the tails of tadpoles, and membranes between the fingers of frogs. Another cause he thinks may be from the librations to one side, and to the other, of the crystalline lens in the eye, by means of involuntary actions of the muscles, which constitute the ciliary process. Both these theories lie under the same objection as that of Dr.

Wells before mentioned; namely, that the apparent motions of objects, after the observer has revolved for some time, should appear to vibrate this way and that; and not to circulate uniformly in a direction contrary to that, in which the observer had revolved.

M. Sauvage has, lastly, mentioned the theory of colours left in the eye, which he has termed impressions on the retina. He says, "Experience teaches us, that impressions made on the retina by a visible object remain some seconds after the object is removed; as appears from the circle of fire which we see, when a fire-stick is whirled round in the dark; therefore when we are carried round our own axis in a circle, we undergo a temporary vertigo, when we stop; because the impressions of the circumjacent objects remain for a time afterwards on the retina." Nosolog. Method. Clas. VIII. I. 1. We have before observed, that the changes of these colours remaining in the eye, evinces them to be motions of the fine terminations of the retina, and not impressions on it; as impressions on a passive substance must either remain, or cease intirely. See an additional note at the end of the second volume.

Any one, who stands alone on the top of a high tower, if he has not been accustomed to balance himself by objects placed at such distances and with such inclinations, begins to stagger, and endeavours to recover himself by his muscular feelings. During this time the apparent motion of objects at a distance below him is very great, and the spectra of these apparent motions continue a little time after he has experienced them; and he is persuaded to incline the contrary way to counteract their effects; and either immediately falls, or applying his hands to the building, uses his muscular feelings to preserve his perpendicular attitude, contrary to the erroneous persuasions of his eyes. Whilst the person, who walks in the dark, staggers, but without dizziness; for he neither has the sensation of moving objects to take off his attention from his muscular feelings, nor has he the spectra of those motions continued on his retina to add to his confusion. It happens indeed sometimes to one landing on a tower, that the idea of his not having room to extend his base by moving one of his feet outwards, when he begins to incline, superadds fears to his other inconveniences; which like surprise, joy, or any great degree of sensation, enervates him in a moment, by employing the whole sensorial power, and by thus breaking all the associated trains and tribes of motion.

7. The irritative ideas of objects, whilst we are awake, are perpetually present to our sense of sight; as we view the furniture of our rooms, or the

ground, we tread upon, throughout the whole day without attending to it. And as our bodies are never at perfect rest during our waking hours, these irritative ideas of objects are attended perpetually with irritative ideas of their apparent motions. The ideas of apparent motions are always irritative ideas, because we never attend to them, whether we attend to the objects themselves, or to their real motions, or to neither. Hence the ideas of the apparent motions of objects are a complete circle of irritative ideas, which continue throughout the day.

Also during all our waking hours, there is a perpetual confused sound of various bodies, as of the wind in our rooms, the fire, distant conversations, mechanic business; this continued buzz, as we are seldom quite motionless, changes its loudness perpetually, like the sound of a bell; which rises and falls as long as it continues, and seems to pulsate on the ear. This any one may experience by turning himself round near a waterfall; or by striking a glass bell, and then moving the direction of its mouth towards the ears, or from them, as long as its vibrations continue. Hence this undulation of indistinct sound makes another concomitant circle of irritative ideas, which continues throughout the day.

We hear this undulating sound, when we are perfectly at rest ourselves, from other sonorous bodies besides bells; as from two organ-pipes, which are nearly but not quite in unison, when they are sounded together. When a bell is struck, the circular form is changed into an eliptic one; the longest axis of which, as the vibrations continue, moves round the periphery of the bell; and when either axis of this elipse is pointed towards our ears, the sound is louder; and less when the intermediate parts of the elipse are opposite to us. The vibrations of the two organ-pipes may be compared to Nonius's rule; the sound is louder, when they coincide, and less at the intermediate times. But, as the sound of bells is the most familiar of those sounds, which have a considerable battement, the vertiginous patients, who attend to the irritative circles of sounds above described, generally compare it to the noise of bells.

The peristaltic motions of our stomach and intestines, and the secretions of the various glands, are other circles of irritative motions, some of them more or less complete, according to our abstinence or satiety.

So that the irritative ideas of the apparent motions of objects, the irritative battements of sounds, and the movements of our bowels and glands compose a great circle of irritative tribes of motion: and when one considerable part of this circle of motions becomes interrupted, the whole proceeds in confusion, as described in Section XVII. 1. 7. on Catenation of Motions.

8. Hence a violent vertigo, from whatever cause it happens, is generally attended with undulating noise in the head, perversions of the motions of

the stomach and duodenum, unusual excretion of bile and gastric juice, with much pale urine, sometimes with yellowness of the skin, and a disordered secretion of almost every gland of the body, till at length the arterial system is affected, and fever succeeds.

Thus bilious vomitings accompany the vertigo occasioned by the motion of a ship; and when the brain is rendered vertiginous by a paralytic affection of any part of the body, a vomiting generally ensues, and a great discharge of bile: and hence great injuries of the head from external violence are succeeded with bilious vomitings, and sometimes with abscesses of the liver. And hence, when a patient is inclined to vomit from other causes, as in some fevers, any motions of the attendants in his room, or of himself when he is raised or turned in his bed, presently induces the vomiting by superadding a degree of vertigo.

9. And conversely it is very usual with those, whose stomachs are affected from internal causes, to be afflicted with vertigo, and noise in the head; such is the vertigo of drunken people, which continues, when their eyes are closed, and themselves in a recumbent posture, as well as when they are in an erect posture, and have their eyes open. And thus the irritation of a stone in the bile-duct, or in the ureter, or an inflammation of any of the intestines, are accompanied with vomitings and vertigo.

In these cases the irritative motions of the stomach, which are in general not attended to, become so changed by some unnatural stimulus, as to become uneasy, and excite our sensation or attention. And thus the other irritative trains of motions, which are associated with it, become disordered by their sympathy. The same happens, when a piece of gravel sticks in the ureter, or when some part of the intestinal canal becomes inflamed. In these cases the irritative muscular motions are first disturbed by unusual stimulus, and a disordered action of the sensual motions, or dizziness ensues. While in sea-sickness the irritative sensual motions, as vertigo, precedes; and the disordered irritative muscular motions, as those of the stomach in vomiting, follow.

10. When these irritative motions are disturbed, if the degree be not very great, the exertion of voluntary attention to any other object, or any sudden sensation, will disjoin these new habits of motion. Thus some drunken people have become sober immediately, when any accident has strongly excited their attention; and sea-sickness has vanished, when the ship has been in danger. Hence when our attention to other objects is most relaxed, as just before we fall asleep, or between our reveries when awake, these irritative ideas of motion and sound are most liable to be perceived; as those, who have been at sea, or have travelled long in a coach, seem to perceive the vibrations of the ship, or the rattling of the wheels, at these intervals; which

cease again, as soon as they exert their attention. That is, at those intervals they attend to the apparent motions, and to the battement of sounds of the bodies around them, and for a moment mistake them for those real motions of the ship, and noise of wheels, which they had lately been accustomed to: or at these intervals of reverie, or on the approach of sleep, these supposed motions or sounds may be produced entirely by imagination.

We may conclude from this account of vertigo, that sea-sickness is not an effort of nature to relieve herself, but a necessary consequence of the associations or catenations of animal motions. And may thence infer, that the vomiting, which attends the gravel in the ureter, inflammations of the bowels, and the commencement of some fevers, has a similar origin, and is not always an effort of the vis medicatrix naturæ. But where the action of the organ is the immediate consequence of the stimulating cause, it is frequently exerted to dislodge that stimulus, as in vomiting up an emetic drug; at other times, the action of an organ is a general effort to relieve pain, as in convulsions of the locomotive muscles; other actions drink up and carry on the fluids, as in absorption and secretion; all which may be termed efforts of nature to relieve, or to preserve herself.

11. The cure of vertigo will frequently depend on our previously investigating the cause of it, which from what has been delivered above may originate from the disorder of any part of the great tribes of irritative motions, and of the associate motions catenated with them.

Many people, when they arrive at fifty or sixty years of age, are affected with slight vertigo; which is generally but wrongly ascribed to indigestion, but in reality arises from a beginning defect of their sight; as about this time they also find it necessary to begin to use spectacles, when they read small prints, especially in winter, or by candle light, but are yet able to read without them during the summer days, when the light is stronger. These people do not see objects so distinctly as formerly, and by exerting their eyes more than usual, they perceive the apparent motions of objects, and confound them with the real motions of them; and therefore cannot accurately balance themselves so as easily to preserve their perpendicularity by them.

That is, the apparent motions of objects, which are at rest, as we move by them, should only excite irritative ideas: but as these are now become less distinct, owing to the beginning imperfection of our sight, we are induced *voluntarily* to attend to them; and then these apparent motions become succeeded by sensation; and thus the other parts of the trains of irritative ideas, or irritative muscular motions, become disordered, as explained above. In these cases of slight vertigo I have always promised my patients, that they would get free from it in two or three months, as they should acquire the habit of balancing their bodies by less distinct objects, and have seldom been mistaken in my prognostic.

There is an auditory vertigo, which is called a noise in the head, explained in No. 7. of this section, which also is very liable to affect people in the advance of life, and is owing to their hearing less perfectly than before. This is sometimes called a ringing, and sometimes a singing, or buzzing, in the ears, and is occasioned by our first experiencing a disagreeable sensation from our not being able distinctly to hear the sounds, we used formerly to hear distinctly. And this disagreeable sensation excites desire and consequent volition; and when we voluntarily attend to small indistinct sounds, even the whispering of the air in a room, and the pulsations of the arteries of the ear are succeeded by sensation; which minute sounds ought only to have produced irritative sensual motions, or unperceived ideas. See Section XVII. 3. 6. These patients after a while lose this auditory vertigo, by acquiring a new habit of not attending voluntarily to these indistinct sounds, but contenting themselves with the less accuracy of their sense of hearing.

Another kind of vertigo begins with the disordered action of some irritative muscular motions, as those of the stomach from intoxication, or from emetics; or those of the ureter, from the stimulus of a stone lodged in it; and it is probable, that the disordered motions of some of the great congeries of glands, as of those which form the liver, or of the intestinal canal, may occasion vertigo in consequence of their motions being associated or catenated with the great circles of irritative motions; and from hence it appears, that the means of cure must be adapted to the cause.

To prevent sea-sickness it is probable, that the habit of swinging for a week or two before going on shipboard might be of service. For the vertigo from failure of sight, spectacles may be used. For the auditory vertigo, æther may be dropt into the ear to stimulate the part, or to dissolve ear-wax, if such be a part of the cause. For the vertigo arising from indigestion, the peruvian bark and a blister are recommended. And for that owing to a stone in the ureter, venesection, cathartics, opiates, sal soda aerated.

12. Definition of vertigo. 1. Some of the irritative sensual, or muscular motions, which were usually not succeeded by sensation, are in this disease succeeded by sensation; and the trains or circles of motions, which were usually catenated with them, are interrupted, or inverted, or proceed in confusion. 2. The sensitive and voluntary motions continue undisturbed. 3. The associate trains or circles of motions continue; but their catenations with some of the irritative motions are disordered, or inverted, or dissevered.

SECT. XXI
OF DRUNKENNESS

1. In the state of nature when the sense of hunger is appeased by the stimulus of agreeable food, the business of the day is over, and the human savage is at peace with the world, he then exerts little attention to external objects, pleasing reveries of imagination succeed, and at length sleep is the result: till the nourishment which he has procured, is carried over every part of the system to repair the injuries of action, and he awakens with fresh vigour, and feels a renewal of his sense of hunger.

The juices of some bitter vegetables, as of the poppy and the laurocerasus, and the ardent spirit produced in the fermentation of the sugar found in vegetable juices, are so agreeable to the nerves of the stomach, that, taken in a small quantity, they instantly pacify the sense of hunger; and the inattention to external stimuli with the reveries of imagination, and sleep, succeeds, in the same manner as when the stomach is filled with other less intoxicating food.

This inattention to the irritative motions occasioned by external stimuli is a very important circumstance in the approach of sleep, and is produced in young children by rocking their cradles: during which all visible objects become indistinct to them. An uniform soft repeated sound, as the murmurs of a gentle current, or of bees, are said to produce the same effect, by presenting indistinct ideas of inconsequential sounds, and by thus stealing our attention from other objects, whilst by their continued reiterations they become familiar themselves, and we cease gradually to attend to any thing, and sleep ensues.

2. After great fatigue or inanition, when the stomach is suddenly filled with flesh and vegetable food, the inattention to external stimuli, and the reveries of imagination, become so conspicuous as to amount to a degree of intoxication. The same is at any time produced by superadding a little wine or opium to our common meals; or by taking these separately in considerable quantity; and this more efficaciously after fatigue or inanition; because a less quantity of any stimulating material will excite an organ into energetic action, after it has lately been torpid from defect of stimulus; as objects appear more luminous, after we have been in the dark; and because the suspension of volition, which is the immediate cause of sleep, is sooner induced, after a continued voluntary exertion has in part exhausted the sensorial power of volition; in the same manner as we cannot contract a single muscle long together without intervals of inaction.

3. In the beginning of intoxication we are inclined to sleep, as mentioned above, but by the excitement of external circumstances, as of noise, light, business, or by the exertion of volition, we prevent the approaches of it, and continue to take into our stomach greater quantities of the inebriating materials. By these means the irritative movements of the stomach are excited into greater action than is natural; and in consequence all the irritative tribes and trains of motion, which are catenated with them, become susceptible of stronger action from their accustomed stimuli; because these motions are excited both by their usual irritation, and by their association with the increased actions of the stomach and lacteals. Hence the skin glows, and the heat of the body is increased, by the more energetic action of the whole glandular system; and pleasure is introduced in consequence of these increased motions from internal stimulus. According to Law 5. Sect. IV. on Animal Causation.

From this great increase of irritative motions from internal stimulus, and the increased sensation introduced into the system in consequence; and secondly, from the increased sensitive motions in consequence of this additional quantity of sensation, so much sensorial power is expended, that the voluntary power becomes feebly exerted, and the irritation from the stimulus of external objects is less forcible; the external parts of the eye are not therefore voluntarily adapted to the distances of objects, whence the apparent motions of those objects either are seen double, or become too indistinct for the purpose of balancing the body, and vertigo is induced.

Hence we become acquainted with that very curious circumstance, why the drunken vertigo is attended with an increase of pleasure; for the irritative ideas and motions occasioned by internal stimulus, that were not attended to in our sober hours, are now just so much increased as to be succeeded by pleasurable sensation, in the same manner as the more violent motions of our organs are succeeded by painful sensation. And hence a greater quantity of pleasurable sensation is introduced into the constitution; which is attended in some people with an increase of benevolence and good humour.

If the apparent motions of objects is much increased, as when we revolve on one foot, or are swung on a rope, the ideas of these apparent motions are also attended to, and are succeeded with pleasureable sensation, till they become familiar to us by frequent use. Hence children are at first delighted with these kinds of exercise, and with riding, and failing, and hence rocking young children inclines them to sleep. For though in the vertigo from intoxication the irritative ideas of the apparent motions of objects are indistinct from their decrease of energy: yet in the vertigo occasioned by rocking or swinging the irritative ideas of the apparent motions of objects are increased in energy, and hence they induce pleasure into the system, but are equally indistinct, and in consequence equally unfit to balance ourselves by. This addition of pleasure precludes desire or aversion, and in consequence the voluntary power is feebly exerted, and on this account rocking young children inclines them to sleep.

In what manner opium and wine act in relieving pain is another article, that well deserves our attention. There are many pains that originate from defect as well as from excess of stimulus; of these are those of the six appetites of hunger, thirst, lust, the want of heat, of distention, and of fresh air. Thus if our cutaneous capillaries cease to act from the diminished stimulus of heat, when we are exposed to cold weather, or our stomach is uneasy for want of food; these are both pains from defect of stimulus, and in consequence opium, which stimulates all the moving system into increased action, must relieve them. But this is not the case in those pains, which arise from excess of stimulus, as in violent inflammations: in these the exhibition of opium is frequently injurious by increasing the action of the system already too great, as in inflammation of the bowels mortification is often produced by the stimulus of opium. Where, however, no such bad consequences follow; the stimulus of opium, by increasing all the motions of the system, expends so much of the sensorial power, that the actions of the whole system soon become feebler, and in consequence those which produced the pain and inflammation.

4. When intoxication proceeds a little further, the quantity of pleasurable sensation is so far increased, that all desire ceases, for there is no pain in the system to excite it. Hence the voluntary exertions are diminished, staggering and stammering succeed; and the trains of ideas become more and more inconsistent from this defect of voluntary exertion, as explained in the sections on sleep and reverie, whilst those passions which are unmixed with volition are more vividly felt, and shewn with less reserve; hence pining love, or superstitious fear, and the maudling tear dropped on the remembrance of the most trifling distress.

5. At length all these circumstances are increased; the quantity of pleasure introduced into the system by the increased irritative muscular motions of the whole sanguiferous, and glandular, and absorbent systems, becomes so great, that the organs of sense are more forcibly excited into action by this internal pleasurable sensation, than by the irritation from the stimulus of external objects. Hence the drunkard ceases to attend to external stimuli, and as volition is now also suspended, the trains of his ideas become totally inconsistent as in dreams, or delirium: and at length a stupor succeeds from the great exhaustion of sensorial power, which probably does not even admit of dreams, and in which, as in apoplexy, no motions continue but those from internal stimuli, from sensation, and from association.

6. In other people a paroxysm of drunkenness has another termination; the inebriate, as soon as he begins to be vertiginous, makes pale urine in great quantities and very frequently, and at length becomes sick, vomits repeatedly, or purges, or has profuse sweats, and a temporary fever ensues with a quick strong pulse. This in some hours is succeeded by sleep; but the unfortunate bacchanalian does not perfectly recover himself till about the same time of the succeeding day, when his course of inebriation began. As shewn in Sect. XVII. 1. 7. on Catenation. The temporary fever with strong pulse is owing to the same cause as the glow on the skin mentioned in the third paragraph of this Section: the flow of urine and sickness arises from the whole system of irritative motions being thrown into confusion by their associations with each other; as in sea-sickness, mentioned in Sect. XX. 4. on Vertigo; and which is more fully explained in Section XXIX. on Diabetes.

7. In this vertigo from internal causes we see objects double, as two candles instead of one, which is thus explained. Two lines drawn through the axes of our two eyes meet at the object we attend to: this angle of the optic axes increases or diminishes with the less or greater distances of objects. All objects before or behind the place where this angle is formed, appear double; as any one may observe by holding up a pen between his eyes and the candle; when he looks attentively at a spot on the pen, and

carelessly at the candle, it will appear double; and the reverse when he looks attentively at the candle and carelessly at the pen; so that in this case the muscles of the eye, like those of the limbs, stagger and are disobedient to the expiring efforts of volition. Numerous objects are indeed sometimes seen by the inebriate, occasioned by the refractions made by the tears, which stand upon his eye-lids.

8. This vertigo also continues, when the inebriate lies in his bed, in the dark, or with his eyes closed; and this more powerfully than when he is erect, and in the light. For the irritative ideas of the apparent motions of objects are now excited by irritation from internal stimulus, or by association with other irritative motions; and the inebriate, like one in a dream, believes the objects of these irritative motions to be present, and feels himself vertiginous. I have observed in this situation, so long as my eyes and mind were intent upon a book, the sickness and vertigo ceased, and were renewed again the moment I discontinued this attention; as was explained in the preceding account of sea-sickness. Some drunken people have been known to become sober instantly from some accident, that has strongly excited their attention, as the pain of a broken bone, or the news of their house being on fire.

9. Sometimes the vertigo from internal causes, as from intoxication, or at the beginning of some fevers, becomes so universal, that the irritative motions which belong to other organs of sense are succeeded by sensation or attention, as well as those of the eye. The vertiginous noise in the ears has been explained in Section XX. on Vertigo. The taste of the saliva, which in general is not attended to, becomes perceptible, and the patients complain of a bad taste in their mouth.

The common smells of the surrounding air sometimes excite the attention of these patients, and bad smells are complained of, which to other people are imperceptible. The irritative motions that belong to the sense of pressure, or of touch, are attended to, and the patient conceives the bed to librate, and is fearful of falling out of it. The irritative motions belonging to the senses of distention, and of heat, like those above mentioned, become attended to at this time: hence we feel the pulsation of our arteries all over us, and complain of heat, or of cold, in parts of the body where there is no accumulation or diminution of actual heat. All which are to be explained, as in the last paragraph, by the irritative ideas belonging to the various senses being now excited by internal stimuli, or by their associations with other irritative motions. And that the inebriate, like one in a dream, believes the external objects, which usually caused these irritative ideas, to be now present.

10. The diseases in consequence of frequent inebriety, or of daily taking much vinous spirit without inebriety, consist in the paralysis, which is liable to succeed violent stimulation. Organs, whose actions are associated with others, are frequently more affected than the organ, which is stimulated into too violent action. See Sect. XXIV. 2. 8. Hence in drunken people it generally happens, that the secretory vessels of the liver become first paralytic, and a torpor with consequent gall-stones or schirrus of this viscus is induced with concomitant jaundice; otherwise it becomes inflamed in consequence of previous torpor, and this inflammation is frequently transferred to a more sensible part, which is associated with it, and produces the gout, or the rosy eruption of the face, or some other leprous eruption on the head, or arms, or legs. Sometimes the stomach is first affected, and paralysis of the lacteal system is induced: whence a total abhorrence from flesh-food, and general emaciation. In others the lymphatic system is affected with paralysis, and dropsy is the consequence. In some inebriates the torpor of the liver produces pain without apparent schirrus, or gall stones, or inflammation, or consequent gout, and in these epilepsy or insanity are often the consequence. All which will be more fully treated of in the course of the work.

I am well aware, that it is a common opinion, that the gout is as frequently owing to gluttony in eating, as to intemperance in drinking fermented or spirituous liquors. To this I answer, that I have seen no person afflicted with the gout, who has not drank freely of fermented liquor, as wine and water, or small beer; though as the disposition to all the diseases, which have originated from intoxication, is in some degree hereditary, a less quantity of spirituous potation will induce the gout in those, who inherit the disposition from their parents. To which I must add, that in young people the rheumatism is frequently mistaken for the gout.

Spice is seldom taken in such quantity as to do any material injury to the system, flesh-meats as well as vegetables are the natural diet of mankind; with these a glutton may be crammed up to the throat, and fed fat like a stalled ox; but he will not be diseased, unless he adds spirituous or fermented liquor to his food. This is well known in the distilleries, where the swine, which are fattened by the spirituous sediments of barrels, acquire diseased livers. But mark what happens to a man, who drinks a quart of wine or of ale, if he has not been habituated to it. He loses the use both of his limbs and of his understanding! He becomes a temporary idiot, and has a temporary stroke of the palsy! And though he slowly recovers after some hours, is it not reasonable to conclude, that a perpetual repetition of so powerful a poison must at length permanently affect him?—If a person accidentally becomes intoxicated by eating a few mushrooms of a peculiar kind, a general alarm is excited, and he is said to be poisoned, and emetics are exhibited; but so familiarised are we to the intoxication from vinous spirit, that it occasions laughter rather than alarm.

There is however considerable danger in too hastily discontinuing the use of so strong a stimulus, lest the torpor of the system, or paralysis, should sooner be induced by the omission than by the continuance of this habit, when unfortunately acquired. A golden rule for determining the quantity, which may with safety be discontinued, is delivered in Sect. XII. 7. 8.

11. Definition of drunkenness. Many of the irritative motions are much increased in energy by internal stimulation.

2. A great additional quantity of pleasurable sensation is occasioned by this increased exertion of the irritative motions. And many sensitive motions are produced in consequence of this increased sensation.

3. The associated trains and tribes of motions, catenated with the increased irritative and sensitive motions, are disturbed, and proceed in confusion.

4. The faculty of volition is gradually impaired, whence proceeds the instability of locomotion, inaccuracy of perception, and inconsistency of ideas; and is at length totally suspended, and a temporary apoplexy succeeds.

SECT. XXII
OF PROPENSITY TO MOTION, REPETITION AND IMITATION

I. Accumulation of sensorial power in hemiplagia, in sleep, in cold fit of fever, in the locomotive muscles, in the organs of sense. Produces propensity to action. II. Repetition by three sensorial powers. In rhimes and alliterations, in music, dancing, architecture, landscape-painting, beauty. III. 1. Perception consists in imitation. Four kinds of imitation. 2. Voluntary. Dogs taught to dance. 3. Sensitive. Hence sympathy, and all our virtues. Contagious matter of venereal ulcers, of hydrophobia, of jail-fever, of small-pox, produced by imitation, and the sex of the embryon. 4. Irritative imitation. 5. Imitations resolvable into associations.

I. 1. In the hemiplagia, when the limbs on one side have lost their power of voluntary motion, the patient is for many days perpetually employed in moving those of the other. 2. When the voluntary power is suspended during sleep, there commences a ceaseless flow of sensitive motions, or ideas of imagination, which compose our dreams. 3. When in the cold fit of an intermittent fever some parts of the system have for a time continued torpid, and have thus expended less than their usual expenditure of sensorial power; a hot fit succeeds, with violent action of those vessels, which had previously been quiescent. All these are explained from an accumulation of sensorial power during the inactivity of some part of the system.

Besides the very great quantity of sensorial power perpetually produced and expended in moving the arterial, venous, and glandular systems, with the various organs or digestion, as described in Section XXXII. 3. 2. there is also a constant expenditure of it by the action of our locomotive muscles and organs of sense. Thus the thickness of the optic nerves, where they enter the eye, and the great expansion of the nerves of touch beneath the whole of the cuticle, evince the great consumption of sensorial power by these senses. And our perpetual muscular actions in the common offices of life, and in constantly preserving the perpendicularity of our bodies during the day, evince a considerable expenditure of the spirit of animation by our locomotive muscles. It follows, that if the exertion of these organs of sense and muscles be for a while intermitted, that some quantity of sensorial power must be accumulated, and a propensity to activity of some kind ensue from the increased excitability of the system. Whence proceeds the irksomeness of a continued attitude, and of an indolent life.

However small this hourly accumulation of the spirit of animation may be, it produces a propensity to some kind of action; but it nevertheless requires either desire or aversion, either pleasure or pain, or some external stimulus, or a previous link of association, to excite the system into activity; thus it frequently happens, when the mind and body are so unemployed as not to possess any of the three first kinds of stimuli, that the last takes place, and consumes the small but perpetual accumulation of sensorial power. Whence some indolent people repeat the same verse for hours together, or hum the same tune. Thus the poet:

> Onward he trudged, not knowing what he sought,

> And whistled, as he went, for want of thought.

II. The repetitions of motions may be at first produced either by volition, or by sensation, or by irritation, but they soon become easier to perform than any other kinds of action, because they soon become associated together, according to Law the seventh, Section IV. on Animal Causation. And because their frequency of repetition, if as much sensorial power be produced during every reiteration as is expended, adds to the facility of their production.

If a stimulus be repeated at uniform intervals of time, as described in Sect. XII. 3. 3. the action, whether of our muscles or organs of sense, is produced with still greater facility or energy; because the sensorial power of association, mentioned above, is combined with the sensorial power of irritation; that is, in common language, the acquired habit assists the power of the stimulus.

This not only obtains in the annual, lunar, and diurnal catenations of animal motions, as explained in Sect. XXXVI. which are thus performed with great facility and energy; but in every less circle of actions or ideas, as in the burthen of a song, or the reiterations of a dance. To the facility and distinctness, with which we hear sounds at repeated intervals, we owe the pleasure, which we receive from musical time, and from poetic time; as described in Botanic Garden, P. 2. Interlude 3. And to this the pleasure we receive from the rhimes and alliterations of modern verification; the source of which without this key would be difficult to discover. And to this likewise should be ascribed the beauty of the duplicature in the perfect tense of the Greek verbs, and of some Latin ones, as tango tetegi, mordeo momordi.

There is no variety of notes referable to the gamut in the beating of the drum, yet if it be performed in musical time, it is agreeable to our ears; and therefore this pleasurable sensation must be owing to the repetition of the divisions of the sounds at certain intervals of time, or musical bars. Whether these times or bars are distinguished by a pause, or by an emphasis, or accent, certain it is, that this distinction is perpetually repeated; otherwise the ear could not determine instantly, whether the successions of sound were in common or in triple time. In common time there is a division between every two crotchets, or other notes of equivalent time; though the bar in written music is put after every fourth crotchet, or notes equivalent in time; in triple time the division or bar is after every three crotchets, or notes equivalent; so that in common time the repetition recurs more frequently than in triple time. The grave or heroic verses of the Greek and Latin poets are written in common time; the French heroic verses, and Mr. Anstie's humorous verses in his Bath Guide, are written in the same time as the Greek and Latin verses, but are one bar shorter. The English grave or heroic verses are measured by triple time, as Mr. Pope's translation of Homer.

But besides these little circles of musical time, there are the greater returning periods, and the still more distant choruses, which, like the rhimes at the ends of verses, owe their beauty to repetition; that is, to the facility and distinctness with which we perceive sounds, which we expect to perceive, or have perceived before; or in the language of this work, to the greater ease and energy with which our organ is excited by the combined sensorial powers of association and irritation, than by the latter singly.

A certain uniformity or repetition of parts enters the very composition of harmony. Thus two octaves nearest to each other in the scale commence their vibrations together after every second vibration of the higher one. And where the first, third, and fifth compose a chord the vibrations concur or coincide frequently, though less to than in the two octaves. It is probable that these chords bear some analogy to a mixture of three alternate colours in the sun's spectrum separated by a prism.

The pleasure we receive from a melodious succession of notes referable to the gamut is derived from another source, viz. to the pandiculation or counteraction of antagonist fibres. See Botanic Garden, P. 2. Interlude 3. If to these be added our early associations of agreeable ideas with certain proportions of sound, I suppose, from these three sources springs all the delight of music, so celebrated by ancient authors, and so enthusiastically cultivated at present. See Sect. XVI. No. 10. on Instinct.

This kind of pleasure arising from repetition, that is from the facility and distinctness, with which we perceive and understand repeated sensations, enters into all the agreeable arts; and when it is carried to excess is termed formality. The art of dancing like that of music depends for a great part of the pleasure, it affords, on repetition; architecture, especially the Grecian, consists of one part being a repetition of another; and hence the beauty of the pyramidal outline in landscape-painting; where one side of the picture may be said in some measure to balance the other. So universally does repetition contribute to our pleasure in the fine arts, that beauty itself has been defined by some writers to consist in a due combination of uniformity and variety. See Sect. XVI. 6.

III. 1. Man is termed by Aristotle an imitative animal; this propensity to imitation not only appears in the actions of children, but in all the customs and fashions of the world: many thousands tread in the beaten paths of others, for one who traverses regions of his own discovery. The origin of this propensity of imitation has not, that I recollect, been deduced from any known principle; when any action presents itself to the view of a child, as of whetting a knife, or threading a needle, the parts of this action in respect of time, motion, figure, is imitated by a part of the retina of his eye; to perform this action therefore with his hands is easier to him than to invent any new action, because it consists in repeating with another set of fibres, viz. with the moving muscles, what he had just performed by some parts of the retina; just as in dancing we transfer the times of motion from the actions of the auditory nerves to the muscles of the limbs. Imitation therefore consists of repetition, which we have shewn above to be the easiest kind of animal action, and which we perpetually fall into, when we possess an accumulation of sensorial power, which is not otherwise called into exertion.

It has been shewn, that our ideas are configurations of the organs of sense, produced originally in consequence of the stimulus of external bodies. And that these ideas, or configurations of the organs of sense, referable in some property a correspondent property of external matter; as the parts of the senses of light and of touch, which are excited into action, resemble in figure the figure of the stimulating body; and probably also the colour, and the quantity of density, which they perceive. As explained in Sect. XIV. 2. 2. Hence it appears, that our perceptions themselves are copies, that is, imitations of some properties of external matter; and the propensity to imitation is thus interwoven with our existence, as it is produced by the stimuli of external bodies, and is afterwards repeated by our volitions and sensations, and thus constitutes all the operations of our minds.

2. Imitations resolve themselves into four kinds, voluntary, sensitive, irritative, and associate. The voluntary imitations are, when we imitate

deliberately the actions of others, either by mimicry, as in acting a play, or in delineating a flower; or in the common actions of our lives, as in our dress, cookery, language, manners, and even in our habits of thinking.

Not only the greatest part of mankind learn all the common arts of life by imitating others, but brute animals seem capable of acquiring knowledge with greater facility by imitating each other, than by any methods by which we can teach them; as dogs and cats, when they are sick, learn of each other to eat grass; and I suppose, that by making an artificial dog perform certain tricks, as in dancing on his hinder legs, a living dog might be easily induced to imitate them; and that the readiest way of instructing dumb animals is by practising them with others of the same species, which have already learned the arts we wish to teach them. The important use of imitation in acquiring natural language is mentioned in Section XVI. 7. and 8. on Instinct.

3. The sensitive imitations are the immediate consequences of pleasure or pain, and these are often produced even contrary to the efforts of the will. Thus many young men on seeing cruel surgical operations become sick, and some even feel pain in the parts of their own bodies, which they see tortured or wounded in others; that is, they in some measure imitate by the exertions of their own fibres the violent actions, which they witnessed in those of others. In this case a double imitation takes place, first the observer imitates with the extremities of the optic nerve the mangled limbs, which are present before his eyes; then by a second imitation he excites to violent action of the fibres of his own limbs as to produce pain in those parts of his own body, which he saw wounded in another. In these pains produced by imitation the effect has some similarity to the cause, which distinguishes them from those produced by association; as the pains of the teeth, called tooth-edge, which are produced by association with disagreeable sounds, as explained in Sect. XVI. 10.

The effect of this powerful agent, imitation, in the moral world, is mentioned in Sect. XVI. 7. as it is the foundation of all our intellectual sympathies with the pains and pleasures of others, and is in consequence the source of all our virtues. For in what consists our sympathy with the miseries, or with the joys, of our fellow creatures, but in an involuntary excitation of ideas in some measure similar or imitative of those, which we believe to exist in the minds of the persons, whom we commiserate or congratulate?

There are certain concurrent or successive actions of some of the glands, or other parts of the body, which are possessed of sensation, which become intelligible from this propensity to imitation. Of these are the production of matter by the membranes of the fauces, or by the skin, in consequence of the venereal disease previously affecting the parts of generation. Since as no fever is excited, and as neither the blood of such patients, nor even the matter from ulcers of the throat, or from cutaneous ulcers, will by inoculation produce the venereal disease in others, as observed by Mr. Hunter, there is reason to conclude, that no contagious matter is conveyed thither by the blood-vessels, but that a milder matter is formed by the actions of the fine vessels in those membranes imitating each other. See Section XXXIII. 2. 9. In this disease the actions of these vessels producing ulcers on the throat and skin are imperfect imitations of those producing chanker, or gonorrhœa; since the matter produced by them is not infectious, while the imitative actions in the hydrophobia appear to be perfect resemblances, as they produce a material equally infectious with the original one, which induced them.

The contagion from the bite of a mad dog differs from other contagious materials, from its being communicable from other animals to mankind, and from many animals to each other; the phenomena attending the hydrophobia are in some degree explicable on the foregoing theory. The infectious matter does not appear to enter the circulation, as it cannot be traced along the course of the lymphatics from the wound, nor is there any swelling of the lymphatic glands, nor does any fever attend, as occurs in the small-pox, and in many other contagious diseases; yet by some unknown process the disease is communicated from the wound to the throat, and that many months after the injury, so as to produce pain and hydrophobia, with a secretion of infectious saliva of the same kind, as that of the mad dog, which inflicted the wound.

This subject is very intricate.—It would appear, that by certain morbid actions of the salivary glands of the mad dog, a peculiar kind of saliva is produced; which being instilled into a wound of another animal stimulates the cutaneous or mucous glands into morbid actions, but which are ineffectual in respect to the production of a similar contagious material; but the salivary glands by irritative sympathy are thrown into similar action, and produce an infectious saliva similar to that instilled into the wound.

Though in many contagious fevers a material similar to that which produced the disease, is thus generated by imitation; yet there are other infectious materials, which do not thus propagate themselves, but which seem to act like slow poisons. Of this kind was the contagious matter, which produced the jail-fever at the assizes at Oxford about a century ago. Which, though fatal to so many, was not communicated to their nurses or attendants. In these cases, the imitations of the fine vessels, as above described, appear to be imperfect, and do not therefore produce a matter similar to that, which stimulates them; in this circumstance resembling the venereal matter in ulcers of the throat or skin, according to the curious discovery of Mr. Hunter above related, who found, by repeated inoculations, that it would not infect. Hunter on Venereal Disease, Part vi. ch. 1.

Another example of morbid imitation is in the production of a great quantity of contagious matter, as in the inoculated small-pox, from a small quantity of it inserted into the arm, and probably diffused in the blood. These particles of contagious matter stimulate the extremities of the fine arteries of the skin, and cause them to imitate some properties of those particles of contagious matter, so as to produce a thousandfold of a similar material. See Sect. XXXIII. 2. 6. Other instances are mentioned in the Section on Generation, which shew the probability that the extremities of the seminal glands may imitate certain ideas of the mind, or actions of the organs of sense, and thus occasion the male or female sex of the embryon. See Sect. XXXIX. 6.

4. We come now to those imitations, which are not attended with sensation. Of these are all the irritative ideas already explained, as when the retina of the eye imitates by its action or configuration the tree or the bench, which I shun in walking past without attending to them. Other examples of these irritative imitations are daily observable in common life; thus one yawning person shall set a whole company a yawning; and some have acquired winking of the eyes or impediments of speech by imitating their companions without being conscious of it.

5. Besides the three species of imitations above described there may be some associate motions, which may imitate each other in the kind as well as in the quantity of their action; but it is difficult to distinguish them from the associations of motions treated of in Section XXXV. Where the actions of other persons are imitated there can be no doubt, or where we imitate a preconceived idea by exertion of our locomotive muscles, as in painting a dragon; all these imitations may aptly be referred to the sources above described of the propensity to activity, and the facility of repetition; at the same time I do not affirm, that all those other apparent sensitive and irritative imitations may not be resolvable into associations of a peculiar kind, in which certain distant parts of similar irritability or sensibility, and which have habitually acted together, may affect each other exactly with the same kinds of motion; as many parts are known to sympathise in the quantity of their motions. And that therefore they may be ultimately resolvable into associations of action, as described in Sect. XXXV.

SECT. XXIII
OF THE CIRCULATORY SYSTEM

I. The heart and arteries have no antagonist muscles. Veins absorb the blood, propel it forwards, and distend the heart; contraction of the heart distends the arteries. Vena portarum. II. Glands which take their fluids from the blood. With long necks, with short necks. III. Absorbent system. IV. Heat given out from glandular secretions. Blood changes colour in the lungs and in the glands and capillaries. V. Blood is absorbed by veins, as chyle by lacteal vessels, otherwise they could not join their streams. VI. Two kinds of stimulus, agreeable and disagreeable. Glandular appetency. Glands originally possessed sensation.

I. We now step forwards to illustrate some of the phenomena of diseases, and to trace out their most efficacious methods of cure; and shall commence this subject with a short description of the circulatory system.

As the nerves, whose extremities form our various organs of sense and muscles, are all joined, or communicate, by means of the brain, for the convenience perhaps of the distribution of a subtile ethereal fluid for the purpose of motion; so all those vessels of the body, which carry the grosser fluids for the purposes of nutrition, communicate with each other by the heart.

The heart and arteries are hollow muscles, and are therefore indued with power of contraction in consequence of stimulus, like all other muscular fibres; but, as they have no antagonist muscles, the cavities of the vessels, which they form, would remain for ever closed, after they have contracted themselves, unless some extraneous power be applied to again distend them. This extraneous power in respect to the heart is the current of blood, which is perpetually absorbed by the veins from the various glands and capillaries, and pushed into the heart by a power probably very similar to that, which raises the sap in vegetables in the spring, which, according to Dr. Hale's experiment on the stump of a vine, exerted a force equal to a column of water above twenty feet high. This force of the current of blood in the veins is partly produced by their absorbent power, exerted at the beginning of every fine ramification; which may be conceived to be a mouth absorbing blood, as the mouths of the lacteals and lymphatics absorb chyle and lymph. And partly by their intermitted compression by the pulsations of their generally concomitant arteries; by which the blood is perpetually propelled towards the heart, as the valves in many veins, and the absorbent mouths in them all, will not suffer it to return.

The blood, thus forcibly injected into the chambers of the heart, distends this combination of hollow muscles; till by the stimulus of distention they contract themselves; and, pushing forwards the blood into the arteries, exert sufficient force to overcome in less than a second of time the vis inertiæ, and perhaps some elasticity, of the very extensive ramifications of the two great systems of the aortal and pulmonary arteries. The power necessary to do this in so short a time must be considerable, and has been variously estimated by different physiologists.

The muscular coats of the arterial system are then brought into action by the stimulus of distention, and propel the blood to the mouths, or through the convolutions, which precede the secretory apertures of the various glands and capillaries.

In the vessels of the liver there is no intervention of the heart; but the vena portarum, which does the office of an artery, is distended by the blood poured into it from the mesenteric veins, and is by this distention stimulated to contract itself, and propel the blood to the mouths of the numerous glands, which compose that viscus.

II. The glandular system of vessels may be divided into those, which take some fluid from the circulation; and those, which give something to it. Those, which take their fluid from the circulation are the various glands, by which the tears, bile, urine, perspiration, and many other secretions are produced; these glands probably consist of a mouth to select, a belly to digest, and an excretory aperture to emit their appropriated fluids; the blood is conveyed by the power of the heart and arteries to the mouths of these glands, it is there taken up by the living power of the gland, and carried forwards to its belly, and excretory aperture, where a part is separated, and the remainder absorbed by the veins for further purposes.

Some of these glands are furnished with long convoluted necks or tubes, as the seminal ones, which are curiously seen when injected with quicksilver. Others seem to consist of shorter tubes, as that great congeries of glands, which constitute the liver, and those of the kidneys. Some have their excretory apertures opening into reservoirs, as the urinary and gall-bladders. And others on the external body, as those which secrete the tears, and perspirable matter.

Another great system of glands, which have very short necks, are the capillary vessels; by which the insensible perspiration is secreted on the skin; and the mucus of various consistences, which lubricates the interstices of the cellular membrane, of the muscular fibres, and of all the larger cavities of the body. From the want of a long convolution of vessels some have doubted, whether these capillaries should be considered as glands, and have been led to conclude, that the perspirable matter rather exuded than was secreted. But the fluid of perspiration is not simple water, though that part of it, which exhales into the air may be such; for there is another part of it, which in a state of health is absorbed again; but which, when the absorbents are diseased, remains on the surface of the skin, in the form of scurf, or indurated mucus. Another thing, which shews their similitude to other glands, is their sensibility to certain affections of the mind; as is seen in the deeper colour of the skin in the blush of shame, or the greater paleness of it from fear.

III. Another series of glandular vessels is called the absorbent system; these open their mouths into all the cavities, and upon all those surfaces of the body, where the excretory apertures of the other glands pour out their fluids. The mouths of the absorbent system drink up a part or the whole of these fluids, and carry them forwards by their living power to their respective glands, which are called conglobate glands. There these fluids undergo some change, before they pass on into the circulation; but if they are very acrid, the conglobate gland swells, and sometimes suppurates, as in inoculation of the small-pox, in the plague, and in venereal absorptions; at other times the fluid may perhaps continue there, till it undergoes some chemical change, that renders it less noxious; or, what is more likely, till it is regurgitated by the retrograde motion of the gland in spontaneous sweats or diarrhœas, as disagreeing food is vomited from the stomach.

IV. As all the fluids, that pass through these glands, and capillary vessels, undergo a chemical change, acquiring new combinations, the matter of heat is at the same time given out; this is apparent, since whatever increases insensible perspiration, increases the heat of the skin; and when the action of these vessels is much increased but for a moment, as in blushing, a vivid heat on the skin is the immediate consequence. So when great bilious secretions, or those of any other gland, are produced, heat is generated in the part in proportion to the quantity of the secretion.

The heat produced on the skin by blushing may be thought by some too sudden to be pronounced a chemical effect, as the fermentations or new combinations taking place in a fluid is in general a slower process. Yet are there many chemical mixtures in which heat is given out as instantaneously; as in solutions of metals in acids, or in mixtures of essential oils and acids, as of oil of cloves and acid of nitre. So the bruised parts of an unripe apple become almost instantaneously sweet; and if the chemico-animal process of digestion be stopped for but a moment, as by fear, or even by voluntary eructation, a great quantity of air is generated, by the fermentation, which instantly succeeds the stop of digestion. By the experiments of Dr. Hales it appears, that an apple during fermentation gave up above six hundred times its bulk of air; and the materials in the stomach are such, and in such a situation, as immediately to run into fermentation, when digestion is impeded.

As the blood passes through the small vessels of the lungs, which connect the pulmonary artery and vein, it undergoes a change of colour from a dark to a light red; which may be termed a chemical change, as it is known to be effected by an admixture of oxygene, or vital air; which, according to a discovery of Dr. Priestley, passes through the moist membranes, which constitute the sides of these vessels. As the blood passes through the capillary vessels, and glands, which connect the aorta and its various branches with their correspondent veins in the extremities of the body, it again loses the bright red colour, and undergoes some new combinations in the glands or capillaries, in which the matter of heat is given out from the secreted fluids. This process therefore, as well as the process of respiration, has some analogy to combustion, as the vital air or oxygene seems to become united to some inflammable base, and the matter of heat escapes from the new acid, which is thus produced.

V. After the blood has passed these glands and capillaries, and parted with whatever they chose to take from it, the remainder is received by the veins, which are a set of blood-absorbing vessels in general corresponding with the ramifications of the arterial system. At the extremity of the fine convolutions of the glands the arterial force ceases; this in respect to the capillary vessels, which unite the extremities of the arteries with the commencement of the veins, is evident to the eye, on viewing the tail of a tadpole by means of a solar, or even by a common microscope, for globules of blood are seen to endeavour to pass, and to return again and again, before they become absorbed by the mouths of the veins; which returning of these globules evinces, that the arterial force behind them has ceased. The veins are furnished with valves like the lymphatic absorbents; and the great trunks of the veins, and of the lacteals and lymphatics, join together before the ingress of their fluids into the left chamber of the heart; both which evince, that the blood in the veins, and the lymph and chyle in the lacteals and lymphatics, are carried on by a similar force; otherwise the stream, which was propelled with a less power, could not enter the vessels, which contained the stream propelled with a greater power. From whence it appears, that the veins are a system of vessels absorbing blood, as the lacteals and lymphatics are a system of vessels absorbing chyle and lymph. See Sect. XXVII. 1.

VI. The movements of their adapted fluids in the various vessels of the body are carried forwards by the actions of those vessels in consequence of two kinds of stimulus, one of which may be compared to a pleasurable sensation or desire inducing the vessel to seize, and, as it were, to swallow the particles thus selected from the blood; as is done by the mouths of the various glands, veins, and other absorbents, which may be called glandular appetency. The other kind of stimulus may be compared to disagreeable sensation, or aversion, as when the heart has received the blood, and is stimulated by it to push it forwards into the arteries; the same again stimulates the arteries to contract, and carry forwards the blood to their extremities, the glands and capillaries. Thus the mesenteric veins absorb the blood from the intestines by glandular appetency, and carry it forward to the vena portarum; which acting as an artery contracts itself by disagreeable stimulus, and pushes it to its ramified extremities, the various glands, which constitute the liver.

It seems probable, that at the beginning of the formation of these vessels in the embryon, an agreeable sensation was in reality felt by the glands during secretion, as is now felt in the act of swallowing palatable food; and that a disagreeable sensation was originally felt by the heart from the distention occasioned by the blood, or by its chemical stimulus; but that by habit these are all become irritative motions; that is, such motions as do not affect the whole system, except when the vessels are diseased by inflammation.

SECT. XXIV
OF THE SECRETIONS OF SALIVA, AND OF TEARS, AND OF THE LACRYMAL SACK

I. Secretion of saliva increased by mercury in the blood. 1. By the food in the mouth. Dryness of the mouth not from a deficiency of saliva. 2. By Sensitive ideas. 3. By volition. 4. By distasteful substances. It is secreted in a dilute and saline state. It then becomes more viscid. 5. By ideas of distasteful substances. 6. By nausea. 7. By aversion. 8. By catenation with stimulating substances in the ear. II. 1. Secretion of tears less in sleep. From stimulation of their excretory duct. 2. Lacrymal sack is a gland. 3. Its uses. 4. Tears are secreted, when the nasal duct is stimulated. 5. Or when it is excited by sensation. 6. Or by volition. 7. The lacrymal sack can regurgitate its contents into the eye. 8. More tears are secreted by association with the irritation of the nasal duct of the lacrymal sack, than the puncta lacrymalia can imbibe. Of the gout in the liver and stomach.

I. The salival glands drink up a certain fluid from the circumfluent blood, and pour it into the mouth. They are sometimes stimulated into action by the blood, that surrounds their origin, or by some part of that heterogeneous fluid: for when mercurial salts, or oxydes, are mixed with the blood, they stimulate these glands into unnatural exertions; and then an unusual quantity of saliva is separated.

1. As the saliva secreted by these glands is most wanted during the mastication of our food, it happens, when the terminations of their ducts in the mouth are stimulated into action, the salival glands themselves are brought into increased action at the same time by association, and separate a greater quantity of their juices from the blood; in the same manner as tears are produced in greater abundance during the stimulus of the vapour of onions, or of any other acrid material in the eye.

The saliva is thus naturally poured into the mouth only during the stimulus of our food in mastication; for when there is too great an exhalation of the mucilaginous secretion from the membranes, which line the mouth, or too great an absorption of it, the mouth becomes dry, though there is no deficiency in the quantity of saliva; as in those who sleep with their mouths open, and in some fevers.

2. Though during the mastication of our natural food the salival glands are excited into action by the stimulus on their excretory ducts, and a due quantity of saliva is separated from the blood, and poured into the mouth; yet as this mastication of our food is always attended with a degree of pleasure; and that pleasurable sensation is also connected with our ideas of certain kinds of aliment; it follows, that when these ideas are reproduced, the pleasurable sensation arises along with them, and the salival glands are excited into action, and fill the mouth with saliva from this sensitive association, as is frequently seen in dogs, who slaver at the sight of food.

3. We have also a voluntary power over the action of these salival glands, for we can at any time produce a flow of saliva into our mouth, and spit out, or swallow it at will.

4. If any very acrid material be held in the mouth, as the root of pyrethrum, or the leaves of tobacco, the salival glands are stimulated into stronger action than is natural, and thence secrete a much larger quantity of saliva; which is at the same time more viscid than in its natural state; because the lymphatics, that open their mouths into the ducts of the salival glands, and on the membranes, which line the mouth, are likewise stimulated into stronger action, and absorb the more liquid parts of the saliva with greater avidity; and the remainder is left both in greater quantity and more viscid.

The increased absorption in the mouth by some stimulating substances, which are called astringents, as crab juice, is evident from the instant dryness produced in the mouth by a small quantity of them.

As the extremities of the glands are of exquisite tenuity, as appears by their difficulty of injection, it was necessary for them to secrete their fluids in a very dilute state; and, probably for the purpose of stimulating them into action, a quantity of neutral salt is likewise secreted or formed by the gland. This aqueous and saline part of all secreted fluids is again reabsorbed into the habit. More than half of some secreted fluids is thus imbibed from the reservoirs, into which they are poured; as in the urinary bladder much more than half of what is secreted by the kidneys becomes reabsorbed by the lymphatics, which are thickly dispersed around the neck of the bladder. This seems to be the purpose of the urinary bladders of fish, as otherwise such a receptacle for the urine could have been of no use to an animal immersed in water.

5. The idea of substances disagreeably acrid will also produce a quantity of saliva in the mouth; as when we smell very putrid vapours, we

are induced to spit out our saliva, as if something disagreeable was actually upon our palates.

6. When disagreeable food in the stomach produces nausea, a flow of saliva is excited in the mouth by association; as efforts to vomit are frequently produced by disagreeable drugs in the mouth by the same kind of association.

7. A preternatural flow of saliva is likewise sometimes occasioned by a disease of the voluntary power; for if we think about our saliva, and determine not to swallow it, or not to spit it out, an exertion is produced by the will, and more saliva is secreted against our wish; that is, by our aversion, which bears the same analogy to desire, as pain does to pleasure; as they are only modifications of the same disposition of the sensorium. See Class IV. 3. 2. 1.

8. The quantity of saliva may also be increased beyond what is natural, by the catenation of the motions of these glands with other motions, or sensations, as by an extraneous body in the ear; of which I have known an instance; or by the application of stizolobium, siliqua hirsuta, cowhage, to the seat of the parotis, as some writers have affirmed.

II. 1. The lacrymal gland drinks up a certain fluid from the circumfluent blood, and pours it on the ball of the eye, on the upper part of the external corner of the eyelids. Though it may perhaps be stimulated into the performance of its natural action by the blood, which surrounds its origin, or by some part of that heterogeneous fluid; yet as the tears secreted by this gland are more wanted at some times than at others, its secretion is variable, like that of the saliva above mentioned, and is chiefly produced when its excretory duct is stimulated; for in our common sleep there seems to be little or no secretion of tears; though they are occasionally produced by our sensations in dreams.

Thus when any extraneous material on the eye-ball, or the dryness of the external covering of it, or the coldness of the air, or the acrimony of some vapours, as of onions, stimulates the excretory duct of the lacrymal gland, it discharges its contents upon the ball; a quicker secretion takes place in the gland, and abundant tears succeed, to moisten, clean, and lubricate the eye. These by frequent nictitation are diffused over the whole ball, and as the external angle of the eye in winking is closed sooner than the internal angle, the tears are gradually driven forwards, and downwards from the lacrymal gland to the puncta lacrymalia.

2. The lacrymal sack, with its puncta lacrymalia, and its nasal duct, is a complete gland; and is singular in this respect, that it neither derives its fluid from, nor disgorges it into the circulation. The simplicity of the

structure of this gland, and both the extremities of it being on the surface of the body, makes it well worthy our minuter observation; as the actions of more intricate and concealed glands may be better understood from their analogy to this.

3. This simple gland consists of two absorbing mouths, a belly, and an excretory duct. As the tears are brought to the internal angle of the eye, these two mouths drink them up, being stimulated into action by this fluid, which they absorb. The belly of the gland, or lacrymal sack, is thus filled, in which the saline part of the tears is absorbed, and when the other end of the gland, or nasal duct, is stimulated by the dryness, or pained by the coldness of the air, or affected by any acrimonious dust or vapour in the nostrils, it is excited into action together with the sack, and the tears are disgorged upon the membrane, which lines the nostrils; where they serve a second purpose to moisten, clean, and lubricate, the organ of smell.

4. When the nasal duct of this gland is stimulated by any very acrid material, as the powder of tobacco, or volatile spirits, it not only disgorges the contents of its belly or receptacle (the lacrymal sack), and absorbs hastily all the fluid, that is ready for it in the corner of the eye; but by the association of its motions with those of the lacrymal gland, it excites that also into increased action, and a large flow of tears is poured into the eye.

5. This nasal duct is likewise excited into strong action by sensitive ideas, as in grief, or joy, and then also by its associations with the lacrymal gland it produces a great flow of tears without any external stimulus; as is more fully explained in Sect. XVI. 8. on Instinct.

6. There are some, famous in the arts of exciting compassion, who are said to have acquired a voluntary power of producing a flow of tears in the eye; which, from what has been said in the section on Instinct above mentioned, I should suspect, is performed by acquiring a voluntary power over the action of this nasal duct.

7. There is another circumstance well worthy our attention, that when by any accident this nasal duct is obstructed, the lacrymal sack, which is the belly or receptacle of this gland, by slight pressure of the finger is enabled to disgorge its contents again into the eye; perhaps the bile in the same manner, when the biliary ducts are obstructed, is returned into the blood by the vessels which secrete it?

8. A very important though minute occurrence must here be observed, that though the lacrymal gland is only excited into action, when we weep at a distressful tale, by its association with this nasal duct, as is more fully explained in Sect. XVI. 8; yet the quantity of tears secreted at once is more than the puncta lacrymalia can readily absorb; which shews *that the motions*

occasioned by associations are frequently more energetic than the original motions, by which they were occasioned. Which we shall have occasion to mention hereafter, to illustrate, why pains frequently exist in a part distant from the cause of them, as in the other end of the urethra, when a stone stimulates the neck of the bladder. And why inflammations frequently arise in parts distant from their cause, as the gutta rosea of drinking people, from an inflamed liver.

The inflammation of a part is generally preceded by a torpor or quiescence of it; if this exists in any large congeries of glands, as in the liver, or any membranous part, as the stomach, pain is produced and chilliness in consequence of the torpor of the vessels. In this situation sometimes an inflammation of the parts succeeds the torpor; at other times a distant more sensible part becomes inflamed; whose actions have previously been associated with it; and the torpor of the first part ceases. This I apprehend happens, when the gout of the foot succeeds a pain of the biliary duct, or of the stomach. Lastly, it sometimes happens, that the pain of torpor exists without any consequent inflammation of the affected part, or of any distant part associated with it, as in the membranes about the temple and eye-brows in hemicrania, and in those pains, which occasion convulsions; if this happens to gouty people, when it affects the liver, I suppose epileptic fits are produced; and, when it affects the stomach, death is the consequence. In these cases the pulse is weak, and the extremities cold, and such medicines as stimulate the quiescent parts into action, or which induce inflammation in them, or in any distant part, which is associated with them, cures the present pain of torpor, and saves the patient.

I have twice seen a gouty inflammation of the liver, attended with jaundice; the patients after a few days were both of them affected with cold fits, like ague-fits, and their feet became affected with gout, and the inflammation of their livers ceased. It is probable, that the uneasy sensations about the stomach, and indigestion, which precedes gouty paroxysms, are generally owing to torpor or slight inflammation of the liver, and biliary ducts; but where great pain with continued sickness, with feeble pulse, and sensation of cold, affect the stomach in patients debilitated by the gout, that it is a torpor of the stomach itself, and destroys the patient from the great connexion of that viscus with the vital organs. See Sect. XXV. 17.

SECT. XXV
OF THE STOMACH AND INTESTINES

1. Of swallowing our food. Ruminating animals. 2. Action of the stomach. 3. Action of the intestines. Irritative motions connected with these. 4. Effects of repletion. 5. Stronger action of the stomach and intestines from more stimulating food. 6. Their action inverted by still greater stimuli. Or by disgustful ideas. Or by volition. 7. Other glands strengthen or invert their motions by sympathy. 8. Vomiting performed by intervals. 9. Inversion of the cutaneous absorbents. 10. Increased secretion of bile and pancreatic juice. 11. Inversion of the lacteals. 12. And of the bile-ducts. 13. Case of a cholera. 14. Further account of the inversion of lacteals. 15. Iliac passions. Valve of the colon. 16. Cure of the iliac passion. 17. Pain of gall-stone distinguished from pain of the stomach. Gout of the stomach from torpor, from inflammation. Intermitting pulse owing to indigestion. To overdose of foxglove. Weak pulse from emetics. Death from a blow on the stomach. From gout of the stomach.

1. The throat, stomach, and intestines, may be considered as one great gland; which like the lacrymal sack above mentioned, neither begins nor ends in the circulation. Though the act of masticating our aliment belongs to the sensitive class of motions, for the pleasure of its taste induces the muscles of the jaw into action; yet the deglutition of it when masticated is generally, if not always, an irritative motion, occasioned by the application of the food already masticated to the origin of the pharinx; in the same manner as we often swallow our spittle without attending to it.

The ruminating class of animals have the power to invert the motion of their gullet, and of their first stomach, from the stimulus of this aliment, when it is a little further prepared; as is their daily practice in chewing the cud; and appears to the eye of any one, who attends to them, whilst they are employed in this second mastication of their food.

2. When our natural aliment arrives into the stomach, this organ is simulated into its proper vermicular action; which beginning at the upper

orifice of it, and terminating at the lower one, gradually mixes together and pushes forwards the digesting materials into the intestine beneath it.

At the same time the glands, that supply the gastric juices, which are necessary to promote the chemical part of the process of digestion, are stimulated to discharge their contained fluids, and to separate a further supply from the blood-vessels: and the lacteals or lymphatics, which open their mouths into the stomach, are stimulated into action, and take up some part of the digesting materials.

3. The remainder of these digesting materials is carried forwards into the upper intestines, and stimulates them into their peristaltic motion similar to that of the stomach; which continues gradually to mix the changing materials, and pass them along through the valve of the colon to the excretory end of this great gland, the sphincter ani.

The digesting materials produce a flow of bile, and of pancreatic juice, as they pass along the duodenum, by stimulating the excretory ducts of the liver and pancreas, which terminate in that intestine: and other branches of the absorbent or lymphatic system, called lacteals, are excited to drink up, as it passes, those parts of the digesting materials, that are proper for their purpose, by its stimulus on their mouths.

4. When the stomach and intestines are thus filled with their proper food, not only the motions of the gastric glands, the pancreas, liver, and lacteal vessels, are excited into action; but at the same time the whole tribe of irritative motions are exerted with greater energy, a greater degree of warmth, colour, plumpness, and moisture, is given to the skin from the increased action of those glands called capillary vessels; pleasurable sensation is excited, the voluntary motions are less easily exerted, and at length suspended; and sleep succeeds, unless it be prevented by the stimulus of surrounding objects, or by voluntary exertion, or by an acquired habit, which was originally produced by one or other of these circumstances, as is explained in Sect. XXI. on Drunkenness.

At this time also, as the blood-vessels become replete with chyle, more urine is separated into the bladder, and less of it is reabsorbed; more mucus poured into the cellular membranes, and less of it reabsorbed; the pulse becomes fuller, and softer, and in general quicker. The reason why less urine and cellular mucus is absorbed after a full meal with sufficient drink is owing to the blood-vessels being fuller: hence one means to promote absorption is to decrease the resistance by emptying the vessels by venesection. From this decreased absorption the urine becomes pale as well as copious, and the skin appears plump as well as florid.

By daily repetition of these movements they all become connected together, and make a diurnal circle of irritative action, and if one of this chain be disturbed, the whole is liable to be put into disorder. See Sect. XX. on Vertigo.

5. When the stomach and intestines receive a quantity of food, whose stimulus is greater than usual, all their motions, and those of the glands and lymphatics, are stimulated into stronger action than usual, and perform their offices with greater vigour and in less time: such are the effects of certain quantities of spice or of vinous spirit.

6. But if the quantity or duration of these stimuli are still further increased, the stomach and throat are stimulated into a motion, whose direction is contrary to the natural one above described; and they regurgitate the materials, which they contain, instead of carrying them forwards. This retrograde motion of the stomach may be compared to the stretchings of wearied limbs the contrary way, and is well elucidated by the following experiment. Look earnestly for a minute or two on an area an inch square of pink silk, placed in a strong light, the eye becomes fatigued, the colour becomes faint, and at length vanishes, for the fatigued eye can no longer be stimulated into direct motions; then on closing the eye a green spectrum will appear in it, which is a colour directly contrary to pink, and which will appear and disappear repeatedly, like the efforts in vomiting. See Section XXIX. 11.

Hence all those drugs, which by their bitter or astringent stimulus increase the action of the stomach, as camomile and white vitriol, if their quantity is increased above a certain dose become emetics.

These inverted motions of the stomach and throat are generally produced from the stimulus of unnatural food, and are attended with the sensation of nausea or sickness: but as this sensation is again connected with an idea of the distasteful food, which induced it; so an idea of nauseous food will also sometimes excite the action of nausea; and that give rise by association to the inversion of the motions of the stomach and throat. As some, who have had horse-flesh or dogs-flesh given them for beef or mutton, are said to have vomited many hours afterwards, when they have been told of the imposition.

I have been told of a person, who had gained a voluntary command over these inverted motions of the stomach and throat, and supported himself by exhibiting this curiosity to the public. At these exhibitions he swallowed a pint of red rough gooseberries, and a pint of white smooth ones, brought them up in small parcels into his mouth, and restored them separately to the spectators, who called for red or white as they pleased, till the whole were redelivered.

7. At the same time that these motions of the stomach and throat are stimulated into inversion, some of the other irritative motions, that had acquired more immediate connexions with the stomach, as those of the gastric glands, are excited into stronger action by this association; and some other of these motions, which are more easily excited, as those of the gastric lymphatics, are inverted by their association with the retrograde motions of the stomach, and regurgitate their contents, and thus a greater quantity of mucus, and of lymph, or chyle, is poured into the stomach, and thrown up along with its contents.

8. These inversions of the motion of the stomach in vomiting are performed by intervals, for the same reason that many other motions are reciprocally exerted and relaxed; for during the time of exertion the stimulus, or sensation, which caused this exertion, is not perceived; but begins to be perceived again, as soon as the exertion ceases, and is some time in again producing its effect. As explained in Sect. XXXIV. on Volition, where it is shewn, that the contractions of the fibres, and the sensation of pain, which occasioned that exertion, cannot exist at the same time. The exertion ceases from another cause also, which is the exhaustion of the sensorial power of the part, and these two causes frequently operate together.

9. At the times of these inverted efforts of the stomach not only the lymphatics, which open their mouths into the stomach, but those of the skin also, are for a time inverted; for sweats are sometimes pushed out during the efforts of vomiting without an increase of heat.

10. But if by a greater stimulus the motions of the stomach are inverted still more violently or more permanently, the duodenum has its peristaltic motions inverted at the same time by their association with those of the stomach; and the bile and pancreatic juice, which it contains, are by the inverted motions brought up into the stomach, and discharged along with its contents; while a greater quantity of bile and pancreatic juice is poured into this intestine; as the glands, that secrete them, are by their association with the motions of the intestine excited into stronger action than usual.

11. The other intestines are by association excited into more powerful action, while the lymphatics, that open their mouths into them, suffer an inversion of their motions corresponding with the lymphatics of the stomach, and duodenum; which with a part of the abundant secretion of bile is carried downwards, and contributes both to stimulate the bowels, and to increase the quantity of the evacuations. This inversion of the motion of the lymphatics appears from the quantity of chyle, which comes away

by stools; which is otherwise absorbed as soon as produced, and by the immense quantity of thin fluid, which is evacuated along with it.

12. But if the stimulus, which inverts the stomach, be still more powerful, or more permanent, it sometimes happens, that the motions of the biliary glands, and of their excretory ducts, are at the same time inverted, and regurgitate their contained bile into the blood-vessels, as appears by the yellow colour of the skin, and of the urine; and it is probable the pancreatic secretion may suffer an inversion at the same time, though we have yet no mark by which this can be ascertained.

13. Mr. — — eat two putrid pigeons out of a cold pigeon-pye, and drank about a pint of beer and ale along with them, and immediately rode about five miles. He was then seized with vomiting, which was after a few periods succeeded by purging; these continued alternately for two hours; and the purging continued by intervals for six or eight hours longer. During this time he could not force himself to drink more than one pint in the whole; this great inability to drink was owing to the nausea, or inverted motions of the stomach, which the voluntary exertion of swallowing could seldom and with difficulty overcome; yet he discharged in the whole at least six quarts; whence came this quantity of liquid? First, the contents of the stomach were emitted, then of the duodenum, gall-bladder, and pancreas, by vomiting. After this the contents of the lower bowels, then the chyle, that was in the lacteal vessels, and in the receptacle of chyle, was regurgitated into the intestines by a retrograde motion of these vessels. And afterwards the mucus deposited in the cellular membrane, and on the surface of all the other membranes, seems to have been absorbed; and with the fluid absorbed from the air to have been carried up their respective lymphatic branches by the increased energy of their natural motions, and down the visceral lymphatics, or lacteals, by the inversion of their motions.

14. It may be difficult to invent experiments to demonstrate the truth of this inversion of some branches of the absorbent system, and increased absorption of others, but the analogy of these vessels to the intestinal canal, and the symptoms of many diseases, render this opinion more probable than many other received opinions of the animal œconomy.

In the above instance, after the yellow excrement was voided, the fluid ceased to have any smell, and appeared like curdled milk, and then a thinner fluid, and some mucus, were evacuated; did not these seem to partake of the chyle, of the mucous fluid from all the cells of the body, and lastly, of the atmospheric moisture? All these facts may be easily observed by any one, who takes a brisk purge.

15. Where the stimulus on the stomach, or on some other part of the intestinal canal, is still more permanent, not only the lacteal vessels, but the whole canal itself, becomes inverted from its associations: this is the iliac passion, in which all the fluids mentioned above are thrown up by the mouth. At this time the valve in the colon, from the inverted motions of that bowel, and the inverted action of this living valve, does not prevent the regurgitation of its contents.

The structure of this valve may be represented by a flexile leathern pipe standing up from the bottom of a vessel of water: its sides collapse by the pressure of the ambient fluid, as a small part of that fluid passes through it; but if it has a living power, and by its inverted action keeps itself open, it becomes like a rigid pipe, and will admit the whole liquid to pass. See Sect. XXIX. 2. 5.

In this case the patient is averse to drink, from the constant inversion of the motions of the stomach, and yet many quarts are daily ejected from the stomach, which at length smell of excrement, and at last seem to be only a thin mucilaginous or aqueous liquor.

From whence is it possible, that this great quantity of fluid for many successive days can be supplied, after the cells of the body have given up their fluids, but from the atmosphere? When the cutaneous branch of absorbents acts with unnatural strength, it is probable the intestinal branch has its motions inverted, and thus a fluid is supplied without entering the arterial system. Could oiling or painting the skin give a check to this disease?

So when the stomach has its motions inverted, the lymphatics of the stomach, which are most strictly associated with it, invert their motions at the same time. But the more distant branches of lymphatics, which are less strictly associated with it, act with increased energy; as the cutaneous lymphatics in the cholera, or iliac passion, above described. And other irritative motions become decreased, as the pulsations of the arteries, from the extra-derivation or exhaustion of the sensorial power.

Sometimes when stronger vomiting takes place the more distant branches of the lymphatic system invert their motions with those of the stomach, and loose stools are produced, and cold sweats.

So when the lacteals have their motions inverted, as during the operation of strong purges, the urinary and cutaneous absorbents have their motions increased to supply the want of fluid in the blood, as in great thirst; but after a meal with sufficient potation the urine is pale, that is, the urinary absorbents act weakly, no supply of water being wanted for the blood. And when the intestinal absorbents act too violently, as when too great quantities of fluid have been drank, the urinary absorbents invert their motions to carry off the superfluity, which is a new circumstance of association, and a temporary diabetes supervenes.

16. I have had the opportunity of seeing four patients in the iliac passion, where the ejected material smelled and looked like excrement. Two of these were so exhausted at the time I saw them, that more blood could not be taken from them, and as their pain had ceased, and they continued to vomit up every thing which they drank, I suspected that a mortification of the bowel had already taken place, and as they were both women advanced in life, and a mortification is produced with less preceding pain in old and weak people, these both died. The other two, who were both young men, had still pain and strength sufficient for further venesection, and they neither of them had any appearance of hernia, both recovered by repeated bleeding, and a scruple of calomel given to one, and half a dram to the other, in very small pills: the usual means of clysters, and purges joined with opiates, had been in vain attempted. I have thought an ounce or two of crude mercury in less violent diseases of this kind has been of use, by contributing to restore its natural motion to some part of the intestinal canal, either by its weight or stimulus; and that hence the whole tube recovered its usual associations of progressive peristaltic motion. I have in three cases seen crude mercury given in small doses, as one or two ounces twice a day, have great effect in stopping pertinacious vomitings.

17. Besides the affections above described, the stomach is liable, like many other membranes of the body, to torpor without consequent inflammation: as happens to the membranes about the head in some cases of hemicrania, or in general head-ach. This torpor of the stomach is attended with indigestion, and consequent flatulency, and with pain, which is usually called the cramp of the stomach, and is relievable by aromatics, essential oils, alcohol, or opium.

The intrusion of a gall-stone into the common bile-duct from the gall-bladder is sometimes mistaken for a pain of the stomach, as neither of them are attended with fever; but in the passage of a gall-stone, the pain is confined to a less space, which is exactly where the common bile-duct enters the duodenum, as explained in Section XXX. 1. 3. Whereas in this gastrodynia the pain is diffused over the whole stomach; and, like other diseases from torpor, the pulse is weaker, and the extremities colder, and the general debility greater, than in the passage of a gall-stone; for in the former the debility is the consequence of the pain, in the latter it is the cause of it.

Though the first fits of the gout, I believe, commence with a torpor of the liver; and the ball of the toe becomes inflamed instead of the membranes of the liver in consequence of this torpor, as a coryza or catarrh frequently succeeds a long exposure of the feet to cold, as in snow, or on a moist brick-floor; yet in old or exhausted constitutions, which have been long habituated to its attacks, it sometimes commences with a torpor of the stomach, and is transferable to every membrane of the body. When the gout begins with torpor of the stomach, a painful sensation of cold occurs, which the patient compares to ice, with weak pulse, cold extremities, and sickness; this in its slighter degree is relievable by spice, wine, or opium; in its greater degree it is succeeded by sudden death, which is owing to the sympathy of the stomach with the heart, as explained below.

If the stomach becomes inflamed in consequence of this gouty torpor of it, or in consequence of its sympathy with some other part, the danger is less. A sickness and vomiting continues many days, or even weeks, the stomach rejecting every thing stimulant, even opium or alcohol, together with much viscid mucus; till the inflammation at length ceases, as happens when other membranes, as those of the joints, are the seat of gouty inflammation; as observed in Sect. XXIV. 2. 8.

The sympathy, or association of motions, between those of the stomach and those of the heart, are evinced in many diseases. First, many people are occasionally affected with an intermission of their pulse for a few days, which then ceases again. In this case there is a stop of the motion of the heart, and at the same time a tendency to eructation from the stomach. As soon as the patient feels a tendency to the intermission of the motion of his heart, if he voluntarily brings up wind from his stomach, the stop of the heart does not occur. From hence I conclude that the stop of digestion is the primary disease; and that air is instantly generated from the aliment, which begins to ferment, if the digestive process is impeded for a moment, (see Sect. XXIII. 4.); and that the stop of the heart is in consequence of the association of the motions of these viscera, as explained in Sect. XXXV. 1. 4.; but if the little air, which is instantly generated during the temporary torpor of the stomach, be evacuated, the digestion recommences, and the temporary torpor of the heart does not follow. One patient, whom I lately saw, and who had been five or six days much troubled with this intermission of a pulsation of his heart, and who had hemicrania with some fever, was immediately relieved from them all by losing ten ounces of blood, which had what is termed an inflammatory crust on it.

Another instance of this association between the motions of the stomach and heart is evinced by the exhibition of an over dose of foxglove, which induces an incessant vomiting, which is attended with very slow, and sometimes intermitting pulse.—Which continues in spite of the exhibition of wine and opium for two or three days. To the same association must be ascribed the weak pulse, which constantly attends the exhibition of emetics during their operation. And also the sudden deaths, which have been occasioned in boxing by a blow on the stomach; and lastly, the sudden death of those, who have been long debilitated by the gout, from the torpor of the stomach. See Sect. XXXV. 1. 4.

SECT. XXVI
OF THE CAPILLARY GLANDS
AND MEMBRANES

I. 1. The capillary vessels are glands. 2. Their excretory ducts. Experiments on the mucus of the intestines, abdomen, cellular membrane, and on the humours of the eye. 3. Scurf on the head, cough, catarrh, diarrhœa, gonorrhœa. 4. Rheumatism. Gout. Leprosy. II. 1. The most minute membranes are unorganized. 2. Larger membranes are composed of the ducts of the capillaries, and the mouths of the absorbents. 3. Mucilaginous fluid is secreted on their surfaces. III. Three kinds of rheumatism.

I. 1. The capillary-vessels are like all the other glands except the absorbent system, inasmuch as they receive blood from the arteries, separate a fluid from it, and return the remainder by the veins.

2. This series of glands is of the most extensive use, as their excretory ducts open on the whole external skin forming its perspirative pores, and on the internal surfaces of every cavity of the body. Their secretion on the skin is termed insensible perspiration, which in health is in part reabsorbed by the mouths of the lymphatics, and in part evaporated in the air; the secretion on the membranes, which line the larger cavities of the body, which have external openings, as the mouth and intestinal canal, is termed mucus, but is not however coagulable by heat; and the secretion on the membranes of those cavities of the body, which have no external openings, is called lymph or water, as in the cavities of the cellular membrane, and of the abdomen; this lymph however is coagulable by the heat of boiling water. Some mucus nearly as viscid as the white of egg, which was discharged by stool, did not coagulate, though I evaporated it to one fourth of the quantity, nor did the aqueous and vitreous humours of a sheep's eye coagulate by the like experiment: but the serosity from an anasarcous leg, and that from the abdomen of a dropsical person, and the crystalline humour of a sheep's eye, coagulated in the same heat.

3. When any of these capillary glands are stimulated into greater irritative actions, than is natural, they secrete a more copious material;

and as the mouths of the absorbent system, which open in their vicinity, are at the same time stimulated into greater action, the thinner and more saline part of the secreted fluid is taken up again; and the remainder is not only more copious but also more viscid than natural. This is more or less troublesome or noxious according to the importance of the functions of the part affected: on the skin and bronchiæ, where this secretion ought naturally to evaporate, it becomes so viscid as to adhere to the membrane; on the tongue it forms a pellicle, which can with difficulty be scraped off; produces the scurf on the heads of many people; and the mucus, which is spit up by others in coughing. On the nostrils and fauces, when the secretion of these capillary glands is increased, it is termed simple catarrh; when in the intestines, a mucous diarrhœa; and in the urethra, or vagina, it has the name of gonorrhœa, or fluor albus.

4. When these capillary glands become inflamed, a still more viscid or even cretaceous humour is produced upon the surfaces of the membranes, which is the cause or the effect of rheumatism, gout, leprosy, and of hard tumours of the legs, which are generally termed scorbutic; all which will be treated of hereafter.

II. 1. The whole surface of the body, with all its cavities and contents, are covered with membrane. It lines every vessel, forms every cell, and binds together all the muscular and perhaps the osseous fibres of the body; and is itself therefore probably a simpler substance than those fibres. And as the containing vessels of the body from the largest to the least are thus lined and connected with membranes, it follows that these membranes themselves consisted of unorganized materials.

For however small we may conceive the diameters of the minutest vessels of the body, which escape our eyes and glasses, yet these vessels must consist of coats or sides, which are made up of an unorganized material, and which are probably produced from a gluten, which hardens after its production, like the silk or web of caterpillars and spiders. Of this material consist the membranes, which line the shells of eggs, and the shell itself, both which are unorganized, and are formed from mucus, which hardens after it is formed, either by the absorption of its more fluid part, or by its uniting with some part of the atmosphere. Such is also the production of the shells of snails, and of shell-fish, and I suppose of the enamel of the teeth.

2. But though the membranes, that compose the sides of the most minute vessels, are in truth unorganized materials, yet the larger membranes, which are perceptible to the eye, seem to be composed of an intertexture of the mouths of the absorbent system, and of the excretory ducts of the capillaries, with their concomitant arteries, veins, and nerves: and from

this construction it is evident, that these membranes must possess great irritability to peculiar stimuli, though they are incapable of any motions, that are visible to the naked eye: and daily experience shews us, that in their inflamed state they have the greatest sensibility to pain, as in the pleurisy and paronychia.

3. On all these membranes a mucilaginous or aqueous fluid is secreted, which moistens and lubricates their surfaces, as was explained in Section XXIII. 2. Some have doubted, whether this mucus is separated from the blood by an appropriated set of glands, or exudes through the membranes, or is an abrasion or destruction of the surface of the membrane itself, which is continually repaired on the other side of it, but the great analogy between the capillary vessels, and the other glands, countenances the former opinion; and evinces, that these capillaries are the glands, that secrete it; to which we must add, that the blood in passing these capillary vessels undergoes a change in its colour from florid to purple, and gives out a quantity of heat; from whence, as in other glands, we must conclude that something is secreted from it.

III. The seat of rheumatism is in the membranes, or upon them; but there are three very distinct diseases, which commonly are confounded under this name. First, when a membrane becomes affected with torpor, or inactivity of the vessels which compose it, pain and coldness succeed, as in the hemicrania, and other head-achs, which are generally termed nervous rheumatism; they exist whether the part be at rest or in motion, and are generally attended with other marks of debility.

Another rheumatism is said to exist, when inflammation and swelling, as well as pain, affect some of the membranes of the joints, as of the ancles, wrists, knees, elbows, and sometimes of the ribs. This is accompanied with fever, is analogous to pleurisy and other inflammations, and is termed the acute rheumatism.

A third disease is called chronic rheumatism, which is distinguished from that first mentioned, as in this the pain only affects the patient during the motion of the part, and from the second kind of rheumatism above described, as it is not attended with quick pulse or inflammation. It is generally believed to succeed the acute rheumatism of the same part, and that some coagulable lymph, or cretaceous, or calculous material, has been left on the membrane; which gives pain, when the muscles move over it, as some extraneous body would do, which was too insoluble to be absorbed. Hence there is an analogy between this chronic rheumatism and the diseases which produce gravel or gout-stones; and it may perhaps receive relief from the same remedies, such as aerated sal soda.

SECT. XXVII
OF HÆMORRHAGES

I. *The veins are absorbent vessels. 1. Hæmorrhages from inflummation. Case of hæmorrhage from the kidney cured by cold bathing. Case of hæmorrhage from the nose cured by cold immersion. II. Hæmorrhage from venous paralysis. Of Piles. Black stools. Petechiæ. Consumption. Scurvy of the lungs. Blackness of the face and eyes in epileptic fits. Cure of hæmorrhages from venous inability.*

I. As the imbibing mouths of the absorbent system already described open on the surface, and into the larger cavities of the body, so there is another system of absorbent vessels, which are not commonly esteemed such, I mean the veins, which take up the blood from the various glands and capillaries, after their proper fluids or secretions have been separated from it.

The veins resemble the other absorbent vessels; as the progression of their contents is carried on in the same manner in both, they alike absorb their appropriated fluids, and have valves to prevent its regurgitation by the accidents of mechanical violence. This appears first, because there is no pulsation in the very beginnings of the veins, as is seen by microscopes; which must happen, if the blood was carried into them by the actions of the arteries. For though the concurrence of various venous streams of blood from different distances must prevent any pulsation in the larger branches, yet in the very beginnings of all these branches a pulsation must unavoidably exist, if the circulation in them was owing to the intermitted force of the arteries. Secondly, the venous absorption of blood from the penis, and from the teats of female animals after their erection, is still more similar to the lymphatic absorption, as it is previously poured into cells, where all arterial impulse must cease.

There is an experiment, which seems to evince this venous absorption, which consists in the external application of a stimulus to the lips, as of vinegar, by which they become instantly pale; that is, the bibulous mouths of the veins by this stimulus are excited to absorb the blood faster, than it can be supplied by the usual arterial exertion. See Sect. XXIII. 5.

There are two kinds of hæmorrhages frequent in diseases, one is where the glandular or capillary action is too powerfully exerted, and propels the blood forwards more hastily, than the veins can absorb it; and the other is, where the absorbent power of the veins is diminished, or a branch of them is become totally paralytic.

1. The former of these cases is known by the heat of the part, and the general fever or inflammation that accompanies the hæmorrhage. An hæmorrhage from the nose or from the lungs is sometimes a crisis of inflammatory diseases, as of the hepatitis and gout, and generally ceases spontaneously, when the vessels are considerably emptied. Sometimes the hæmorrhage recurs by daily periods accompanying the hot fits of fever, and ceasing in the cold fits, or in the intermissions; this is to be cured by removing the febrile paroxysms, which will be treated of in their place. Otherwise it is cured by venesection, by the internal or external preparations of lead, or by the application of cold, with an abstemious diet, and diluting liquids, like other inflammations. Which by inducing a quiescence on those glandular parts, that are affected, prevents a greater quantity of blood from being protruded forwards, than the veins are capable of absorbing.

Mr. B—— had an hæmorrhage from his kidney, and parted with not less than a pint of blood a day (by conjecture) along with his urine for above a fortnight: venesections, mucilages, balsams, preparations of lead, the bark, alum, and dragon's blood, opiates, with a large blister on his loins, were separately tried, in large doses, to no purpose. He was then directed to bathe in a cold spring up to the middle of his body only, the upper part being covered, and the hæmorrhage diminished at the first, and ceased at the second immersion.

In this case the external capillaries were rendered quiescent by the coldness of the water, and thence a less quantity of blood was circulated through them; and the internal capillaries, or other glands, became quiescent from their irritative associations with the external ones; and the hæmorrhage was stopped a sufficient time for the ruptured vessels to contract their apertures, or for the blood in those apertures to coagulate.

Mrs. K—— had a continued haemorrhage from her nose for some days; the ruptured vessel was not to be reached by plugs up the nostrils, and the sensibility of her fauces was such that nothing could be born behind the uvula. After repeated venesection, and other common applications, she was directed to immerse her whole head into a pail of water, which was made colder by the addition of several handfuls of salt, and the hæmorrhage immediately ceased, and returned no more; but her pulse continued hard, and she was necessitated to lose blood from the arm on the succeeding day.

Query, might not the cold bath instantly stop hæmorrhages from the lungs in inflammatory cases?—for the shortness of breath of those, who go suddenly into cold water, is not owing to the accumulation of blood in the lungs, but to the quiescence of the pulmonary capillaries from association, as explained in Section XXXII. 3. 2.

II. The other kind of hæmorrhage is known from its being attended with a weak pulse, and other symptoms of general debility, and very frequently occurs in those, who have diseased livers, owing to intemperance in the use of fermented liquors. These constitutions are shewn to be liable to paralysis of the lymphatic absorbents, producing the various kinds of dropsies in Section XXIX. 5. Now if any branch of the venous system loses its power of absorption, the part swells, and at length bursts and discharges the blood, which the capillaries or other glands circulate through them.

It sometimes happens that the large external veins of the legs burst, and effuse their blood; but this occurs most frequently in the veins of the intestines, as the vena portarum is liable to suffer from a schirrus of the liver opposing the progression of the blood, which is absorbed from the intestines. Hence the piles are a symptom of hepatic obstruction, and hence the copious discharges downwards or upwards of a black material, which has been called melancholia, or black bile; but is no other than the blood, which is probably discharged from the veins of the intestines.

J.F. Meckel, in his Experimenta de Finibus Vasorum, published at Berlin, 1772, mentions his discovery of a communication of a lymphatic vessel with the gastric branch of the vena portarum. It is possible, that when the motion of the lymphatic becomes retrograde in some diseases, that blood may obtain a passage into it, where it anastomoses with the vein, and thus be poured into the intestines. A discharge of blood with the urine sometimes attends diabetes, and may have its source in the same manner.

Mr. A——, who had been a hard drinker, and had the gutta rosacea on his face and breast, after a stroke of the palsy voided near a quart of a black viscid material by stool: on diluting it with water it did not become yellow, as it must have done if it had been inspissated bile, but continued black like the grounds of coffee.

But any other part of the venous system may become quiescent or totally paralytic as well as the veins of the intestines: all which occur more frequently in those who have diseased livers, than in any others. Hence troublesome bleedings of the nose, or from the lungs with a weak pulse; hence hæmorrhages from the kidneys, too great menstruation; and hence the oozing of blood from every part of the body, and the petechiæ in those fevers, which are termed putrid, and which is erroneously ascribed to the thinness of the blood: for the blood in inflammatory diseases is equally fluid before it coagulates in the cold air.

Is not that hereditary consumption, which occurs chiefly in dark-eyed people about the age of twenty, and commences with slight pulmonary hæmorrhages without fever, a disease of this kind?—These hæmorrhages frequently begin during sleep, when the irritability of the lungs is not sufficient in these patients to carry on the circulation without the assistance of volition; for in our waking hours, the motions of the lungs are in part voluntary, especially if any difficulty of breathing renders the efforts of volition necessary. See Class I. 2. 1. 3. and Class III. 2. 1. 12. Another species of pulmonary consumption which seems more certainly of scrophulous origin is described in the next Section, No. 2.

I have seen two cases of women, of about forty years of age, both of whom were seized with quick weak pulse, with difficult respiration, and who spit up by coughing much viscid mucus mixed with dark coloured blood. They had both large vibices on their limbs, and petechiæ; in one the feet were in danger of mortification, in the other the legs were œdematous. To relieve the difficult respiration, about six ounces of blood were taken from one of them, which to my surprise was sizy, like inflamed blood: they had both palpitations or unequal pulsations of the heart. They continued four or five weeks with pale and bloated countenances, and did not cease spitting phlegm mixed with black blood, and the pulse seldom slower than 130 or 135 in a minute. This blood, from its dark colour, and from the many vibices and petechiæ, seems to have been venous blood; the quickness of the pulse, and the irregularity of the motion of the heart, are to be ascribed to debility of that part of the system; as the extravasation of blood originated from the defect of venous absorption. The approximation of these two cases to sea-scurvy is peculiar, and may allow them to be called scorbutus pulmonalis. Had these been younger subjects, and the paralysis of the veins had only affected the lungs, it is probable the disease would have been a pulmonary consumption.

Last week I saw a gentleman of Birmingham, who had for ten days laboured under great palpitation of his heart, which was so distinctly felt by the hand, as to discountenance the idea of there being a fluid in the pericardium. He frequently spit up mucus stained with dark coloured blood, his pulse very unequal and very weak, with cold hands and nose. He could not lie down at all, and for about ten days past could not sleep a minute together, but waked perpetually with great uneasiness. Could those symptoms be owing to very extensive adhesions of the lungs? or is this a scorbutus pulmonalis? After a few days he suddenly got so much better as to be able to sleep many hours at a time by the use of one grain of powder of foxglove twice a day, and a grain of opium at night. After a few days longer, the bark was exhibited, and the opium continued with some wine; and the palpitations of his heart became much relieved, and he recovered his usual degree of health, but died suddenly some months afterwards.

In epileptic fits the patients frequently become black in the face, from the temporary paralysis of the venous system of this part. I have known two instances where the blackness has continued many days. M. P——, who had drank intemperately, was seized with the epilepsy when he was in his fortieth year; in one of these fits the white part of his eyes was left totally black with effused blood; which was attended with no pain or heat, and was in a few weeks gradually absorbed, changing colour as is usual with vibices from bruises.

The hæmorrhages produced from the inability of the veins to absorb the refluent blood, is cured by opium, the preparations of steel, lead, the bark, vitriolic acid, and blisters; but these have the effect with much more certainty, if a venesection to a few ounces, and a moderate cathartic with four or six grains of calomel be premised, where the patient is not already too much debilitated; as one great means of promoting the absorption of any fluid consists in previously emptying the vessels, which are to receive it.

SECT. XXVIII
OF THE PARALYSIS OF THE ABSORBENT SYSTEM

I. Paralysis of the lacteals, atrophy. Distaste to animal food. II. Cause of dropsy. Cause of herpes. Scrophula. Mesenteric consumption. Pulmonary consumption. Why ulcers in the lungs are so difficult to heal.

The term paralysis has generally been used to express the loss of voluntary motion, as in the hemiplagia, but may with equal propriety be applied to express the disobediency of the muscular fibres to the other kinds of stimulus; as to those of irritation or sensation.

I. There is a species of atrophy, which has not been well understood; when the absorbent vessels of the stomach and intestines have been long inured to the stimulus of too much spirituous liquor, they at length, either by the too sudden omission of fermented or spirituous potation, or from the gradual decay of nature, become in a certain degree paralytic; now it is observed in the larger muscles of the body, when one side is paralytic, the other is more frequently in motion, owing to the less expenditure of sensorial power in the paralytic limbs; so in this case the other part of the absorbent system acts with greater force, or with greater perseverance, in consequence of the paralysis of the lacteals; and the body becomes greatly emaciated in a small time.

I have seen several patients in this disease, of which the following are the circumstances. 1. They were men about fifty years of age, and had lived freely in respect to fermented liquors. 2. They lost their appetite to animal food. 3. They became suddenly emaciated to a great degree. 4. Their skins were dry and rough. 5. They coughed and expectorated with difficulty a viscid phlegm. 6. The membrane of the tongue was dry and red, and liable to become ulcerous.

The inability to digest animal food, and the consequent distaste to it, generally precedes the dropsy, and other diseases, which originate from spirituous potation. I suppose when the stomach becomes inirritable, that there is at the same time a deficiency of gastric acid; hence milk seldom agrees with these patients, unless it be previously curdled, as they have not sufficient gastric acid to curdle it; and hence vegetable food, which is itself acescent, will agree with their stomachs longer than animal food, which requires more of the gastric acid for its digestion.

In this disease the skin is dry from the increased absorption of the cutaneous lymphatics, the fat is absorbed from the increased absorption of the cellular lymphatics, the mucus of the lungs is too viscid to be easily spit up by the increased absorption of the thinner parts of it, the membrana sneideriana becomes dry, covered with hardened mucus, and at length becomes inflamed and full of aphthæ, and either these sloughs, or pulmonary ulcers, terminate the scene.

II. The immediate cause of dropsy is the paralysis of some other branches of the absorbent system, which are called lymphatics, and which open into the larger cavities of the body, or into the cells of the cellular membrane; whence those cavities or cells become distended with the fluid, which is hourly secreted into them for the purpose of lubricating their surfaces. As is more fully explained in No. 5. of the next Section.

As those lymphatic vessels consist generally of a long neck or mouth, which drinks up its appropriated fluid, and of a conglobate gland, in which this fluid undergoes some change, it happens, that sometimes the mouth of the lymphatic, and sometimes the belly or glandular part of it, becomes totally or partially paralytic. In the former case, where the mouths of the cutaneous lymphatics become torpid or quiescent, the fluid secreted on the skin ceases to be absorbed, and erodes the skin by its saline acrimony, and produces eruptions termed herpes, the discharge from which is as salt, as the tears, which are secreted too fast to be reabsorbed, as in grief, or when the puncta lacrymalia are obstructed, and which running down the cheek redden and inflame the skin.

When the mouths of the lymphatics, which open on the mucous membrane of the nostrils, become torpid, as on walking into the air in a frosty morning; the mucus, which continues to be secreted, has not its aqueous and saline part reabsorbed, which running over the upper lip inflames it, and has a salt taste, if it falls on the tongue.

When the belly, or glandular part of these lymphatics, becomes torpid, the fluid absorbed by its mouth stagnates, and forms a tumour in the gland. This disease is called the scrophula. If these glands suppurate externally, they gradually heal, as those of the neck; if they suppurate without an opening on the external habit, as the mesenteric glands, a hectic fever ensues, which destroys the patient; if they suppurate in the lungs, a pulmonary consumption ensues, which is believed thus to differ from that described in the preceding Section, in respect to its seat or proximate cause.

It is remarkable, that matter produced by suppuration will lie concealed in the body many weeks, or even months, without producing hectic fever; but as soon as the wound is opened, so as to admit air to the surface of the ulcer, a hectic fever supervenes, even in very few hours, which is probably owing to the azotic part of the atmosphere rather than to the oxygene; because those medicines, which contain much oxygene, as the calces or oxydes of metals, externally applied, greatly contribute to heal ulcers, of these are the solutions of lead and mercury, and copper in acids, or their precipitates.

Hence when wounds are to be healed by the first intention, as it is called, it is necessary carefully to exclude the air from them. Hence we have one cause, which prevents pulmonary ulcers from healing, which is their being perpetually exposed to the air.

Both the dark-eyed patients, which are affected with pulmonary ulcers from deficient venous absorption, as described in Section. XXVII. 2. and the light-eyed patients from deficient lymphatic absorption, which we are now treating of, have generally large apertures of the iris; these large pupils of the eyes are a common mark of want of irritability; and it generally happens, that an increase of sensibility, that is, of motions in consequence of sensation, attends these constitutions. See Sect. XXXI. 2. Whence inflammations may occur in these from stagnated fluids more frequently than in those constitutions, which possess more irritability and less sensibility.

Great expectations in respect to the cure of consumptions, as well as of many other diseases, are produced by the very ingenious exertions of DR. BEDDOES; who has established an apparatus for breathing various mixtures of airs or gasses, at the hot-wells near Bristol, which well deserves the attention of the public.

DR. BEDDOES very ingeniously concludes, from the florid colour of the blood of consumptive patients, that it abounds in oxygene; and that the redness of their tongues, and lips, and the fine blush of their cheeks shew the presence of the same principle, like flesh reddened by nitre. And adds, that the circumstance of the consumptions of pregnant women being stopped in their progress during pregnancy, at which time their blood may be supposed to be in part deprived of its oxygene, by oxygenating the blood of the fœtus, is a forceable argument in favour of this theory; which must soon be confirmed or confuted by his experiments. See Essay on Scurvy, Consumption, &c. by Dr. Beddoes. Murray. London. Also Letter to Dr. Darwin, by the same. Murray. London.

SECT. XXIX
ON THE RETROGRADE MOTIONS
OF THE ABSORBENT SYSTEM

I. *Account of the absorbent system.* II. *The valves of the absorbent vessels may suffer their fluids to regurgitate in some diseases.* III. *Communication from the alimentary canal to the bladder by means of the absorbent vessels.* IV. *The phenomena of diabetes explained.* V. 1. *The phenomena of dropsies explained.* 2. *Cases of the use of foxglove.* VI. *Of cold sweats.* VII. *Translations of matter, of chyle, of milk, of urine, operation of purging drugs applied externally.* VIII. *Circumstances by which the fluids, that are effused by the retrograde motions of the absorbent vessels, are distinguished.* IX. *Retrograde motions of vegetable juices.* X. *Objections answered.* XI. *The causes, which induce the retrograde motions of animal vessels, and the medicines by which the natural motions are restored.*

N.B. The following Section is a translation of a part of a Latin thesis written by the late Mr. Charles Darwin, which was printed with his prize-dissertation on a criterion between matter and mucus in 1780. Sold by Cadell, London.

I. *Account of the Absorbent System.*

1. The absorbent system of vessels in animal bodies consists of several branches, differing in respect to their situations, and to the fluids, which they absorb.

The intestinal absorbents open their mouths on the internal surfaces of the intestines; their office is to drink up the chyle and the other fluids from the alimentary canal; and they are termed lacteals, to distinguish them from the other absorbent vessels, which have been termed lymphatics.

Those, whose mouths are dispersed on the external skin, imbibe a great quantity of water from the atmosphere, and a part of the perspirable matter, which does not evaporate, and are termed cutaneous absorbents.

Those, which arise from the internal surface of the bronchia, and which imbibe moisture from the atmosphere, and a part of the bronchial mucus, are called pulmonary absorbents.

Those, which open their innumerable mouths into the cells of the whole cellular membrane; and whose use is to take up the fluid, which is poured into those cells, after it has done its office there; may be called cellular absorbents.

Those, which arise from the internal surfaces of the membranes, which line the larger cavities of the body, as the thorax, abdomen, scrotum, pericardium, take up the mucus poured into those cavities; and are distinguished by the names of their respective cavities.

Whilst those, which arise from the internal surfaces of the urinary bladder, gall-bladder, salivary ducts, or other receptacles of secreted fluids, may take their names from those fluids; the thinner parts of which it is their office to absorb: as urinary, bilious, or salivary absorbents.

2. Many of these absorbent vessels, both lacteals and lymphatics, like some of the veins, are replete with valves: which seem designed to assist the progress of their fluids, or at least to prevent their regurgitation; where they are subjected to the intermitted pressure of the muscular, or arterial actions in their neighbourhood.

These valves do not however appear to be necessary to all the absorbents, any more than to all the veins; since they are not found to exist in the absorbent system of fish; according to the discoveries of the ingenious, and much lamented Mr. Hewson. Philos. Trans. v. 59, Enquiries into the Lymph. Syst. p. 94.

3. These absorbent vessels are also furnished with glands, which are called conglobate glands; whose use is not at present sufficiently investigated; but it is probable that they resemble the conglomerate glands both in structure and in use, except that their absorbent mouths are for the conveniency of situation placed at a greater distance from the body of the gland. The conglomerate glands open their mouths immediately into the sanguiferous vessels, which bring the blood, from whence they absorb their respective fluids, quite up to the gland: but these conglobate glands collect their adapted fluids from very distant membranes, or cysts, by means of mouths furnished with long necks for this purpose; and which are called lacteals, or lymphatics.

4. The fluids, thus collected from various parts of the body, pass by means of the thoracic duct into the left subclavian near the jugular vein; except indeed that those collected from the right side of the head and neck,

and from the right arm, are carried into the right subclavian vein: and sometimes even the lymphatics from the right side of the lungs are inserted into the right subclavian vein; whilst those of the left side of the head open but just into the summit of the thoracic duct.

5. In the absorbent system there are many anastomoses of the vessels, which seem of great consequence to the preservation of health. These anastomoses are discovered by dissection to be very frequent between the intestinal and urinary lymphatics, as mentioned by Mr. Hewson, (Phil. Trans. v. 58.)

6. Nor do all the intestinal absorbents seem to terminate in the thoracic duct, as appears from some curious experiments of D. Munro, who gave madder to some animals, having previously put a ligature on the thoracic duct, and found their bones, and the serum of their blood, coloured red.

II. The Valves of the Absorbent System may suffer
their Fluids to regurgitate in some Diseases.

1. The many valves, which occur in the progress of the lymphatic and lacteal vessels, would seem insuperable obstacles to the regurgitation of their contents. But as these valves are placed in vessels, which are indued with life, and are themselves indued with life also; and are very irritable into those natural motions, which absorb, or propel the fluids they contain; it is possible, in some diseases, where these valves or vessels are stimulated into unnatural exertions, or are become paralytic, that during the diastole of the part of the vessel to which the valve is attached, the valve may not so completely close, as to prevent the relapse of the lymph or chyle. This is rendered more probable, by the experiments of injecting mercury, or water, or suet, or by blowing air down these vessels: all which pass the valves very easily, contrary to the natural course of their fluids, when the vessels are thus a little forcibly dilated, as mentioned by Dr. Haller, Elem. Physiol. t. iii. s. 4.

"The valves of the thoracic duct are few, some assert they are not more than twelve, and that they do not very accurately perform their office, as they do not close the whole area of the duct, and thence may permit chyle to repass them downwards. In living animals, however, though not always, yet more frequently than in the dead, they prevent the chyle from returning. The principal of these valves is that, which presides over the insertion of the thoracic duct, into the subclavian vein; many have believed this also to perform the office of a valve, both to admit the chyle into the vein, and to preclude the blood from entering the duct; but in my opinion it is scarcely sufficient for this purpose." Haller, Elem. Phys. t. vii. p. 226.

2. The mouths of the lymphatics seem to admit water to pass through them after death, the inverted way, easier than the natural one; since an inverted bladder readily lets out the water with which it is filled; whence it may be inferred, that there is no obstacle at the mouths of these vessels to prevent the regurgitation of their contained fluids.

I was induced to repeat this experiment, and having accurately tied the ureters and neck of a fresh ox's bladder, I made an opening at the fundus of it; and then, having turned it inside outwards, filled it half full with water, and was surprised to see it empty itself so hastily. I thought the experiment more apposite to my purpose by suspending the bladder with its neck downwards, as the lymphatics are chiefly spread upon this part of it, as shewn by Dr. Watson, Philos. Trans. v. 59. p. 392.

3. In some diseases, as in the diabetes and scrophula, it is probable the valves themselves are diseased, and are thence incapable of preventing the return of the fluids they should support. Thus the valves of the aorta itself have frequently been found schirrous, according to the dissections of Mons. Lieutaud, and have given rise to an interrupted pulse, and laborious palpitations, by suffering a return of part of the blood into the heart. Nor are any parts of the body so liable to schirrosity as the lymphatic glands and vessels, insomuch that their schirrosities have acquired a distinct name, and been termed scrophula.

4. There are valves in other parts of the body, analogous to those of the absorbent system, and which are liable, when diseased, to regurgitate their contents: thus the upper and lower orifices of the stomach are closed by valves, which, when too great quantities of warm water have been drank with a design to promote vomiting, have sometimes resisted the utmost efforts of the abdominal muscles, and diaphragm: yet, at other times, the upper valve, or cardia, easily permits the evacuation of the contents of the stomach; whilst the inferior valve, or pylorus, permits the bile, and other contents of the duodenum, to regurgitate into the stomach.

5. The valve of the colon is well adapted to prevent the retrograde motion of the excrements; yet, as this valve is possessed of a living power, in the iliac passion, either from spasm, or other unnatural exertions, it keeps itself open, and either suffers or promotes the retrograde movements of the contents of the intestines below; as in ruminating animals the mouth of the first stomach seems to be so constructed, as to facilitate or assist the regurgitation of the food; the rings of the œsophagus afterwards contracting themselves in inverted order. De Haeu, by means of a syringe, forced so much water into the rectum intestinum of a dog, that he vomited it in a full stream from his mouth; and in the iliac passion above mentioned,

excrements and clyster are often evacuated by the mouth. See Section XXV. 15.

6. The puncta lacrymalia, with the lacrymal sack and nasal duct, compose a complete gland, and much resemble the intestinal canal: the puncta lacrymalia are absorbent mouths, that take up the tears from the eye, when they have done their office there, and convey them into the nostrils; but when the nasal duct is obstructed, and the lacrymal sack distended with its fluid, on pressure with the finger the mouths of this gland (puncta lacrymalia) will readily disgorge the fluid, they had previously absorbed, back into the eye.

7. As the capillary vessels receive blood from the arteries, and separating the mucus, or perspirable matter from it, convey the remainder back by the veins; these capillary vessels are a set of glands, in every respect similar to the secretory vessels of the liver, or other large congeries of glands. The beginnings of these capillary vessels have frequent anastomoses into each other, in which circumstance they are resembled by the lacteals; and like the mouths or beginnings of other glands, they are a set of absorbent vessels, which drink up the blood which is brought to them by the arteries, as the chyle is drank up by the lacteals: for the circulation of the blood through the capillaries is proved to be independent of arterial impulse; since in the blush of shame, and in partial inflammations, their action is increased, without any increase of the motion of the heart.

8. Yet not only the mouths, or beginnings of these anastomosing capillaries are frequently seen by microscopes, to regurgitate some particles of blood, during the struggles of the animal; but retrograde motion of the blood, in the veins of those animals, from the very heart of the extremity of the limbs, is observable, by intervals, during the distresses of the dying creature. Haller, Elem. Physiol. t. i. p. 216. Now, as the veins have perhaps all of them a valve somewhere between their extremities and the heart, here is ocular demonstration of the fluids in this diseased condition of the animal, repassing through venous valves: and it is hence highly probable, from the strictest analogy, that if the course of the fluids, in the lymphatic vessels, could be subjected to microscopic observation, they would also, in the diseased state of the animal, be seen to repass the valves, and the mouths of those vessels, which had previously absorbed them, or promoted their progression.

III. Communication from the Alimentary Canal to
the Bladder, by means of the Absorbent Vessels.

Many medical philosophers, both ancient and modern, have suspected that there was a nearer communication between the stomach and the urinary bladder, than that of the circulation: they were led into this opinion from the great expedition with which cold water, when drank to excess, passes off by the bladder; and from the similarity of the urine, when produced in this hasty manner, with the material that was drank.

The former of these circumstances happens perpetually to those who drink abundance of cold water, when they are much heated by exercise, and to many at the beginning of intoxication.

Of the latter, many instances are recorded by Etmuller, t. xi. p. 716. where simple water, wine, and wine with sugar, and emulsions, were returned by urine unchanged.

There are other experiments, that seem to demonstrate the existence of another passage to the bladder, besides that through the kidneys. Thus Dr. Kratzenstein put ligatures on the ureters of a dog, and then emptied the bladder by a catheter; yet in a little time the dog drank greedily, and made a quantity of water, (Disputat. Morbor. Halleri. t. iv. p. 63.) A similar experiment is related in the Philosophical Transactions, with the same event, (No. 65, 67, for the year 1670.)

Add to this, that in some morbid cases the urine has continued to pass, after the suppuration or total destruction of the kidneys; of which many instances are referred to in the Elem. Physiol. t. vii. p. 379. of Dr. Haller.

From all which it must be concluded, that some fluids have passed from the stomach or abdomen, without having gone through the sanguiferous circulation: and as the bladder is supplied with many lymphatics, as described by Dr. Watson, in the Philos. Trans. v. 59. p. 392. and as no other vessels open into it besides these and the ureters, it seems evident, that the unnatural urine, produced as above described, when the ureters were tied, or the kidneys obliterated, was carried into the bladder by the retrograde motions of the urinary branch of the lymphatic system.

The more certainly to ascertain the existence of another communication between the stomach and bladder, besides that of the circulation, the following experiment was made, to which I must beg your patient attention:—A friend of mine (June 14, 1772) on drinking repeatedly of cold small punch, till he began to be intoxicated, made a quantity of colourless urine. He then drank about two drams of nitre dissolved in some of the punch, and eat about twenty stalks of boiled asparagus: on continuing to drink more of the punch, the next urine that he made was quite clear, and without smell; but in a little time another quantity was made, which was not quite so colourless, and had a strong smell of the asparagus: he then lost about four ounces of blood from the arm.

The smell of asparagus was not at all perceptible in the blood, neither when fresh taken, nor the next morning, as myself and two others accurately attended to; yet this smell was strongly perceived in the urine, which was made just before the blood was taken from his arm.

Some bibulous paper, moistened in the serum of this blood, and suffered to dry, shewed no signs of nitre by its manner of burning. But some of the same paper, moistened in the urine, and dried, on being ignited, evidently shewed the presence of nitre. This blood and the urine stood some days exposed to the sun in the open air, till they were evaporated to about a fourth of their original quantity, and began to stink: the paper, which was then moistened with the concentrated urine, shewed the presence of much nitre by its manner of burning; whilst that moistened with the blood shewed no such appearance at all.

Hence it appears, that certain fluids at the beginning of intoxication, find another passage to the bladder besides the long course of the arterial circulation; and as the intestinal absorbents are joined with the urinary lymphatics by frequent anastomoses, as Hewson has demonstrated; and as there is no other road, we may justly conclude, that these fluids pass into the bladder by the urinary branch of the lymphatics, which has its motions inverted during the diseased state of the animal.

A gentleman, who had been some weeks affected with jaundice, and whose urine was in consequence of a very deep yellow, took some cold small punch, in which was dissolved about a dram of nitre; he then took repeated draughts of the punch, and kept himself in a cool room, till on the approach of slight intoxication he made a large quantity of water; this water had a slight yellow tinge, as might be expected from a small admixture of bile secreted from the kidneys; but if the whole of it had passed through the sanguiferous vessels, which were now replete with bile (his whole skin being as yellow as gold) would not this urine also, as well as that he had made for weeks before, have been of a deep yellow? Paper dipped in this water, and dryed, and ignited, shewed evident marks of the presence of nitre, when the flame was blown out.

IV. *The Phænomena of the Diabetes*
explained, and of some Diarrhœas.

The phenomena of many diseases are only explicable from the retrograde motions of some of the branches of the lymphatic system; as the great and immediate flow of pale urine in the beginning of drunkenness; in hysteric paroxysms; from being exposed to cold air; or to the influence of fear or anxiety.

Before we endeavour to illustrate this doctrine, by describing the phænomena of these diseases, we must premise one circumstance; that all the branches of the lymphatic system have a certain sympathy with each other, insomuch that when one branch is stimulated into unusual kinds or quantities of motion, some other branch has its motions either increased, or decreased, or inverted at the same time. This kind of sympathy can only be proved by the concurrent testimony of numerous facts, which will be related in the course of the work. I shall only add here, that it is probable, that this sympathy does not depend on any communication of nervous filaments, but on habit; owing to the various branches of this system having frequently been stimulated into action at the same time.

There are a thousand instances of involuntary motions associated in this manner; as in the act of vomiting, while the motions of the stomach and œsophagus are inverted, the pulsations of the arterial system by a certain sympathy become weaker; and when the bowels or kidneys are stimulated by poison, a stone, or inflammation, into more violent action; the stomach and œsophagus by sympathy invert their motions.

1. When any one drinks a moderate quantity of vinous spirit, the whole system acts with more energy by consent with the stomach and intestines, as is seen from the glow on the skin, and the increase of strength and activity; but when a greater quantity of this inebriating material is drank, at the same time that the lacteals are excited into greater action to absorb it; it frequently happens, that the urinary branch of absorbents, which is connected with the lacteals by many anastomoses, inverts its motions, and a great quantity of pale unanimalized urine is discharged. By this wise contrivance too much of an unnecessary fluid is prevented from entering the circulation—This may be called the drunken diabetes, to distinguish it from the other temporary diabetes, which occur in hysteric diseases, and from continued fear or anxiety.

2. If this idle ingurgitation of too much vinous spirit be daily practised, the urinary branch of absorbents at length gains an habit of inverting its motions, whenever the lacteals are much stimulated; and the whole or a great part of the chyle is thus daily carried to the bladder without entering the circulation, and the body becomes emaciated. This is one kind of chronic diabetes, and may be distinguished from the others by the taste and appearance of the urine; which is sweet, and the colour of whey, and may be termed the chyliferous diabetes.

3. Many children have a similar deposition of chyle in their urine, from the irritation of worms in their intestines, which stimulating the mouths of the lacteals into unnatural action, the urinary branch of the absorbents

becomes inverted, and carries part of the chyle to the bladder: part of
the chyle also has been carried to the iliac and lumbar glands, of which
instances are recorded by Haller, t. vii. 225. and which can be explained on
no other theory: but the dissections of the lymphatic system of the human
body, which have yet been published, are not sufficiently extensive for our
purpose; yet if we may reason from comparative anatomy, this translation of
chyle to the bladder is much illustrated by the account given of this system
of vessels in a turtle, by Mr. Hewson, who observed, "That the lacteals near
the root of the mesentery anastomose, so as to form a net-work, from which
several large branches go into some considerable lymphatics lying near the
spine; and which can be traced almost to the anus, and particularly to the
kidneys." Philos. Trans. v. 59. p. 199—Enquiries, p. 74.

4. At the same time that the urinary branch of absorbents, in the
beginning of diabetes, is excited into inverted action, the cellular branch is
excited by the sympathy above mentioned, into more energetic action; and
the fat, that was before deposited, is reabsorbed and thrown into the blood
vessels; where it floats, and was mistaken for chyle, till the late experiments
of the ingenious Mr. Hewson demonstrated it to be fat.

This appearance of what was mistaken for chyle in the blood, which
was drawn from these patients, and the obstructed liver, which very
frequently accompanies this disease, seems to have led Dr. Mead to suspect
the diabetes was owing to a defect of sanguification; and that the schirrosity
of the liver was the original cause of it: but as the schirrhus of the liver is
most frequently owing to the same causes, that produce the diabetes and
dropsies; namely, the great use of fermented liquors; there is no wonder
they should exist together, without being the consequence of each other.

5. If the cutaneous branch of absorbents gains a habit of being excited
into stronger action, and imbibes greater quantities of moisture from the
atmosphere, at the same time that the urinary branch has its motions
inverted, another kind of diabetes is formed, which may be termed the
aqueous diabetes. In this diabetes the cutaneous absorbents frequently
imbibe an amazing quantity of atmospheric moisture; insomuch that there
are authentic histories, where many gallons a day, for many weeks together,
above the quantity that has been drank, have been discharged by urine.

Dr. Keil, in his Medicina Statica, found that he gained eighteen ounces
from the moist air of one night; and Dr. Percival affirms, that one of his
hands imbibed, after being well chafed, near an ounce and half of water, in a
quarter of an hour. (Transact. of the College, London, vol. ii. p. 102.) Home's
Medic. Facts, p. 2. sect. 3.

The pale urine in hysterical women, or which is produced by fear or anxiety, is a temporary complaint of this kind; and it would in reality be the same disease, if it was confirmed by habit.

6. The purging stools, and pale urine, occasioned by exposing the naked body to cold air, or sprinkling it with cold water, originate from a similar cause; for the mouths of the cutaneous lymphatics being suddenly exposed to cold become torpid, and cease, or nearly cease, to act; whilst, by the sympathy above described, not only the lymphatics of the bladder and intestines cease also to absorb the more aqueous and saline part of the fluids secreted into them; but it is probable that these lymphatics invert their motions, and return the fluids, which were previously absorbed, into the intestines and bladder. At the very instant that the body is exposed naked to the cold air, an unusual movement is felt in the bowels; as is experienced by boys going into the cold bath: this could not occur from an obstruction of the perspirable matter, since there is not time, for that to be returned to the bowels by the course of the circulation.

There is also a chronic aqueous diarrhœa, in which the atmospheric moisture, drank up by the cutaneous and pulmonary lymphatics, is poured into the intestines, by the retrograde motions of the lacteals. This disease is most similar to the aqueous diabetes, and is frequently exchanged for it: a distinct instance of this is recorded by Benningerus, Cent. v. Obs. 98. in which an aqueous diarrhœa succeeded an aqueous diabetes, and destroyed the patient. There is a curious example of this, described by Sympson (De Re Medica)—"A young man (says he) was seized with a fever, upon which a diarrhœa came on, with great stupor; and he refused to drink any thing, though he was parched up with excessive heat: the better to supply him with moisture, I directed his feet to be immersed in cold water; immediately I observed a wonderful decrease of water in the vessel, and then an impetuous stream of a fluid, scarcely coloured, was discharged by stool, like a cataract."

7. There is another kind of diarrhœa, which has been called cæliaca; in this disease the chyle, drank up by the lacteals of the small intestines, is probably poured into the large intestines, by the retrograde motions of their lacteals: as in the chyliferous diabetes, the chyle is poured into the bladder, by the retrograde motions of the urinary branch of absorbents.

The chyliferous diabetes, like this chyliferous diarrhœa, produces sudden atrophy; since the nourishment, which ought to supply the hourly waste of the body, is expelled by the bladder, or rectum: whilst the aqueous diabetes, and the aqueous diarrhœa produce excessive thirst; because the moisture, which is obtained from the atmosphere, is not conveyed to the thoracic receptacle, as it ought to be, but to the bladder, or lower intestines; whence the chyle, blood, and whole system of glands, are robbed of their proportion of humidity.

8. There is a third species of diabetes, in which the urine is mucilaginous, and appears ropy in pouring it from one vessel into another; and will sometimes coagulate over the fire. This disease appears by intervals, and ceases again, and seems to be occasioned by a previous dropsy in some part of the body. When such a collection is reabsorbed, it is not always returned into the circulation; but the same irritation that stimulates one lymphatic branch to reabsorb the deposited fluid, inverts the urinary branch, and pours it into the bladder. Hence this mucilaginous diabetes is a cure, or the consequence of a cure, of a worse disease, rather than a disease itself.

Dr. Cotunnius gave half an ounce of cream of tartar, every morning, to a patient, who had the anasarca; and he voided a great quantity of urine; a part of which, put over the fire, coagulated, on the evaporation of half of it, so as to look like the white of an egg. De Ischiade Nervos.

This kind of diabetes frequently precedes a dropsy; and has this remarkable circumstance attending it, that it generally happens in the night; as during the recumbent state of the body, the fluid, that was accumulated in the cellular membrane, or in the lungs, is more readily absorbed, as it is less impeded by its gravity. I have seen more than one instance of this disease. Mr. D. a man in the decline of life, who had long accustomed himself to spirituous liquor, had swelled legs, and other symptoms of approaching anasarca; about once in a week, or ten days, for several months, he was seized, on going to bed, with great general uneasiness, which his attendants resembled to an hysteric fit; and which terminated in a great discharge of viscid urine; his legs became less swelled, and he continued in better health for some days afterwards. I had not the opportunity to try if this urine would coagulate over the fire, when part of it was evaporated, which I imagine would be the criterion of this kind of diabetes; as the mucilaginous fluid deposited in the cells and cysts of the body, which have no communication with the external air, seems to acquire, by stagnation, this property of coagulation by heat, which the secreted mucus of the intestines and bladder do not appear to possess; as I have found by experiment: and if any one should suppose this coagulable urine was separated from the blood by the kidneys, he may recollect, that in the most inflammatory diseases, in which the blood is most replete or most ready to part with the coagulable lymph, none of this appears in the urine.

9. Different kinds of diabetes require different methods of cure. For the first kind, or chyliferous diabetes, after clearing the stomach and intestines,

by ipecacuanha and rhubarb, to evacuate any acid material, which may too powerfully stimulate the mouths of the lacteals, repeated and large doses of tincture of cantharides have been much recommended. The specific stimulus of this medicine, on the neck of the bladder, is likely to excite the numerous absorbent vessels, which are spread on that part, into stronger natural actions, and by that means prevent their retrograde ones; till, by persisting in the use of the medicine, their natural habits of motions might again be established. Another indication of cure, requires such medicines, as by lining the intestines with mucilaginous substances, or with such as consist of smooth particles, or which chemically destroy the acrimony of their contents, may prevent the too great action of the intestinal absorbents. For this purpose, I have found the earth precipitated from a solution of alum, by means of fixed alcali, given in the dose of half a dram every six hours, of great advantage, with a few grains of rhubarb, so as to produce a daily evacuation.

The food should consist of materials that have the least stimulus, with calcareous water, as of Bristol and Matlock; that the mouths of the lacteals may be as little stimulated as is necessary for their proper absorption; lest with their greater exertions, should be connected by sympathy, the inverted motions of the urinary lymphatics.

The same method may be employed with equal advantage in the aqueous diabetes, so great is the sympathy between the skin and the stomach. To which, however, some application to the skin might be usefully added; as rubbing the patient all over with oil, to prevent the too great action of the cutaneous absorbents. I knew an experiment of this kind made upon one patient with apparent advantage.

The mucilaginous diabetes will require the same treatment, which is most efficacious in the dropsy, and will be described below. I must add, that the diet and medicines above mentioned, are strongly recommended by various authors, as by Morgan, Willis, Harris, and Etmuller; but more histories of the successful treatment of these diseases are wanting to fully ascertain the most efficacious methods of cure.

In a letter from Mr. Charles Darwin, dated April 24, 1778, Edinburgh, is the subsequent passage:—"A man who had long laboured under a diabetes died yesterday in the clinical ward. He had for some time drank four, and passed twelve pounds of fluid daily; each pound of urine contained an ounce of sugar. He took, without considerable relief, gum kino, sanguis diaconis melted with alum, tincture of cantharides, isinglass, gum arabic, crabs eyes, spirit of hartshorn, and eat ten or fifteen oysters thrice a day. Dr. Home, having read my thesis, bled him, and found that neither the fresh blood nor the serum tasted sweet. His body was opened this morning—every viscus appeared in a sound and natural state, except that the left kidney had a very small pelvis, and that there was a considerable enlargement of most of the mesenteric lymphatic glands. I intend to insert this in my thesis, as it coincides with the experiment, where some asparagus was eaten at the beginning of intoxication, and its smell perceived in the urine, though not in the blood."

The following case of chyliferous diabetes is extracted from some letters of Mr. Hughes, to whose unremitted care the infirmary at Stafford for many years was much indebted. Dated October 10, 1778.

Richard Davis, aged 33, a whitesmith by trade, had drank hard by intervals; was much troubled with sweating of his hands, which incommoded him in his occupation, but which ceased on his frequently dipping them in lime. About seven months ago he began to make large quantities of water; his legs are œdematous, his belly tense, and he complains of a rising in his throat, like the globus hystericus: he eats twice as much as other people, drinks about fourteen pints of small beer a day, besides a pint of ale, some milk-porridge, and a bason of broth, and he makes about eighteen pints of water a day.

He tried alum, dragon's blood, steel, blue vitriol, and cantharides in large quantities, and duly repeated, under the care of Dr. Underhill, but without any effect; except that on the day after he omitted the cantharides, he made but twelve pints of water, but on the next day this good effect ceased again.

November 21.—He made eighteen pints of water, and he now, at Dr. Darwin's request, took a grain of opium every four hours, and five grains of aloes at night; and had a flannel shirt given him.

22.—Made sixteen pints. 23.—Thirteen pints: drinks less.

24.—Increased the opium to a grain and quarter every four hours: he made twelve pints.

25.—Increased the opium to a grain and half: he now makes ten pints; and drinks eight pints in a day.

The opium was gradually increased during the next fortnight, till he took three grains every four hours, but without any further diminution of his water. During the use of the opium he sweat much in the nights, so as to have large drops stand on his face and all over him. The quantity of opium was then gradually decreased, but not totally omitted, as he continued to take about a grain morning and evening.

January 17.—He makes fourteen pints of water a day. Dr. Underhill now directed him two scruples of common rosin triturated with as much sugar, every six hours; and three grains of opium every night.

19.—Makes fifteen pints of water: sweats at night.

21.—Makes seventeen pints of water; has twitchings of his limbs in a morning, and pains of his legs: he now takes a dram of rosin for a dose, and continues the opium.

23.—Water more coloured, and reduced to sixteen pints, and he thinks has a brackish taste.

26.—Water reduced to fourteen pints.

28.—Water thirteen pints: he continues the opium, and takes four scruples of the rosin for a dose.

February 1.—Water twelve pints.

4.—Water eleven pints: twitchings less; takes five scruples for a dose.

8.—Water ten pints: has had many stools.

12.—Appetite less: purges very much.

After this the rosin either purged him, or would not stay on his stomach; and he gradually relapsed nearly to his former condition, and in a few months sunk under the disease.

October 3, Mr. Hughes evaporated two quarts of the water, and obtained from it four ounces and half of a hard and brittle saccharine mass, like treacle which had been some time boiled. Four ounces of blood, which he took from his arm with design to examine it, had the common appearances, except that the serum resembled cheese-whey; and that on the evidence of four persons, two of whom did not know what it was they tasted, *the serum had a saltish taste.*

From hence it appears, that the saccharine matter, with which the urine of these patients so much abounds, does not enter the blood-vessels like the nitre and asparagus mentioned above; but that the process of digestion resembles the process of the germination of vegetables, or of making barley into malt; as the vast quantity of sugar found in the urine must be made from the food which he took (which was double that taken by others), and from the fourteen pints of small beer which he drank. And, secondly, as the serum of the blood was not sweet, the chyle appears to have been conveyed to the bladder without entering the circulation of the blood, since so large a quantity of sugar, as was found in the urine, namely, twenty ounces a day, could not have previously existed in the blood without being perceptible to the taste.

November 1. Mr. Hughes dissolved two drams of nitre in a pint of a decoction of the roots of asparagus, and added to it two ounces of tincture of rhubarb: the patient took a fourth part of this mixture every five minutes, till he had taken the whole. — In about half an hour he made eighteen ounces of water, which was very manifestly tinged with the rhubarb; the smell of asparagus was doubtful.

He then lost four ounces of blood, the serum of which was not so opake as that drawn before, but of a yellowish cast, as the serum of thc blood usually appears.

Paper, dipped three or four times in the tinged urine and dried again, did not scintillate when it was set on fire; but when the flame was blown out, the fire ran along the paper for half an inch; which, when the same paper was unimpregnated, it would not do; nor when the same paper was dipped in urine made before he took the nitre, and dried in the same manner.

Paper, dipped in the serum of the blood and dried in the same manner as in the urine, did not scintillate when the flame was blown out, but burnt exactly in the same manner as the same paper dipped in the serum of blood drawn from another person.

This experiment, which is copied from a letter of Mr. Hughes, as well as the former, seems to evince the existence of another passage from the intestines to the bladder, in this disease, besides that of the sanguiferous system; and coincides with the curious experiment related in section the third, except that the smell of the asparagus was not here perceived, owing perhaps to the roots having been made use of instead of the heads.

The rising in the throat of this patient, and the twitchings of his limbs, seem to indicate some similarity between the diabetes and the hysteric disease, besides the great flow of pale urine, which is common to them both.

Perhaps if the mesenteric glands were nicely inspected in the dissections of these patients; and if the thoracic duct, and the larger branches of the lacteals, and if the lymphatics, which arise from the bladder, were well examined by injection, or by the knife, the cause of diabetes might be more certainly understood.

The opium alone, and the opium with the rosin, seem much to have served this patient, and might probably have effected a cure, if the disease had been slighter, or the medicine had been exhibited, before it had been confirmed by habit during the seven months it had continued. The increase of the quantity of water on beginning the large doses of rosin was probably owing to his omitting the morning doses of opium.

V. *The Phænomena of Dropsies explained.*

I. Some inebriates have their paroxysms of inebriety terminated by much pale urine, or profuse sweats, or vomiting, or stools; others have their paroxysms terminated by stupor, or sleep, without the above evacuations.

The former kind of these inebriates have been observed to be more liable to diabetes and dropsy; and the latter to gout, gravel, and leprosy. Evoe! attend ye bacchanalians! start at this dark train of evils, and, amid your immodest jests, and idiot laughter, recollect,

Quem Deus vult perdere, prius dementat.

In those who are subject to diabetes and dropsy, the absorbent vessels are naturally more irritable than in the latter; and by being frequently disturbed or inverted by violent stimulus, and by their too great sympathy with each other, they become at length either entirely paralytic, or are only susceptible of motion from the stimulus of very acrid materials; as every part of the body, after having been used to great irritations, becomes less affected by smaller ones. Thus we cannot distinguish objects in the night, for some time after we come out of a strong light, though the iris is presently dilated; and the air of a summer evening appears cold, after we have been exposed to the heat of the day.

There are no cells in the body, where dropsy may not be produced, if the lymphatics cease to absorb that mucilaginous fluid, which is perpetually deposited in them, for the purpose of lubricating their surfaces.

If the lymphatic branch, which opens into the cellular membrane, either does its office imperfectly, or not at all; these cells become replete with a mucilaginous fluid, which, after it has stagnated some time in the cells, will coagulate over the fire; and is erroneously called water. Wherever the seat of this disease is, (unless in the lungs or other pendent viscera) the mucilaginous liquid above mentioned will subside to the most depending parts of the body, as the feet and legs, when those are lower than the head and trunk; for all these cells have communications with each other.

When the cellular absorbents are become insensible to their usual irritations, it most frequently happens, but not always, that the cutaneous branch of absorbents, which is strictly associated with them, suffers the like inability. And then, as no pwater is absorbed from the atmosphere, the urine is not only less diluted at the time of its secretion, and consequently in less quantity and higher coloured: but great thirst is at the same time induced, for as no water is absorbed from the atmosphere to dilute the chyle and blood, the lacteals and other absorbent vessels, which have not lost their powers, are excited into more constant or more violent action, to supply this deficiency; whence the urine becomes still less in quantity, and of a deeper colour, and turbid like the yolk of an egg, owing to a greater absorption of its thinner parts. From this stronger action of those absorbents, which still retain their irritability, the fat is also absorbed, and the whole body becomes emaciated. This increased exertion of some branches of the lymphatics, while others are totally or partially paralytic, is resembled by what constantly occurs in the hemiplagia; when the patient has lost the use of the limbs on one side, he is incessantly moving those of the other; for the moving power, not having access to the paralytic limbs, becomes redundant in those which are not diseased.

The paucity of urine and thirst cannot be explained from a greater quantity of mucilaginous fluid being deposited in the cellular membrane: for though these symptoms have continued many weeks, or even months, this collection frequently does not amount to more than very few pints. Hence also the difficulty of promoting copious sweats in anasarca is accounted for, as well as the great thirst, paucity of urine, and loss of fat; since, when the cutaneous branch of absorbents is paralytic, or nearly so, there is already too small a quantity of aqueous fluid in the blood: nor can these torpid cutaneous lymphatics be readily excited into retrograde motions.

Hence likewise we understand, why in the ascites, and some other dropsies, there is often no thirst, and no paucity of urine; in these cases the cutaneous absorbents continue to do their office.

Some have believed, that dropsies were occasioned by the inability of the kidneys, from having only observed the paucity of urine; and have thence laboured much to obtain diuretic medicines; but it is daily observable, that those who die of a total inability to make water, do not become dropsical in consequence of it: Fernelius mentions one, who laboured under a perfect suppression of urine during twenty days before his death, and yet had no symptoms of dropsy. Pathol. 1. vi. c. 8. From the same idea many physicians have restrained their patients from drinking, though their thirst has been very urgent; and some cases have been published, where this cruel regimen has been thought advantageous: but others of nicer observation are of opinion, that it has always aggravated the distresses of the patient; and though it has abated his swellings, yet by inducing a fever it has hastened his dissolution. See Transactions of the College, London, vol. ii. p. 235. Cases of Dropsy by Dr. G. Baker.

The cure of anasarca, so far as respects the evacuation of the accumulated fluid, coincides with the idea of the retrograde action of the lymphatic system. It is well known that vomits, and other drugs, which induce sickness or nausea; at the same time that they evacuate the stomach, produce a great absorption of the lymph accumulated in the cellular membrane. In the operation of a vomit, not only the motions of the stomach and duodenum become inverted, but also those of the lymphatics and lacteals, which belong to them; whence a great quantity of chyle and lymph is perpetually poured into the stomach and intestines, during the operation, and evacuated by the mouth. Now at the same time, other branches of the lymphatic system, viz. those which open on the cellular membrane, are brought into more energetic action, by the sympathy above mentioned, and an increase of their absorption is produced.

Hence repeated vomits, and cupreous salts, and small doses of squill or foxglove, are so efficacious in this disease. And as drastic purges act also by inverting the motions of the lacteals; and thence the other branches of lymphatics are induced into more powerful natural action, by sympathy, and drink up the fluids from all the cells of the body; and by their anastomoses, pour them into the lacteal branches; which, by their inverted actions, return them into the intestines; and they are thus evacuated from the body:—these purges also are used with success in discharging the accumulated fluid in anasarca.

II. The following cases are related with design to ascertain the particular kinds of dropsy in which the digitalis purpurea, or common foxglove, is preferable to squill, or other evacuants, and were first published in 1780, in a pamphlet entitled Experiments on mucilaginous and purulent Matter, &c. Cadell. London. Other cases of dropsy, treated with digitalis, were afterwards published by Dr. Darwin in the Medical Transactions, vol. iii. in which there is a mistake in respect to the dose of the powder of foxglove, which should have been from five grains to one, instead of from five grains to ten.

Anasarca of the Lungs.

1. A lady, between forty and fifty years of age, had been indisposed some time, was then seized with cough and fever, and afterwards expectorated much digested mucus. This expectoration suddenly ceased, and a considerable difficulty of breathing supervened, with a pulse very irregular both in velocity and strength; she was much distressed at first lying down, and at first rising; but after a minute or two bore either of those attitudes with ease. She had no pain or numbness in her arms; she had no hectic fever, nor any cold shiverings, and the urine was in due quantity, and of the natural colour.

The difficulty of breathing was twice considerably relieved by small doses of ipecacuanha, which operated upwards and downwards, but recurred in a few days: she was then directed a decoction of foxglove, (digitalis purpurea) prepared by boiling four ounces of the fresh leaves from two pints of water to one pint; to which was added two ounces of vinous spirit: she took three large spoonfuls of this mixture every two hours, till she had taken it four times; a continued sickness supervened, with frequent vomiting, and a copious flow of urine: these evacuations continued at intervals for two or three days, and relieved the difficulty of breathing—She had some relapses afterwards, which were again relieved by the repetition of the decoction of foxglove.

2. A gentleman, about sixty years of age, who had been addicted to an immoderate use of fermented liquors, and had been very corpulent, gradually lost his strength and flesh, had great difficulty of breathing, with legs somewhat swelled, and a very irregular pulse. He was very much distressed at first lying down, and at first rising from his bed, yet in a minute or two was easy in both those attitudes. He made straw-coloured urine in due quantity, and had no pain or numbness of his arms.

He took a large spoonful of the decoction of foxglove, as above, every hour, for ten or twelve successive hours, had incessant sickness for about two days, and passed a large quantity of urine; upon which his breath became quite easy, and the swelling of his legs subsided; but as his whole constitution was already sinking from the previous intemperance of his life, he did not survive more than three or four months.

Hydrops Pericardii.

3. A gentleman of temperate life and sedulous application to business, between thirty and forty years of age, had long been subject, at intervals, to an irregular pulse: a few months ago he became weak, with difficulty of breathing, and dry cough. In this situation a physician of eminence directed him to abstain from all animal food and fermented liquor, during which regimen all his complaints increased; he now became emaciated, and totally lost his appetite; his pulse very irregular both in velocity and strength; with great difficulty of breathing, and some swelling of his legs; yet he could lie down horizontally in his bed, though he got little sleep, and passed a due quantity of urine, and of the natural colour: no fullness or hardness could be perceived about the region of the liver; and he had no pain or numbness in his arms.

One night he had a most profuse sweat all over his body and limbs, which quite deluged his bed, and for a day or two somewhat relieved his difficulty of breathing, and his pulse became less irregular: this copious sweat recurred three or four times at the intervals of five or six days, and repeatedly alleviated his symptoms.

He was directed one large spoonful of the above decoction of foxglove every hour, till it procured some considerable evacuation: after he had taken it eleven successive hours he had a few liquid stools, attended with a great flow of urine, which last had a dark tinge, as if mixed with a few drops of blood: he continued sick at intervals for two days, but his breath became quite easy, and his pulse quite regular, the swelling of his legs disappeared, and his appetite and sleep returned.

He then took three grains of white vitriol twice a day, with some bitter medicines, and a grain of opium with five grains of rhubarb every night; was advised to eat flesh meat, and spice, as his stomach would bear it, with small beer, and a few glasses of wine; and had issues made in his thighs; and has suffered no relapse.

4. A lady, about fifty years of age, had for some weeks great difficulty of breathing, with very irregular pulse, and considerable general debility: she could lie down in bed, and the urine was in due quantity and of the natural colour, and she had no pain or numbness of her arms.

She took one large spoonful of the above decoction of foxglove every hour, for ten or twelve successive hours; was sick, and made a quantity of pale urine for about two days, and was quite relieved both of the difficulty of breathing, and the irregularity of her pulse. She then took a grain of opium, and five grains of rhubarb, every night, night, for many weeks; with some slight chalybeate and bitter medicines, and has suffered no relapse.

Hydrops Thoracis.

5. A tradesman, about fifty years of age, became weak and short of breath, especially on increase of motion, with pain in one arm, about the insertion of the biceps muscle. He observed he sometimes in the night made an unusual quantity of pale water. He took calomel, alum, and peruvian bark, and all his symptoms increased: his legs began to swell considerably; his breath became more difficult, and he could not lie down in bed; but all this time he made a due quantity of straw-coloured water.

The decoction of foxglove was given as in the preceding cases, which operated chiefly by purging, and seemed to relieve his breath for a day or two; but also seemed to contribute to weaken him.—He became after some weeks universally dropsical, and died comatous.

6. A young lady of delicate constitution, with light eyes and hair, and who had perhaps lived too abstemiously both in respect to the quantity and quality of what she eat and drank, was seized with great difficulty of

breathing, so as to threaten immediate death. Her extremities were quite cold, and her breath felt cold to the back of one's hand. She had no sweat, nor could be down for a single moment; and had previously, and at present, complained of great weakness and pain and numbness of both her arms; had no swelling of her legs, no thirst, water in due quantity and colour. Her sister, about a year before, was afflicted with similar symptoms, was repeatedly blooded, and died universally dropsical.

A grain of opium was given immediately, and repeated every six hours with evident and amazing advantage; afterwards a blister, with chalybeates, bitters, and essential oils, were exhibited, but nothing had such eminent effect in relieving the difficulty of breathing and coldness of her extremities as opium, by the use of which in a few weeks she perfectly regained her health, and has suffered no relapse.

Ascites.

7. A young lady of delicate constitution having been exposed to great fear, cold, and fatigue, by the overturn of a chaise in the night, began with pain and tumour in the right hypochondrium: in a few months a fluctuation was felt throughout the whole abdomen, more distinctly perceptible indeed about the region of the stomach; since the integuments of the lower part of the abdomen generally become thickened in this disease by a degree of anasarca. Her legs were not swelled, no thirst, water in due quantity and colour.—She took the foxglove so as to induce sickness and stools, but without abating the swelling, and was obliged at length to submit to the operation of tapping.

8. A man about sixty-seven, who had long been accustomed to spirituous potation, had some time laboured under ascites; his legs somewhat swelled; his breath easy in all attitudes; no appetite; great thirst; urine in exceedingly small quantity, very deep coloured, and turbid; pulse equal. He took the foxglove in such quantity as vomited him, and induced sickness for two days; but procured no flow of urine, or diminution of his swelling; but was thought to leave him considerably weaker.

9. A corpulent man, accustomed to large potation of fermented liquors, had vehement cough, difficult breathing, anasarca of his legs, thighs, and hands, and considerable tumour, with evident fluctuation of his abdomen; his pulse was equal; his urine in small quantity, of deep colour, and turbid. These swellings had been twice considerably abated by drastic cathartics. He took three ounces of a decoction of foxglove (made by boiling one ounce of the fresh leaves in a pint of water) every three hours, for two whole days; it then began to vomit and purge him violently, and promoted a great flow of urine; he was by these evacuations completely emptied in twelve hours.

After two or three months all these symptoms returned, and were again relieved by the use of the foxglove; and thus in the space of about three years he was about ten times evacuated, and continued all that time his usual potations: excepting at first, the medicine operated only by urine, and did not appear considerably to weaken him—The last time he took it, it had no effect; and a few weeks afterwards he vomited a great quantity of blood, and expired.

QUERIES.

1. As the first six of these patients had a due discharge of urine, and of the natural colour, was not the feat of the disease confined to some part of the thorax, and the swelling of the legs rather a symptom of the obstructed circulation of the blood, than of a paralysis of the cellular lymphatics of those parts?

2. When the original disease is a general anasarca, do not the cutaneous lymphatics always become paralytic at the same time with the cellular ones, by their greater sympathy with each other? and hence the paucity of urine, and the great thirst, distinguish this kind of dropsy?

3. In the anasarca of the lungs, when the disease is not very great, though the patients have considerable difficulty of breathing at their first lying down, yet after a minute or two their breath becomes easy again; and the same occurs at their first rising. Is not this owing to the time necessary for the fluid in the cells of the lungs to change its place, so as the least to incommode respiration in the new attitude?

4. In the dropsy of the pericardium does not the patient bear the horizontal or perpendicular attitude with equal ease? Does this circumstance distinguish the dropsy of the pericardium from that of the lungs and of the thorax?

5. Do the universal sweats distinguish the dropsy of the pericardium, or of the thorax? and those, which cover the upper parts of the body only, the anasarca of the lungs?

6. When in the dropsy of the thorax, the patient endeavours to lie down, does not the extravasated fluid compress the upper parts of the bronchia, and totally preclude the access of air to every part of the lungs; whilst in the perpendicular attitude the inferior parts of the lungs only are compressed? Does not something similar to this occur in the anasarca of the lungs, when the disease is very great, and thus prevent those patients also from lying down?

7. As a principal branch of the fourth cervical nerve of the left side, after having joined a branch of the third and of the second cervical nerves,

descending between the subclavian vein and artery, is received in a groove formed for it in the pericardium, and is obliged to make a considerable turn outwards to go over the prominent part of it, where the point of the heart is lodged, in its course to the diaphragm; and as the other phrenic nerve of the right side has a straight course to the diaphragm; and as many other considerable branches of this fourth pair of cervical nerves are spread on the arms; does not a pain in the left arm distinguish a disease of the pericardium, as in the angina pectoris, or in the dropsy of the pericardium? and does not a pain or weakness in both arms distinguish the dropsy of the thorax?

8. Do not the dropsies of the thorax and pericardium frequently exist together, and thus add to the uncertainty and fatality of the disease?

9. Might not the foxglove be serviceable in hydrocephalus internus, in hydrocele, and in white swellings of the joints?

VI. *Of cold Sweats.*

There have been histories given of chronical immoderate sweatings, which bear some analogy to the diabetes. Dr. Willis mentions a lady then living, whose sweats where for many years so profuse, that all her bed-clothes were not only moistened, but deluged with them every night; and that many ounces, and sometimes pints, of this sweat, were received in vessels properly placed, as it trickled down her body. He adds, that she had great thirst, had taken many medicines, and submitted to various rules of life, and changes of climate, but still continued to have these immoderate sweats. Pharmac. ration. de sudore anglico.

Dr. Willis has also observed, that the sudor anglicanus which appeared in England, in 1483, and continued till 1551, was in some respects similar to the diabetes; and as Dr. Caius, who saw this disease, mentions the viscidity, as well as the quantity of these sweats, and adds, that the extremities were often cold, when the internal parts were burnt up with heat and thirst, with great and speedy emaciation and debility: there is great reason to believe, that the fluids were absorbed from the cells of the body by the cellular and cystic branches of the lymphatics, and poured on the skin by the retrograde motions of the cutaneous ones.

Sydenham has recorded, in the stationary fever of the year 1685, the viscid sweats flowing from the head, which were probably from the same source as those in the sweating plague above mentioned.

It is very common in dropsies of the chest or lungs to have the difficulty of breathing relieved by copious sweats, flowing from the head and neck. Mr. P. about 50 years of age, had for many weeks been afflicted with anasarca of his legs and thighs, attended with difficulty of breathing; and had repeatedly been relieved by squill, other bitters, and chalybeates.—One night the difficulty of breathing became so great, that it was thought he must have expired; but so copious a sweat came out of his head and neck, that in a few hours some pints, by estimation, were wiped off from those parts, and his breath was for a time relieved. This dyspnœa and these sweats recurred at intervals, and after some weeks he ceased to exist. The skin of his head and neck felt cold to the hand, and appeared pale at the time these sweats flowed so abundantly; which is a proof, that they were produced by an inverted motion of the absorbents of those parts: for sweats, which are the consequence of an increased action of the sanguiferous system, are always attended with a warmth of the skin, greater than is natural, and a more florid colour; as the sweats from exercise, or those that succeed the cold fits of agues. Can any one explain how these partial sweats should relieve the difficulty of breathing in anasarca, but by supposing that the pulmonary branch of absorbents drank up the fluid in the cavity of the thorax, or in the cells of the lungs, and threw it on the skin, by the retrograde motions of the cutaneous branch? for, if we could suppose, that the increased action of the cutaneous glands or capillaries poured upon the skin this fluid, previously absorbed from the lungs; why is not the whole surface of the body covered with sweat? why is not the skin warm? Add to this, that the sweats above mentioned were clammy or glutinous, which the condensed perspirable matter is not; whence it would seem to have been a different fluid from that of common perspiration.

Dr. Dobson, of Liverpool, has given a very ingenious explanation of the acid sweats, which he observed in a diabetic patient—he thinks part of the chyle is secreted by the skin, and afterwards undergoes an acetous fermentation.—Can the chyle get thither, but by an inverted motion of the cutaneous lymphatics? in the same manner as it is carried to the bladder, by the inverted motions of the urinary lymphatics. Medic. Observat. and Enq. London, vol. v.

Are not the cold sweats in some fainting fits, and in dying people, owing to an inverted motion of the cutaneous lymphatics? for in these there can be no increased arterial or glandular action.

Is the difficulty of breathing, arising from anasarca of the lungs, relieved by sweats from the head and neck; whilst that difficulty of breathing, which arises from a dropsy of the thorax, or pericardium, is never attended with these sweats of the head? and thence can these diseases be distinguished from each other? Do the periodic returns of nocturnal asthma rise from a temporary dropsy of the lungs, collected during their more torpid state in sound deep, and then re-absorbed by the vehement efforts of the disordered organs of respiration, and carried off by the copious sweats about the head and neck?

More extensive and accurate dissections of the lymphatic system are wanting to enable us to unravel these knots of science.

VII. *Translations of Matter, of Chyle, of Milk, of Urine.*
Operation of purging Drugs applied externally.

1. The translations of matter from one part of the body to another, can only receive an explanation from the doctrine of the occasional retrograde motions of some branches of the lymphatic system: for how can matter, absorbed and mixed with the whole mass of blood, be so hastily collected again in any one part? and is it not an immutable law, in animal bodies, that each gland can secrete no other, but its own proper fluid? which is, in part, fabricated in the very gland by an animal process, which it there undergoes: of these purulent translations innumerable and very remarkable instances are recorded.

2. The chyle, which is seen among the materials thrown up by violent vomiting, or in purging stools, can only come thither by its having been poured into the bowels by the inverted motions of the lacteals: for our aliment is not converted into chyle in the stomach or intestines by a chemical process, but is made in the very mouths of the lacteals; or in the mesenteric glands; in the same manner as other secreted fluids are made by an animal process in their adapted glands.

Here a curious phænomenon in the exhibition of mercury is worth explaining:—If a moderate dose of calomel, as six or ten grains, be swallowed, and within one or two days a cathartic is given, a salivation is prevented: but after three or four days, a salivation having come on, repeated purges every day, for a week or two, are required to eliminate the mercury from the constitution. For this acrid metallic preparation, being absorbed by the mouth of the lacteals, continues, for a time arrested by the mesenteric glands, (as the variolous or venereal poisons swell the subaxillar or inguinal glands): which, during the operation of a cathartic, is returned into the intestines by the inverted action of the lacteals, and thus carried out of the system.

Hence we understand the use of vomits or purges, to those who have swallowed either contagious or poisonous materials, even though exhibited a day or even two days after such accidents; namely, that by the retrograde motions of the lacteals and lymphatics, the material still arrested in the mesenteric, or other glands, may be eliminated from the body.

3. Many instances of milk and chyle found in ulcers are given by Haller, El. Physiol. t. vii. p. 12, 23, which admit of no other explanation than by supposing, that the chyle, imbibed by one branch of the absorbent system, was carried to the ulcer, by the inverted motions of another branch of the same system.

4. Mrs. P. on the second day after delivery, was seized with a violent purging, in which, though opiates, mucilages, the bark, and testacea were profusely used, continued many days, till at length she recovered. During the time of this purging, no milk could be drawn from her breasts; but the stools appeared like the curd of milk broken into small pieces. In this case, was not the milk taken up from the follicles of the pectoral glands, and thrown on the intestines, by a retrogression of the intestinal absorbents? for how can we for a moment suspect that the mucous glands of the intestines could separate pure milk from the blood? Doctor Smelly has observed, that loose stools, mixed with milk, which is curdled in the intestines, frequently relieves the turgescency of the breasts of those who studiously repel their milk. Cases in Midwifery, 43, No. 2. 1.

5. J.F. Meckel observed in a patient, whose urine was in small quantity and high coloured, that a copious sweat under the arm-pits, of a perfectly urinous smell, stained the linen; which ceased again when the usual quantity of urine was discharged by the urethra. Here we must believe from analogy, that the urine was first secreted in the kidneys, then re-absorbed by the increased action of the urinary lymphatics, and lastly carried to the axillae by the retrograde motions of the lymphatic branches of those parts. As in the jaundice it is necessary, that the bile should first be secreted by the liver, and re-absorbed into the circulation, to produce the yellowness of the skin; as was formerly demonstrated by the late Dr. Munro, (Edin. Medical Essays) and if in this patient the urine had been re-absorbed into the mass of blood, as the bile in the jaundice, why was it not detected in other parts of the body, as well as in the arm-pits?

6. Cathartic and vermifuge medicines applied externally to the abdomen, seem to be taken up by the cutaneous branch of lymphatics, and poured on the intestines by the retrograde motions of the lacteals, without having passed the circulation.

For when the drastic purges are taken by the mouth, they excite the lacteals of the intestines into retrograde motions, as appears from the chyle, which is found coagulated among the fæces, as was shewn above, (sect. 2 and 4.) And as the cutaneous lymphatics are joined with the lacteals of the intestines, by frequent anastomoses; it would be more extraordinary, when a strong purging drug, absorbed by the skin, is carried to the anastomosing branches of the lacteals unchanged, if it should not excite them into retrograde action as efficaciously, as if it was taken by the mouth, and mixed with the food of the stomach.

VIII. *Circumstances by which the Fluids, that are effused by the retrograde Motions of the absorbent Vessels, are distinguished.*

1. We frequently observe an unusual quantity of mucus or other fluids in some diseases, although the action of the glands, by which those fluids are separated from the blood, is not unusually increased; but when the power of absorption alone is diminished. Thus the catarrhal humour from the nostrils of some, who ride in frosty weather; and the tears, which run down the cheeks of those, who have an obstruction of the puncta lacrymalia; and the ichor of those phagedenic ulcers, which are not attended with inflammation, are all instances of this circumstance.

These fluids however are easily distinguished from others by their abounding in ammoniacal or muriatic salts; whence they inflame the circumjacent skin: thus in the catarrh the upper lip becomes red and swelled from the acrimony of the mucus, and patients complain of the saltness of its taste. The eyes and cheeks are red with the corrosive tears, and the ichor of some herpetic eruptions erodes far and wide the contiguous parts, and is pungently salt to the taste, as some patients have informed me.

Whilst, on the contrary, those fluids, which are effused by the retrograde action of the lymphatics, are for the most part mild and innocent; as water, chyle, and the natural mucus: or they take their properties from the materials previously absorbed, as in the coloured or vinous urine, or that scented with asparagus, described before.

2. Whenever the secretion of any fluid is increased, there is at the same time an increased heat in the part; for the secreted fluid, as the bile, did not previously exist in the mass of blood, but a new combination is produced in the gland. Now as solutions are attended with cold, so combinations are attended with heat; and it is probable the sum of the heat given out by all the secreted fluids of animal bodies may be the cause of their general heat above that of the atmosphere.

Hence the fluids derived from increased secretions are readily distinguished from those originating from the retrograde motions of the lymphatics: thus an increase of heat either in the diseased parts, or diffused over the whole body, is perceptible, when copious bilious stools are consequent to an inflamed liver; or a copious mucous salivation from the inflammatory angina.

3. When any secreted fluid is produced in an unusual quantity, and at the same time the power of absorption is increased in equal proportion, not only the heat of the gland becomes more intense, but the secreted fluid becomes thicker and milder, its thinner and saline parts being re-absorbed: and these are distinguishable both by their greater consistence, and by their heat, from the fluids, which are effused by the retrograde motions of the lymphatics; as is observable towards the termination of gonorrhœa, catarrh, chincough, and in those ulcers, which are said to abound with laudable pus.

4. When chyle is observed in stools, or among the materials ejected by vomit, we may be confident it must have been brought thither by the retrograde motions of the lacteals; for chyle does not previously exist amid the contents of the intestines, but is made in the very mouths of the lacteals, as was before explained.

5. When chyle, milk, or other extraneous fluids are found in the urinary bladder, or in any other excretory receptacle of a gland; no one can for a moment believe, that these have been collected from the mass of blood by a morbid secretion, as it contradicts all analogy.

— — Aurea duræ

Mala ferant quercus? Narcisco floreat alnus?

Pinguia corticibus sudent electra myricæ?—VIRGIL.

IX. *Retrograde Motions of Vegetable juices.*

There are besides some motions of the sap in vegetables, which bear analogy to our present subject; and as the vegetable tribes are by many philosophers held to be inferior animals, it may be a matter of curiosity at least to observe, that their absorbent vessels seem evidently, at times, to be capable of a retrograde motion. Mr. Perault cut off a forked branch of a tree, with the leaves on; and inverting one of the forks into a vessel of water, observed, that the leaves on the other branch continued green much longer than those of a similar branch, cut off from the same tree; which shews, that the water from the vessel was carried up one part of the forked branch, by the retrograde motion of its vessels, and supplied nutriment some time to the other part of the branch, which was out of the water. And the celebrated Dr. Hales found, by numerous very accurate experiments, that the sap of trees rose upwards during the warmer hours of the day, and in part descended again during the cooler ones. Vegetable Statics.

It is well known that the branches of willows, and of many other trees, will either take root in the earth or engraft on other trees, so as to have their natural direction inverted, and yet flourish with vigour.

Dr. Hope has also made this pleasing experiment, after the manner of Hales—he has placed a forked branch, cut from one tree, erect between two others; then cutting off a part of the bark from one fork applied it to a similar branch of one of the trees in its vicinity; and the same of the other fork; so that a tree is seen to grow suspended in the air, between two other trees; which supply their softer friend with due nourishment.

Miranturque novas frondes, et non sua poma.

All these experiments clearly evince, that the juices of vegetables can occasionally pass either upwards or downwards in their absorbent system of vessels.

X. *Objections answered.*

The following experiment, at first view, would seem to invalidate this opinion of the retrograde motions of the lymphatic vessels, in some diseases.

About a gallon of milk having been giving to an hungry swine, he was suffered to live about an hour, and was then killed by a stroke or two on his head with an axe.—On opening his belly the lacteals were well seen filled with chyle; on irritating many of the branches of them with a knife, they did not appear to empty themselves hastily; but they did however carry forwards their contents in a little time.

I then passed a ligature round several branches of lacteals, and irritated them much with a knife beneath the ligature, but could not make them regurgitate their contained fluid into the bowels.

I am not indeed certain, that the nerve was not at the same time included in the ligature, and thus the lymphatic rendered unirritable or lifeless; but this however is certain, that it is not any quantity of any stimulus, which induces the vessels of animal bodies to revert their motions; but a certain quantity of a certain stimulus, as appears from wounds in the stomach, which do not produce vomiting; and wounds of the intestines, which do not produce the cholera morbus.

At Nottingham, a few years ago, two shoemakers quarrelled, and one of them with a knife, which they use in their occupation, stabbed his companion about the region of the stomach. On opening the abdomen of the wounded man after his death the food and medicines he had taken were in part found in the cavity of the belly, on the outside of the bowels; and there was a wound about half an inch long at the bottom of the stomach; which I suppose was distended with liquor and food at the time of the accident; and thence was more liable to be injured at its bottom: but during the whole time he lived, which was about ten days, he had no efforts to vomit, nor ever even complained of being sick at the stomach! Other cases similar to this are mentioned in the philosophical transactions.

Thus, if you vellicate the throat with a feather, nausea is produced; if you wound it with a penknife, pain is induced, but not sickness. So if the soles of the feet of children or their armpits are tickled, convulsive laughter is excited, which ceases the moment the hand is applied, so as to rub them more forcibly.

The experiment therefore above related upon the lacteals of a dead pig, which were included in a strict ligature, proves nothing; as it is not the quantity, but the kind of stimulus, which excites the lymphatic vessels into retrograde motion.

XI. *The Causes which induce the retrograde Motions of animal Vessels; and the Medicines by which the natural Motions are restored.*

1. Such is the construction of animal bodies, that all their parts, which are subjected to less stimuli than nature designed, perform their functions with less accuracy: thus, when too watery or too acescent food is taken into the stomach, indigestion, and flatulency, and heartburn succeed.

2. Another law of irritation, connate with our existence, is, that all those parts of the body, which have previously been exposed to too great a quantity of such stimuli, as strongly affect them, become for some time afterwards disobedient to the natural quantity of their adapted stimuli.— Thus the eye is incapable of seeing objects in an obscure room, though the iris is quite dilated, after having been exposed to the meridian sun.

3. There is a third law of irritation, that all the parts of our bodies, which have been lately subjected to less stimulus, than they have been accustomed to, when they are exposed to their usual quantity of stimulus, are excited into more energetic motions: thus when we come from a dusky cavern into the glare of daylight, our eyes are dazzled; and after emerging from the cold bath, the skin becomes warm and red.

4. There is a fourth law of irritation, that all the parts of our bodies, which are subjected to still stronger stimuli for a length of time, become torpid, and refuse to obey even these stronger stimuli; and thence do their offices very imperfectly.—Thus, if any one looks earnestly for some minutes on an area, an inch diameter, of red silk, placed on a sheet of white paper, the image of the silk will gradually become pale, and at length totally vanish.

5. Nor is it the nerves of sense alone, as the optic and auditory nerves, that thus become torpid, when the stimulus is withdrawn or their irritability decreased; but the motive muscles, when they are deprived of their natural

stimuli, or of their irritability, become torpid and paralytic; as is seen in the tremulous hand of the drunkard in a morning; and in the awkward step of age.

The hollow muscles also, of which the various vessels of the body are constructed, when they are deprived of their natural stimuli, or of their due degree of irritability, not only become tremulous, as the arterial pulsations of dying people; but also frequently invert their motions, as in vomiting, in hysteric suffocations, and diabetes above described.

I must beg your patient attention, for a few moments whilst I endeavour to explain, how the retrograde actions of our hollow muscles are the consequence of their debility; as the tremulous actions of the solid muscles are the consequence of their debility. When, through fatigue, a muscle can act no longer; the antagonist muscles, either by their inanimate elasticity, or by their animal action, draw the limb into a contrary direction: in the solid muscles, as those of locomotion, their actions are associated in tribes, which have been accustomed to synchronous action only; hence when they are fatigued, only a single contrary effort takes place; which is either tremulous, when the fatigued muscles are again immediately brought into action; or it is a pandiculation, or stretching, where they are not immediately again brought into action.

Now the motions of the hollow muscles, as they in general propel a fluid along their cavities, are associated in trains, which have been accustomed to successive actions: hence when one ring of such a muscle is fatigued from its too great debility, and is brought into retrograde action, the next ring from its association falls successively into retrograde action; and so on throughout the whole canal. See Sect. XXV. 6.

6. But as the retrograde motions of the stomach, œsophagus, and fauces in vomiting are, as it were, apparent to the eye; we shall consider this operation more minutely, that the similar operations in the more recondite parts of our system may be easier understood.

From certain nauseous ideas of the mind, from an ungrateful taste in the mouth, or from fœtid smells, vomiting is sometimes instantly excited; or even from a stroke on the head, or from the vibratory motions of a ship; all which originate from association, or sympathy. See Sect. XX. on Vertigo.

But when the stomach is subjected to a less stimulus than is natural, according to the first law of irritation mentioned above, its motions become disturbed, as in hunger; first pain is produced, then sickness, and at length vain efforts to vomit, as many authors inform us.

But when a great quantity of wine, or of opium, is swallowed, the retrograde motions of the stomach do not occur till after several minutes, or even hours; for when the power of so strong a stimulus ceases, according to the second law of irritation, mentioned above, the peristaltic motions become tremulous, and at length retrograde; as is well known to the drunkard, who on the next morning has sickness and vomitings.

When a still greater quantity of wine, or of opium, or when nauseous vegetables, or strong bitters, or metallic salts, are taken into the stomach, they quickly induce vomiting; though all these in less doses excite the stomach into more energetic action, and strengthen the digestion; as the flowers of chamomile, and the vitriol of zinc: for, according to the fourth law of irritation, the stomach will not long be obedient to a stimulus so much greater than is natural; but its action becomes first tremulous and then retrograde.

7. When the motions of any vessels become retrograde, less heat of the body is produced; for in paroxysms of vomiting, of hysteric affections, of diabetes, of asthma, the extremities of the body are cold: hence we may conclude, that these symptoms arise from the debility of the parts in action; for an increase of muscular action is always attended with increase of heat.

8. But as animal debility is owing to defect of stimulus, or to defect of irritability, as shewn above, the method of cure is easily deduced: when the vascular muscles are not excited into their due action by the natural stimuli, we should exhibit those medicines, which possess a still greater degree of stimulus; amongst these are the fœtids, the volatiles, aromatics, bitters, metallic salts, opiates, wine, which indeed should be given in small doses, and frequently repeated. To these should be added constant, but moderate exercise, cheerfulness of mind, and change of country to a warmer climate; and perhaps occasionally the external stimulus of blisters.

It is also frequently useful to diminish the quantity of natural stimulus for a short time, by which afterwards the irritability of the system becomes increased; according to the third law of irritation above-mentioned, hence the use of baths somewhat colder than animal heat, and of equitation in the open air.

The catalogue of diseases owing to the retrograde motions of lymphatics is here omitted, as it will appear in the second volume of this work. The following is the conclusion to this thesis of Mr. CHARLES DARWIN.

Thus have I endeavoured in a concise manner to explain the numerous diseases, which deduce their origin from the inverted motions of the hollow muscles of our bodies: and it is probable, that Saint Vitus's dance, and the stammering of speech, originate from a similar, inverted order of the associated motions of some of the solid muscles; which, as it is foreign to my present purpose, I shall not here discuss.

I beg, illustrious professors, and ingenious fellow-students, that you will recollect how difficult a talk I have attempted, to evince the retrograde motions of the lymphatic vessels, when the vessels themselves for so many ages escaped the eyes and glasses of philosophers: and if you are not yet convinced of the truth of this theory, hold, I entreat you, your minds in suspense, till ANATOMY draws her sword with happier omens, cuts asunder the knots, which entangle PHYSIOLOGY; and, like an augur inspecting the immolated victim, announces to mankind the wisdom of HEAVEN.

SECT. XXX
PARALYSIS OF THE LIVER AND KIDNEYS

I. 1.Bile-ducts less irritable after having been stimulated much. 2. Jaundice from paralysis of the bile-ducts cured by electric shocks. 3. From bile-stones. Experiments on bile-stones. Oil vomit. 4. Palsy of the liver, two cases. 5. Schirrosity of the liver. 6. Large livers of geese. II. Paralysis of the kidneys. III. Story of Prometheus.

I. 1. From the ingurgitation of spirituous liquors into the stomach and duodenum, the termination of the common bile-duct in that bowel becomes stimulated into unnatural action, and a greater quantity of bile is produced from all the secretory vessels of the liver, by the association of their motions with those of their excretory ducts; as has been explained in Section XXIV. and XXV. but as all parts of the body, that have been affected with stronger stimuli for any length of time, become less susceptible of motion, from their natural weaker stimuli, it follows, that the motions of the secretory vessels, and in consequence the secretion of bile, is less than is natural during the intervals of sobriety. 2. If this ingurgitation of spirituous liquors has been daily continued in considerable quantity, and is then suddenly intermitted, a languor or paralysis of the common bile-duct is induced; the bile is prevented from being poured into the intestines; and as the bilious absorbents are stimulated into stronger action by its accumulation, and by the acrimony or viscidity, which it acquires by delay, it is absorbed, and carried to the receptacle of the chyle; or otherwise the secretory vessels of the liver, by the above-mentioned stimulus, invert their motions, and regurgitate their contents into the blood, as sometimes happens to the tears in the lachrymal sack, see Sect. XXIV. 2. 7. and one kind of jaundice is brought on.

There is reason to believe, that the bile is most frequently returned into the circulation by the inverted motions of these hepatic glands, for the bile does not seem liable to be absorbed by the lymphatics, for it soaks through the gall-ducts, and is frequently found in the cellular membrane. This kind of jaundice is not generally attended with pain, neither at the extremity of the bile-duct, where it enters the duodenum, nor on the region of the gall-bladder.

Mr. S. a gentleman between 40 and 50 years of age, had had the jaundice about six weeks, without pain, sickness, or fever; and had taken emetics, cathartics, mercurials, bitters, chalybeates, essential oil, and ether, without apparent advantage. On a supposition that the obstruction of the bile might be owing to the paralysis, or torpid action of the common bile-duct, and the stimulants taken into the stomach seeming to have no effect, I directed half a score smart electric shocks from a coated bottle, which held about a quart, to be passed through the liver, and along the course of the common gall-duct, as near as could be guessed, and on that very day the stools became yellow; he continued the electric shocks a few days more, and his skin gradually became clear.

3. The bilious vomiting and purging, that affects some people by intervals of a few weeks, is a less degree of this disease; the bile-duct is less irritable than natural, and hence the bile becomes accumulated in the gall-bladder, and hepatic ducts, till by its quantity, acrimony or viscidity, a greater degree of irritation is produced, and it is suddenly evacuated, or lastly from the absorption of the more liquid parts of the bile, the remainder becomes inspissated, and chrystallizes into masses too large to pass, and forms another kind of jaundice, where the bile-duct is not quite paralytic, or has regained its irritability.

This disease is attended with much pain, which at first is felt at the pit of the stomach, exactly in the centre of the body, where the bile-duct enters the duodenum; afterwards, when the size of the bile-stones increase, it is also felt on the right side, where the gall-bladder is situated. The former pain at the pit of the stomach recurs by intervals, as the bile-stone is pushed against the neck of the duct; like the paroxysms of the stone in the urinary bladder, the other is a more dull and constant pain.

Where these bile-stones are too large to pass, and the bile-ducts possess their sensibility, this becomes a very painful and hopeless disease. I made the following experiments with a view to their chemical solution.

Some fragments of the same bile-stone were put into the weak spirit of marine salt, which is sold in the shops, and into solution of mild alcali; and into a solution of caustic alcali; and into oil of turpentine; without their being dissolved. All these mixtures were after some time put into a heat of boiling water, and then the oil of turpentine dissolved its fragments of bile-stone, but no alteration was produced upon those in the other liquids except some change of their colour.

Some fragments of the same bile-stone were put into vitriolic æther, and were quickly dissolved without additional heat. Might not æther mixed with yolk of egg or with honey be given advantageously in bilious concretions?

I have in two instances seen from 30 to 50 bile-stones come away by stool, about the size of large peas, after having given six grains of calomel in the evening, and four ounces of oil of almonds or olives on the succeeding morning. I have also given half a pint of good olive or almond oil as an emetic during the painful fit, and repeated it in half an hour, if the first did not operate, with frequent good effect.

4. Another disease of the liver, which I have several times observed, consists in the inability or paralysis of the secretory vessels. This disease has generally the same cause as the preceding one, the too frequent potation of spirituous liquors, or the too sudden omission of them, after the habit is confined; and is greater or less in proportion, as the whole or a part of the liver is affected, and as the inability or paralysis is more or less complete.

This palsy of the liver is known from these symptoms, the patients have generally passed the meridian of life, have drank fermented liquors daily, but perhaps not been opprobrious drunkards; they lose their appetite, then their flesh and strength diminish in consequence, there appears no bile in their stools, nor in their urine, nor is any hardness or swelling perceptible on the region of the liver. But what is peculiar to this disease, and distinguishes it from all others at the first glance of the eye, is the bombycinous colour of the skin, which, like that of full-grown silkworms, has a degree of transparency with a yellow tint not greater than is natural to the serum of the blood.

Mr. C. and Mr. B. both very strong men, between 50 and 60 years of age, who had drank ale at their meals instead of small beer, but were not reputed hard-drinkers, suddenly became weak, lost their appetite, flesh, and strength, with all the symptoms above enumerated, and died in about two months from the beginning of their malady. Mr. C. became anasarcous a few days before his death, and Mr. B. had frequent and great hæmorrhages from an issue, and some parts of his mouth, a few days before his death. In both these cases calomel, bitters and chalybeates were repeatedly used without effect.

One of the patients described above, Mr. C, was by trade a plumber; both of them could digest no food, and died apparently for want of blood. Might not the transfusion of blood be used in these cases with advantage?

5. When the paralysis of the hepatic glands is less complete, or less universal, a schirrosity of some part of the liver is induced; for the secretory vessels retaining some of their living power take up a fluid from the circulation, without being sufficiently irritable to carry it forwards to their excretory ducts; hence the body, or receptacle of each gland, becomes inflated, and this distension increases, till by its very great stimulus

inflammation is produced, or till those parts of the viscus become totally paralytic. This disease is distinguishable from the foregoing by the palpable hardness or largeness of the liver; and as the hepatic glands are not totally paralytic, or the whole liver not affected, some bile continues to be made. The inflammations of this viscus, consequent to the schirrosity of it, belong to the diseases of the sensitive motions, and will be treated of hereafter.

6. The ancients are said to have possessed an art of increasing the livers of geese to a size greater than the remainder of the goose. Martial. l. 13. epig. 58.—This is said to have been done by fat and figs. Horace, l. 2. sat. 8.—Juvenal sets these large livers before an epicure as a great rarity. Sat. 5. l. 114; and Persius, sat. 6. l. 71. Pliny says these large goose-livers were soaked in mulled milk, that is, I suppose, milk mixed with honey and wine; and adds, "that it is uncertain whether Scipio Metellus, of consular dignity, or M. Sestius, a Roman knight, was the great discoverer of this excellent dish." A modern traveller, I believe Mr. Brydone, asserts that the art of enlarging the livers of geese still exists in Sicily; and it is to be lamented that he did not import it into his native country, as some method of affecting the human liver might perhaps have been collected from it; besides the honour he might have acquired in improving our giblet pies.

Our wiser caupones, I am told, know how to fatten their fowls, as well as their geese, for the London markets, by mixing gin instead of figs and fat with their food; by which they are said to become sleepy, and to fatten apace, and probably acquire enlarged livers; as the swine are asserted to do, which are fed on the sediments of barrels in the distilleries; and which so frequently obtains in those, who ingurgitate much ale, or wine, or drams.

II. The irritative diseases of the kidneys, pancreas, spleen, and other glands, are analogous to those of the liver above described, differing only in the consequences attending their inability to action. For instance, when the secretory vessels of the kidneys become disobedient to the stimulus of the passing current of blood, no urine is separated or produced by them; their excretory mouths become filled with concreted mucus, or calculus matter, and in eight or ten days stupor and death supervenes in consequence of the retention of the feculent part of the blood.

This disease in a slighter degree, or when only a part of the kidney is affected, is succeeded by partial inflammation of the kidney in consequence of previous torpor. In that case greater actions of the secretory vessels occur, and the nucleus of gravel is formed by the inflamed mucous membranes of the tubuli uriniferi, as farther explained in its place.

This torpor, or paralysis of the secretory vessels of the kidneys, like that of the liver, owes its origin to their being previously habituated to too great stimulus; which in this country is generally owing to the alcohol contained in ale or wine; and hence must be registered amongst the diseases owing to inebriety; though it may be caused by whatever occasionally inflames the kidney; as too violent riding on horseback, or the cold from a damp bed, or by sleeping on the cold ground; or perhaps by drinking in general too little aqueous fluids.

III. I shall conclude this section on the diseases of the liver induced by spirituous liquors, with the well known story of Prometheus, which seems indeed to have been invented by physicians in those ancient times, when all things were clothed in hieroglyphic, or in fable. Prometheus was painted as stealing fire from heaven, which might well represent the inflammable spirit produced by fermentation; which may be said to animate or enliven the man of clay: whence the conquests of Bacchus, as well as the temporary mirth and noise of his devotees. But the after punishment of those, who steal this accursed fire, is a vulture gnawing the liver; and well allegorises the poor inebriate lingering for years under painful hepatic diseases. When the expediency of laying a further tax on the distillation of spirituous liquors from grain was canvassed before the House of Commons some years ago, it was said of the distillers, with great truth, *"They take the bread from the people, and convert it into poison!"* Yet is this manufactory of disease permitted to continue, as appears by its paying into the treasury above 900,000*l*. near a million of money annually. And thus, under the names of rum, brandy, gin, whisky, usquebaugh, wine, cyder, beer, and porter, alcohol is become the bane of the Christian world, as opium of the Mahometan.

Evoe! parce, liber?

Parce, gravi metuende thirso! — Hor.

SECT. XXXI
OF TEMPERAMENTS

I. The temperament of decreased irritability known by weak pulse, large pupils of the eyes, cold extremities. Are generally supposed to be too irritable. Bear pain better than labour. Natives of North-America contrasted with those upon the coast of Africa. Narrow and broad shouldered people. Irritable constitutions bear labour better than pain. II. Temperament of increased sensibility. Liable to intoxication, to inflammation, hæmoptoe, gutta serena, enthusiasm, delirium, reverie. These constitutions are indolent to voluntary exertions, and dull to irritations. The natives of South-America, and brute animals of this temperament. III. Of increased voluntarity; these are subject to locked jaw, convulsions, epilepsy, mania. Are very active, bear cold, hunger, fatigue. Are suited to great exertions. This temperament distinguishes mankind from other animals. IV. Of increased association. These have great memories, are liable to quartan agues, and stronger sympathies of parts with each other. V. Change of temperaments into one another.

Antient writers have spoken much of temperaments, but without sufficient precision. By temperament of the system should be meant a permanent predisposition to certain classes of diseases: without this definition a temporary predisposition to every distinct malady might be termed a temperament. There are four kinds of constitution, which permanently deviate from good health, and are perhaps sufficiently marked to be distinguished from each other, and constitute the temperaments or predispositions to the irritative, sensitive, voluntary, and associate classes of diseases.

I. *The Temperament of decreased Irritability.*

The diseases, which are caused by irritation, most frequently originate from the defect of it; for those, which are immediately owing to the excess of it, as the hot fits of fever, are generally occasioned by an accumulation of sensorial power in consequence of a previous defect of irritation, as in the preceding cold fits of fever. Whereas the diseases, which are caused by sensation and volition, most frequently originate from the excess of those sensorial powers, as will be explained below.

The temperament of decreased irritability appears from the following circumstances, which shew that the muscular fibres or organs of sense are liable to become torpid or quiescent from less defect of stimulation than is productive of torpor or quiescence in other constitutions.

1. The first is the weak pulse, which in some constitutions is at the same time quick. 2. The next most marked criterion of this temperament is the largeness of the aperture of the iris, or pupil of the eye, which has been reckoned by some a beautiful feature in the female countenance, as an indication of delicacy, but to an experienced observer it is an indication of debility, and is therefore a defect, not an excellence. The third most marked circumstance in this constitution is, that the extremities, as the hands and feet, or nose and ears, are liable to become cold and pale in situations in respect to warmth, where those of greater strength are not affected. Those of this temperament are subject to hysteric affections, nervous fevers, hydrocephalus, scrophula, and consumption, and to all other diseases of debility.

Those, who possess this kind of constitution, are popularly supposed to be more irritable than is natural, but are in reality less so.

This mistake has arisen from their generally having a greater quickness of pulse, as explained in Sect. XII. 1. 4. XII. 3. 3.; but this frequency of pulse is not necessary to the temperament, like the debility of it.

Persons of this temperament are frequently found amongst the softer sex, and amongst narrow-shouldered men; who are said to bear labour worse, and pain better than others. This last circumstance is supposed to have prevented the natives of North America from having been made slaves by the Europeans. They are a narrow-shouldered race of people, and will rather expire under the lash, than be made to labour. Some nations of Asia have small hands, as may be seen by the handles of their scymetars; which with their narrow shoulders shew, that they have not been accustomed to so great labour with their hands and arms, as the European nations in agriculture, and those on the coasts of Africa in swimming and rowing. Dr. Maningham, a popular accoucheur in the beginning of this century, observes in his aphorisms, that broad-shouldered men procreate broad-shouldered children. Now as labour strengthens the muscles employed, and increases their bulk, it would seem that a few generations of labour or of indolence may in this respect change the form and temperament of the body.

On the contrary, those who are happily possessed of a great degree of irritability, bear labour better than pain; and are strong, active, and ingenious. But there is not properly a temperament of increased irritability tending to disease, because an increased quantity of irritative motions generally induces an increase of pleasure or pain, as in intoxication, or inflammation; and then the new motions are the immediate consequences of increased sensation, not of increased irritation; which have hence been so perpetually confounded with each other.

II. *Temperament of Sensibility.*

There is not properly a temperament, or predisposition to disease, from decreased sensibility, since irritability and not sensibility is immediately necessary to bodily health. Hence it is the excess of sensation alone, as it is the defect of irritation, that most frequently produces disease. This temperament of increased sensibility is known from the increased activity of all those motions of the organs of sense and muscles, which are exerted in consequence of pleasure or pain, as in the beginning of drunkenness, and in inflammatory fever. Hence those of this constitution are liable to inflammatory diseases, as hepatitis; and to that kind of consumption which is hereditary, and commences with slight repeated hæmoptoe. They have high-coloured lips, frequently dark hair and dark eyes with large pupils, and are in that case subject to gutta serena. They are liable to enthusiasm, delirium, and reverie. In this last circumstance they are liable to start at the clapping of a door; because the more intent any one is on the passing current of his ideas, the greater surprise he experiences on their being dissevered by some external violence, as explained in Sect. XIX. on reverie.

As in these constitutions more than the natural quantities of sensitive motions are produced by the increased quantity of sensation existing in the habit, it follows, that the irritative motions will be performed in some degree with less energy, owing to the great expenditure of sensorial power on the sensitive ones. Hence those of this temperament do not attend to slight stimulations, as explained in Sect. XIX. But when a stimulus is so great as to excite sensation, it produces greater sensitive actions of the system than in others; such as delirium or inflammation. Hence they are liable to be absent in company; sit or lie long in one posture; and in winter have the skin of their legs burnt into various colours by the fire. Hence also they are fearful of pain; covet music and sleep; and delight in poetry and romance.

As the motions in consequence of sensation are more than natural, it also happens from the greater expenditure of sensorial power on them, that the voluntary motions are less easily exerted. Hence the subjects of this temperament are indolent in respect to all voluntary exertions, whether of mind or body.

A race of people of this description seems to have been found by the Spaniards in the islands of America, where they first landed, ten of whom are said not to have consumed more food than one Spaniard, nor to have been capable of more than one tenth of the exertion of a Spaniard. Robertson's History.—In a state similar to this the greatest part of the animal world pass their lives, between sleep or inactive reverie, except when they are excited by the call of hunger.

III. *The Temperament of increased Voluntarity.*

Those of this constitution differ from both the last mentioned in this, that the pain, which gradually subsides in the first, and is productive of inflammation or delirium in the second, is in this succeeded by the exertion of the muscles or ideas, which are most frequently connected with volition; and they are thence subject to locked jaw, convulsions, epilepsy, and mania, as explained in Sect. XXXIV. Those of this temperament attend to the slightest irritations or sensations, and immediately exert themselves to obtain or avoid the objects of them; they can at the same time bear cold and hunger better than others, of which Charles the Twelfth of Sweden was an instance. They are suited and generally prompted to all great exertions of genius or labour, as their desires are more extensive and more vehement, and their powers of attention and of labour greater. It is this facility of voluntary exertion, which distinguishes men from brutes, and which has made them lords of the creation.

IV. *The Temperament of increased Association.*

This constitution consists in the too great facility, with which the fibrous motions acquire habits of association, and by which these associations become proportionably stronger than in those of the other temperaments. Those of this temperament are slow in voluntary exertions, or in those dependent on sensation, or on irritation. Hence great memories have been said to be attended with less sense and less imagination from Aristotle down to the present time; for by the word memory these writers only understood the unmeaning repetition of words or numbers in the order they were received, without any voluntary efforts of the mind.

In this temperament those associations of motions, which are commonly termed sympathies, act with greater certainty and energy, as those between disturbed vision and the inversion of the motion of the stomach, as in sea-sickness; and the pains in the shoulder from hepatic inflammation. Add to this, that the catenated circles of actions are of greater extent than in the other constitutions. Thus if a strong vomit or cathartic be exhibited in this temperament, a smaller quantity will produce as great an effect, if it be given some weeks afterwards; whereas in other temperaments this is only to be expected, if it be exhibited in a few days after the first dose. Hence quartan agues are formed in those of this temperament, as explained in Section XXXII. on diseases from irritation, and other intermittents are liable to recur from slight causes many weeks after they have been cured by the bark.

V. The first of these temperaments differs from the standard of health from defect, and the others from excess of sensorial power; but it sometimes happens that the same individual, from the changes introduced into his habit by the different seasons of the year, modes or periods of life, or by accidental discases, passes from one of these temperaments to another. Thus a long use of too much fermented liquor produces the temperament of increased sensibility; great indolence and solitude that of decreased irritability; and want of the necessaries of life that of increased voluntarity.

SECT. XXXII
DISEASES OF IRRITATION

I. *Irritative fevers with strong pulse. With weak pulse. Symptoms of fever, Their source. II. 1. Quick pulse is owing to decreased irritability. 2. Not in sleep or in apoplexy. 3. From inanition. Owing to deficiency of sensorial power. III. 1. Causes of fever. From defect of heat. Heat from secretions. Pain of cold in the loins and forehead. 2. Great expense of sensorial power in the vital motions. Immersion in cold water. Succeeding glow of heat. Difficult respiration in cold bathing explained. Why the cold bath invigorates. Bracing and relaxation are mechanical terms. 3. Uses of cold bathing. Uses of cold air in fevers. 4. Ague fits from cold air. Whence their periodical returns. IV. Defect of distention a cause of fever. Deficiency of blood. Transfusion of blood. V. 1. Defect of momentum of the blood from mechanic stimuli. 2 . Air injected into the blood-vessels. 3. Exercise increases the momentum of the blood. 4. Sometimes bleeding increases the momentum of it. VI. Influence of the sun and moon on diseases. The chemical stimulus of the blood. Menstruation obeys the lunations. Queries. VII. Quiesence of large glands a cause of fever. Swelling of the præcordia. VIII. Other causes of quiescence, as hunger, bad air, fear, anxiety. IX. 1. Symptoms of the cold fit. 2. Of the hot fit. 3. Second cold fit why. 4. Inflammation introduced, or delirium, or stupor. X. Recapitulation. Fever not an effort of nature to relieve herself. Doctrine of spasm.*

I. When the contractile sides of the heart and arteries perform a greater number of pulsations in a given time, and move through a greater area at each pulsation, whether these motions are occasioned by the stimulus of the acrimony or quantity of the blood, or by their association with other irritative motions, or by the increased irritability of the arterial system, that is, by an increased quantity of sensorial power, one kind of fever is produced; which may be called Synocha irritativa, or Febris irritativa pulsu forti, or irritative fever with strong pulse.

When the contractile sides of the heart and arteries perform a greater number of pulsations in a given time, but move through a much less area at each pulsation, whether these motions are occasioned by defect of their natural stimuli, or by the defect of other irritative motions with which they are associated, or from the inirritability of the arterial system, that is, from a decreased quantity of sensorial power, another kind of fever arises; which may be termed, Typhus irritativus, or Febris irritativa pulsu debili, or irritative fever with weak pulse. The former of these fevers is the synocha of nosologists, and the latter the typhus mitior, or nervous fever. In the former there appears to be an increase of sensorial power, in the latter a deficiency of it; which is shewn to be the immediate cause of strength and weakness, as defined in Sect. XII. 1. 3.

It should be added, that a temporary quantity of strength or debility may be induced by the defect or excess of stimulus above what is natural; and that in the same fever *debility always exists during the cold fit, though strength does not always exist during the hot fit.*

These fevers are always connected with, and generally induced by, the disordered irritative motions of the organs of sense, or of the intestinal canal, or of the glandular system, or of the absorbent system; and hence are always complicated with some or many of these disordered motions, which are termed the symptoms of the fever, and which compose the great variety in these diseases.

The irritative fevers both with strong and with weak pulse, as well as the sensitive fevers with strong and with weak pulse, which are to be described in the next section, are liable to periodical remissions, and then they take the name of intermittent fevers, and are distinguished by the periodical times of their access.

II. For the better illustration of the phenomena of irritative fevers we must refer the reader to the circumstances of irritation explained in Sect. XII. and shall commence this intricate subject by speaking of the quick pulse, and proceed by considering many of the causes, which either separately or in combination most frequently produce the cold fits of fevers.

1. If the arteries are dilated but to half their usual diameters, though they contract twice as frequently in a given time, they will circulate only half their usual quantity of blood: for as they are cylinders, the blood which they contain must be as the squares of their diameters. Hence when the pulse becomes quicker and smaller in the same proportion, the heart and arteries act with less energy than in their natural state. See Sect. XII. 1. 4.

That this quick small pulse is owing to want of irritability, appears, first, because it attends other symptoms of want of irritability; and, secondly, because on the application of a stimulus greater than usual, it becomes slower and larger. Thus in cold fits of agues, in hysteric palpitations of the heart, and when the body is much exhausted by hæmorrhages, or by fatigue, as well as in nervous fevers, the pulse becomes quick and small; and secondly, in all those cases if an increase of stimulus be added, by giving a little wine or opium; the quick small pulse becomes slower and larger, as any one may easily experience on himself, by counting his pulse after drinking one or two glasses of wine, when he is faint from hunger or fatigue.

Now nothing can so strongly evince that this quick small pulse is owing to defect of irritability, than that an additional stimulus, above what is natural, makes it become slower and larger immediately: for what is meant by a defect of irritability, but that the arteries and heart are not excited into their usual exertions by their usual quantity of stimulus? but if you increase the quantity of stimulus, and they immediately act with their usual energy, this proves their previous want of their natural degree of irritability. Thus the trembling hands of drunkards in a morning become steady, and acquire strength to perform their usual offices, by the accustomed stimulus of a glass or two of brandy.

2. In sleep and in apoplexy the pulse becomes slower, which is not owing to defect of irritability, for it is at the same time larger; and thence the quantity of the circulation is rather increased than diminished. In these cases the organs of sense are closed, and the voluntary power is suspended, while the motions dependent on internal irritations, as those of digestion and secretion, are carried on with more than their usual vigour; which has led superficial observers to confound these cases with those arising from want of irritability. Thus if you lift up the eyelid of an apoplectic patient, who is not actually dying, the iris will, as usual, contract itself, as this motion is associated with the stimulus of light; but it is not so in the last stages of nervous fevers, where the pupil of the eye continues expanded in the broad day-light: in the former case there is a want of voluntary power, in the latter a want of irritability.

Hence also those constitutions which are deficient in quantity of irritability, and which possess too great sensibility, as during the pain of hunger, of hysteric spasms, or nervous headachs, are generally supposed to have too much irritability; and opium, which in its due dose is a most powerful stimulant, is erroneously called a sedative; because by increasing the irritative motions it decreases the pains arising from defect of them.

Why the pulse should become quicker both from an increase of irritation, as in the synocha irritativa, or irritative fever with strong pulse; and from the decrease of it, as in the typhus irritativus, or irritative fever with weak pulse; seems paradoxical. The former circumstance needs no illustration; since if the stimulus of the blood, or the irritability of the sanguiferous system be increased, and the strength of the patient not diminished, it is plain that the motions must be performed quicker and stronger.

In the latter circumstance the weakness of the muscular power of the heart is soon over-balanced by the elasticity of the coats of the arteries, which they possess besides a muscular power of contraction; and hence the arteries are distended to less than their usual diameters. The heart being thus stopped, when it is but half emptied, begins sooner to dilate again; and the arteries being dilated to less than their usual diameters, begin so much sooner to contract themselves; insomuch, that in the last stages of fevers with weakness the frequency of pulsation of the heart and arteries becomes doubled; which, however, is never the case in fevers with strength, in which they seldom exceed 118 or 120 pulsations in a minute. It must be added, that in these cases, while the pulse is very small and very quick, the heart often feels large, and labouring to one's hand; which coincides with the above explanation, shewing that it does not completely empty itself.

3. In cases however of debility from paucity of blood, as in animals which are bleeding to death in the slaughter-house, the quick pulsations of the heart and arteries may be owing to their not being distended to more than half their usual diastole; and in consequence they must contract sooner, or more frequently, in a given time. As weak people are liable to a deficient quantity of blood, this cause may occasionally contribute to quicken the pulse in fevers with debility, which may be known by applying one's hand upon the heart as above; but the principal cause I suppose to consist in the diminution of sensorial power. When a muscle contains, or is supplied with but little sensorial power, its contraction soon ceases, and in consequence may soon recur, as is seen in the trembling hands of people weakened by age or by drunkenness. See Sect. XII. 1. 4. XII. 3. 4.

It may nevertheless frequently happen, that both the deficiency of stimulus, as where the quantity of blood is lessened (as described in No. 4. of this section), and the deficiency of sensorial power, as in those of the temperament of irritability, described in Sect. XXXI. occur at the same time; which will thus add to the quickness of the pulse and to the danger of the disease.

III. 1. A certain degree of heat is necessary to muscular motion, and is, in consequence, essential to life. This is observed in those animals and insects which pass the cold season in a torpid state, and which revive on being warmed by the fire. This necessary stimulus of heat has two sources; one from the fluid atmosphere of heat, in which all things are immersed, and the other from the internal combinations of the particles, which form the various fluids, which are produced in the extensive systems of the glands. When either the external heat, which surrounds us, or the internal production of it, becomes lessened to a certain degree, the pain of cold is perceived.

This pain of cold is experienced most sensibly by our teeth, when ice is held in the mouth; or by our whole system after having been previously accustomed to much warmth. It is probable, that this pain does not arise from the mechanical or chemical effects of a deficiency of heat; but that, like the organs of sense by which we perceive hunger and thirst, this sense of heat suffers pain, when the stimulus of its object is wanting to excite the irritative motions of the organ; that is, when the sensorial power becomes too much accumulated in the quiescent fibres. See Sect. XII. 5. 3. For as the peristaltic motions of the stomach are lessened, when the pain of hunger is great, so the action of the cutaneous capillaries are lessened during the pain of cold; as appears by the paleness of the skin, as explained in Sect. XIV. 6. on the production of ideas.

The pain in the small of the back and forehead in the cold fits of the ague, in nervous hemicrania, and in hysteric paroxysms, when all the irritative motions are much impaired, seems to arise from this cause; the vessels of these membranes or muscles become torpid by their irritative associations with other parts of the body, and thence produce less of their accustomed secretions, and in consequence less heat is evolved, and they experience the pain of cold; which coldness may often be felt by the hand applied upon the affected part.

2. The importance of a greater or less deduction of heat from the system will be more easy to comprehend, if we first consider the great expense of sensorial power used in carrying on the vital motions; that is, which circulates, absorbs, secretes, aerates, and elaborates the whole mass of fluids with unceasing assiduity. The sensorial power, or spirit of animation, used in giving perpetual and strong motion to the heart, which overcomes the elasticity and vis inertiæ of the whole arterial system; next the expense of sensorial power in moving with great force and velocity the innumerable trunks and ramifications of the arterial system; the expense of sensorial power in circulating the whole mass of blood through the long and intricate intortions of the very fine vessels, which compose the glands and capillaries; then the expense of sensorial power in the exertions of the absorbent extremities of all the lacteals, and of all the lymphatics, which open their mouths on the external surface of the skin, and on the internal surfaces of every cell or interstice of the body; then the expense of sensorial power in the venous absorption, by which the blood is received from the capillary vessels, or glands, where the arterial power ceases, and is drank

up, and returned to the heart; next the expense of sensorial power used by the muscles of respiration in their office of perpetually expanding the bronchia, or air-vessels, of the lungs; and lastly in the unceasing peristaltic motions of the stomach and whole system of intestines, and in all the secretions of bile, gastric juice, mucus, perspirable matter, and the various excretions from the system. If we consider the ceaseless expense of sensorial power thus perpetually employed, it will appear to be much greater in a day than all the voluntary exertions of our muscles and organs of sense consume in a week; and all this without any sensible fatigue! Now, if but a part of these vital motions are impeded, or totally stopped for but a short time, we gain an idea, that there must be a great accumulation of sensorial power; as its production in these organs, which are subject to perpetual activity, is continued during their quiescence, and is in consequence accumulated.

While, on the contrary, where those vital organs act too forcibly by increase of stimulus without a proportionally-increased production of sensorial power in the brain, it is evident, that a great deficiency of action, that is torpor, must soon follow, as in fevers; whereas the locomotive muscles, which act only by intervals, are neither liable to so great accumulation of sensorial power during their times of inactivity, nor to so great an exhaustion of it during their times of action.

Thus, on going into a very cold bath, suppose at 33 degrees of heat on Fahrenheit's scale, the action of the subcutaneous capillaries, or glands, and of the mouths of the cutaneous absorbents is diminished, or ceases for a time. Hence less or no blood passes these capillaries, and paleness succeeds. But soon after emerging from the bath, a more florid colour and a greater degree of heat is generated on the skin than was possessed before immersion; for the capillary glands, after this quiescent state, occasioned by the want of stimulus, become more irritable than usual to their natural stimuli, owing to the accumulation of sensorial power, and hence a greater quantity of blood is transmitted through them, and a greater secretion of perspirable matter; and, in consequence, a greater degree of heat succeeds. During the continuance in cold water the breath is cold, and the act of respiration quick and laborious; which have generally been ascribed to the obstruction of the circulating fluid by a spasm of the cutaneous vessels, and by a consequent accumulation of blood in the lungs, occasioned by the pressure as well as by the coldness of the water. This is not a satisfactory account of this curious phænomenon, since at this time the whole circulation is less, as appears from the smallness of the pulse and coldness of the breath; which shew that less blood passes through the lungs in a given time; the same laborious breathing immediately occurs when the paleness of the skin is produced by fear, where no external cold or pressure are applied.

The minute vessels of the bronchia, through which the blood passes from the arterial to the venal system, and which correspond with the cutaneous capillaries, have frequently been exposed to cold air, and become quiescent along with those of the skin; and hence their motions are so associated together, that when one is affected either with quiescence or exertion, the other sympathizes with it, according to the laws of irritative association. See Sect. XXVII. 1. on hæmorrhages.

Besides the quiescence of the minute vessels of the lungs, there are many other systems of vessels which become torpid from their irritative associations with those of the skin, as the absorbents of the bladder and intestines; whence an evacuation of pale urine occurs, when the naked skin is exposed only to the coldness of the atmosphere; and sprinkling the naked body with cold water is known to remove even pertinacious constipation of the bowels. From the quiescence of such extensive systems of vessels as the glands and capillaries of the skin, and the minute vessels of the lungs, with their various absorbent series of vessels, a great accumulation of sensorial powers is occasioned; part of which is again expended in the increased exertion of all these vessels, with an universal glow of heat in consequence of this exertion, and the remainder of it adds vigour to both the vital and voluntary exertions of the whole day.

If the activity of the subcutaneous vessels, and of those with which their actions are associated, was too great before cold immersion, as in the hot days of summer, and by that means the sensorial power was previously diminished, we see the cause why the cold bath gives such present strength; namely, by stopping the unnecessary activity of the subcutaneous vessels, and thus preventing the too great exhaustion of sensorial power; which, in metaphorical language, has been called *bracing* the system: which is, however, a mechanical term, only applicable to drums, or musical strings: as on the contrary the word *relaxation*, when applied to living animal bodies, can only mean too small a quantity of stimulus, or too small a quantity of sensorial power; as explained in Sect. XII. 1.

3. This experiment of cold bathing presents us with a simple fever-fit; for the pulse is weak, small, and quick during the cold immersion; and becomes strong, full, and quick during the subsequent glow of heat; till in a few minutes these symptoms subside, and the temporary fever ceases.

In those constitutions where the degree of inirritability, or of debility, is greater than natural, the coldness and paleness of the skin with the quick and weak pulse continue a long time after the patient leaves the bath; and the subsequent heat approaches by unequal flushings, and he feels himself disordered for many hours. Hence the bathing in a cold spring of water, where the heat is but forty-eight degrees on Fahrenheit's thermometer, much disagrees with those of weak or inirritable habits of body; who possess so little sensorial power, that they cannot without injury bear to have it diminished even for a short time; but who can nevertheless bear the more temperate coldness of Buxton bath, which is about eighty degrees of heat, and which strengthens them, and makes them by habit less liable to great quiescence from small variations of cold, and thence less liable to be disordered by the unavoidable accidents of life. Hence it appears, why people of these inirritable constitutions, which is another expression for sensorial deficiency, are often much injured by bathing in a cold spring of water; and why they should continue but a very short time in baths, which are colder than their bodies; and should gradually increase both the degree of coldness of the water, and the time of their continuance in it, if they would obtain salutary effects from cold immersions. See Sect. XII. 2. 1.

On the other hand, in all cases where the heat of the external surface of the body, or of the internal surface of the lungs, is greater than natural, the use of exposure to cool air may be deduced. In fever-fits attended with strength, that is with great quantity of sensorial power, it removes the additional stimulus of heat from the surfaces above mentioned, and thus prevents their excess of useless motion; and in fever-fits attended with debility, that is with a deficiency of the quantity of sensorial power, it prevents the great and dangerous waste of sensorial power expended in the unnecessary increase of the actions of the glands and capillaries of the skin and lungs.

4. In the same manner, when any one is long exposed to very cold air, a quiescence is produced of the cutaneous and pulmonary capillaries and absorbents, owing to the deficiency of their usual stimulus of heat; and this quiescence of so great a quantity of vessels affects, by irritative association, the whole absorbent and glandular system, which becomes in a greater or less degree quiescent, and a cold fit of fever is produced.

If the deficiency of the stimulus of heat is very great, the quiescence becomes so general as to extinguish life, as in those who are frozen to death.

If the deficiency of heat be in less degree, but yet so great as in some measure to disorder the system, and should occur the succeeding day, it will induce a greater degree of quiescence than before, from its acting in concurrence with the period of the diurnal circle of actions, explained in Sect. XXXVI. Hence from a small beginning a greater and greater degree of quiescence may be induced, till a complete fever-fit is formed; and which will continue to recur at the periods by which it was produced. See Sect. XVII. 3. 6.

If the degree of quiescence occasioned by defect of the stimulus of heat be very great, it will recur a second time by a slighter cause, than that which first induced it. If the cause, which induces the second fit of quiescence, recurs the succeeding day, the quotidian fever is produced; if not till the alternate day, the tertian fever; and if not till after seventy-two hours from the first fit of quiescence, the quartan fever is formed. This last kind of fever recurs less frequently than the other, as it is a disease only of those of the temperament of associability, as mentioned in Sect. XXXI.; for in other constitutions the capability of forming a habit ceases, before the new cause of quiescence is again applied, if that does not occur sooner than in seventy-two hours.

And hence those fevers, whose cause is from cold air of the night or morning, are more liable to observe the solar day in their periods; while those from other causes frequently observe the lunar day in their periods, their paroxysms returning near an hour later every day, as explained in Sect. XXXVI.

IV. Another frequent cause of the cold fits of fever is the defect of the stimulus of distention. The whole arterial system would appear, by the experiments of Haller, to be irritable by no other stimulus, and the motions of the heart and alimentary canal are certainly in some measure dependant on the same cause. See Sect. XIV. 7. Hence there can be no wonder, that the diminution of distention should frequently induce the quiescence, which constitutes the beginning of fever-fits.

Monsieur Leiutaud has judiciously mentioned the deficiency of the quantity of blood amongst the causes of diseases, which he says is frequently evident in dissections: fevers are hence brought on by great hæmorrhages, diarrhœas, or other evacuations; or from the continued use of diet, which contains but little nourishment; or from the exhaustion occasioned by violent fatigue, or by those chronic diseases in which the digestion is much impaired; as where the stomach has been long affected with the gout or schirrus; or in the paralysis of the liver, as described in Sect. XXX. Hence a paroxysm of gout is liable to recur on bleeding or purging; as the torpor of some viscus, which precedes the inflammation of the foot, is thus induced by the want of the stimulus of distention. And hence the extremities of the body, as the nose and fingers, are more liable to become cold, when we have long abstained from food; and hence the pulse is increased both in strength and velocity above the natural standard after a full meal by the stimulus of distention.

However, this stimulus of distention, like the stimulus of heat above described, though it contributes much to the due action not only of the heart, arteries, and alimentary canal, but seems necessary to the proper secretion of all the various glands; yet perhaps it is not the sole cause of any of these numerous motions: for as the lacteals, cutaneous absorbents, and the various glands appear to be stimulated into action by the peculiar pungency of the fluids they absorb, so in the intestinal canal the pungency of the digesting aliment, or the acrimony of the fæces, seem to contribute, as well as their bulk, to promote the peristaltic motions; and in the arterial system, the momentum of the particles of the circulating blood, and their acrimony, stimulate the arteries, as well as the distention occasioned by it. Where the pulse is small this defect of distention is present, and contributes much to produce the febris irritativa pulsu debili, or irritative fever with weak pulse, called by modern writers nervous fever, as a predisponent cause. See Sect. XII. 1. 4. Might not the transfusion of blood, suppose of four ounces daily from a strong man, or other healthful animal, as a sheep or an ass, be used in the early state of nervous or putrid fevers with great prospect of success?

V. 1. The defect of the momentum of the particles of the circulating blood is another cause of the quiescence, with which the cold fits of fever commence. This stimulus of the momentum of the progressive particles of the blood does not act over the whole body like those of heat and distention above described, but is confined to the arterial system; and differs from the stimulus of the distention of the blood, as much as the vibration of the air does from the currents of it. Thus are the different organs of our bodies stimulated by four different mechanic properties of the external world: the sense of touch by the pressure of solid bodies so as to distinguish their figure; the muscular system by the distention, which they occasion; the internal surface of the arteries, by the momentum of their moving particles; and the auditory nerves, by the vibration of them: and these four mechanic properties are as different from each other as the various chemical ones, which are adapted to the numerous glands, and to the other organs of sense.

2. The momentum of the progressive particles of blood is compounded of their velocity and their quantity of matter: hence whatever circumstances diminish either of these without proportionally increasing the other, and without superadding either of the general stimuli of heat or distention, will tend to produce a quiescence of the arterial system, and from thence of all the other irritative motions, which are connected with it.

Hence in all those constitutions or diseases where the blood contains a greater proportion of serum, which is the lightest part of its composition, the pulsations of the arteries are weaker, as in nervous fevers, chlorosis, and hysteric complaints; for in these cases the momentum of the progressive particles of blood is less: and hence, where the denser parts of its composition abound, as the red part of it, or the coagulable lymph, the arterial pulsations are stronger; as in those of robust health, and in inflammatory diseases.

That this stimulus of the momentum of the particles of the circulating fluid is of the greatest consequence to the arterial action, appears from the experiment of injecting air into the blood vessels, which seems to destroy animal life from the want of this stimulus of momentum; for the distention of the arteries is not diminished by it, it possesses no corrosive acrimony, and is less liable to repass the valves than the blood itself; since air-valves in all machinery require much less accuracy of construction than those which are opposed to water.

3. One method of increasing the velocity of the blood, and in consequence the momentum of its particles, is by the exercise of the body, or by the friction of its surface: so, on the contrary, too great indolence contributes to decrease this stimulus of the momentum of the particles of the circulating blood, and thus tends to induce quiescence; as is seen in hysteric cases, and chlorosis, and the other diseases of sedentary people.

4. The velocity of the particles of the blood in certain circumstances is increased by venesection, which, by removing a part of it, diminishes the resistance to the motion of the other part, and hence the momentum of the particles of it is increased. This may be easily understood by considering it in the extreme, since, if the resistance was greatly increased, so as to overcome the propelling power, there could be no velocity, and in consequence no momentum at all. From this circumstance arises that curious phænomenon, the truth of which I have been more than once witness to, that venesection will often instantaneously relieve those nervous pains, which attend the cold periods of hysteric, asthmatic, or epileptic diseases; and that even where large doses of opium have been in vain exhibited. In these cases the pulse becomes stronger after the bleeding, and the extremities regain their natural warmth; and an opiate then given acts with much more certain effect.

VI. There is another cause, which seems occasionally to induce quiescence into some part of our system, I mean the influence of the sun and moon; the attraction of these luminaries, by decreasing the gravity of the particles of the blood, cannot affect their momentum, as their vis inertiæ remains the same; but it may nevertheless produce some chemical change in them, because whatever affects the general attractions of the particles of matter may be supposed from analogy to affect their specific attractions or affinities: and thus the stimulus of the particles of blood may be diminished, though not their momentum. As the tides of the sea obey the southing and northing of the moon (allowing for the time necessary for their motion, and the obstructions of the shores), it is probable, that there are also atmospheric tides on both sides of the earth, which to the inhabitants of another planet might so deflect the light as to resemble the ring of Saturn. Now as these tides of water, or of air, are raised by the diminution of their gravity, it follows, that their pressure on the surface of the earth is no greater than the pressure of the other parts of the ocean, or of the atmosphere, where no such tides exist; and therefore that they cannot affect the mercury in the barometer. In the same manner, the gravity of all other terrestrial bodies is diminished at the times of the southing and northing of the moon, and that in a greater degree when this coincides with the southing and northing of the sun, and this in a still greater degree about the times of the equinoxes. This decrease of the gravity of all bodies during the time the moon passes our zenith or nadir might possibly be shewn by the slower vibrations of a pendulum, compared with a spring clock, or with astronomical observation. Since a pendulum of a certain length moves slower at the line than near the poles, because the gravity being diminished and the vis inertiæ continuing the same, the motive power is less, but the resistance to be overcome continues the same. The combined powers of the lunar and solar attraction is estimated by Sir Isaac Newton not to exceed one 7,868,850th part of the power of gravitation, which seems indeed but a small circumstance to produce any considerable effect on the weight of sublunary bodies, and yet this is sufficient to raise the tides at the equator above ten feet high; and if it be considered, what small impulses of other bodies produce their effects on the organs of sense adapted to the perception of them, as of vibration on the auditory nerves, we shall cease to to be surprised, that so minute a diminution in the gravity of the particles of blood should so far affect their chemical changes, or their stimulating quality, as, joined with other causes, sometimes to produce the beginnings of diseases.

Add to this, that if the lunar influence produces a very small degree of quiescence at first, and if that recurs at certain periods even with less power to produce quiescence than at first, yet the quiescence will daily increase by the acquired habit acting at the same time, till at length so great a degree of quiescence is induced as to produce phrensy, canine madness, epilepsy, hysteric pains or cold fits of fever, instances of many of which are to be found in Dr. Mead's work on this subject. The solar influence also appears daily in several diseases; but as darkness, silence, sleep, and our periodical meals mark the parts of the solar circle of actions, it is sometimes dubious to which of these the periodical returns of these diseases are to be ascribed.

As far as I have been able to observe, the periods of inflammatory diseases observe the solar day; as the gout and rheumatism have their greatest quiescence about noon and midnight, and their exacerbations some hours after; as they have more frequently their immediate cause from cold air, inanition, or fatigue, than from the effects of lunations: whilst the cold fits of hysteric patients, and those in nervous fevers, more frequently occur twice a day, later by near half an hour each time, according to the lunar day; whilst some fits of intermittents, which are undisturbed by medicines, return at regular solar periods, and others at lunar ones; which may, probably, be owing to the difference of the periods of those external circumstances of cold, inanition, or lunation, which immediately caused them.

We must, however, observe, that the periods of quiescence and exacerbation in diseases do not always commence at the times of the syzygies or quadratures of the moon and sun, or at the times of their passing the zenith or nadir; but as it is probable, that the stimulus of the particles of the circumfluent blood is gradually diminished from the time of the quadratures to that of the syzygies, the quiescence may commence at any hour, when co-operating with other causes of quiescence, it becomes great enough to produce a disease: afterwards it will continue to recur at the same period of the lunar or solar influence; the same cause operating conjointly with the acquired habit, that is with the catenation of this new motion with the dissevered links of the lunar or solar circles of animal action.

In this manner the periods of menstruation obey the lunar month with great exactness in healthy patients (and perhaps the venereal orgasm in brute animals does the same), yet these periods do not commence either at the syzygies or quadratures of the lunations, but at whatever time of the lunar periods they begin, they observe the same in their returns till some greater cause disturbs them.

Hence, though the best way to calculate the time of the expected returns of the paroxysms of periodical diseases is to count the number of hours between the commencement of the two preceding fits, yet the following observations may be worth attending to, when we endeavour to prevent the returns of maniacal or epileptic diseases; whose periods (at the beginning of them especially) frequently observe the syzygies of the moon and sun, and particularly about the equinox.

The greatest of the two tides happening in every revolution of the moon, is that when the moon approaches nearest to the zenith or nadir; for this reason, while the sun is in the northern signs, that is during the vernal and summer months, the greater of the two diurnal tides in our latitude is that, when the moon is above the horizon; and when the sun is in the southern signs, or during the autumnal and winter months, the greater tide is that, which arises when the moon is below the horizon: and as the sun approaches somewhat nearer the earth in winter than in summer, the greatest equinoctial tides are observed to be a little before the vernal equinox, and a little after the autumnal one.

Do not the cold periods of lunar diseases commence a few hours before the southing of the moon during the vernal and summer months, and before the northing of the moon during the autumnal and winter months? Do not palsies and apoplexies, which occur about the equinoxes, happen a few days before the vernal equinoctial lunation, and after the autumnal one? Are not the periods of those diurnal diseases more obstinate, that commence many hours before the southing or northing of the moon, than of those which commence at those times? Are not those palsies and apoplexies more dangerous which commence many days before the syzygies of the moon, than those which happen at those times? See Sect. XXXVI. on the periods of diseases.

VII. Another very frequent cause of the cold fit of fever is the quiescence of some of those large congeries of glands, which compose the liver, spleen, or pancreas; one or more of which are frequently so enlarged in the autumnal intermittents as to be perceptible to the touch externally, and are called by the vulgar ague-cakes. As these glands are stimulated into action by the specific pungency of the fluids, which they absorb, the general cause of their quiescence seems to be the too great insipidity of the fluids of the body, co-operating perhaps at the same time with other general causes of quiescence.

Hence, in marshy countries at cold seasons, which have succeeded hot ones, and amongst those, who have lived on innutritious and unstimulating diet, these agues are most frequent. The enlargement of these quiescent viscera, and the swelling of the præcordia in many other fevers, is, most probably, owing to the same cause; which may consist in a general deficiency of the production of sensorial power, as well as in the diminished stimulation of the fluids; and when the quiescence of so great a number of glands, as constitute one of those large viscera, commences, all the other irritative motions are affected by their connection with it, and the cold fit of fever is produced.

VIII. There are many other causes, which produce quiescence of some part of the animal system, as fatigue, hunger, thirst, bad diet, disappointed love, unwholesome air, exhaustion from evacuations, and many others; but the last cause, that we shall mention, as frequently productive of cold fits of fever, is fear or anxiety of mind. The pains, which we are first and most generally acquainted with, have been produced by defect of some stimulus; thus, soon after our nativity we become acquainted with the pain from the coldness of the air, from the want of respiration, and from the want of food. Now all these pains occasioned by defect of stimulus are attended with quiescence of the organ, and at the same time with a greater or less degree of quiescence of other parts of the system: thus, if we even endure the pain of hunger so as to miss one meal instead of our daily habit of repletion, not only the peristaltic motions of the stomach and bowels are diminished, but we are more liable to coldness of our extremities, as of our noses, and ears, and feet, than at other times.

Now, as fear is originally excited by our having experienced pain, and is itself a painful affection, the same quiescence of other fibrous motions accompany it, as have been most frequently connected with this kind of pain, as explained in Sect. XVI. 8. 1. as the coldness and paleness of the skin, trembling, difficult respiration, indigestion, and other symptoms, which contribute to form the cold fit of fevers. Anxiety is fear continued through a longer time, and, by producing chronical torpor of the system, extinguishes life slowly, by what is commonly termed a broken heart.

IX. 1. We now step forwards to consider the other symptoms in consequence of the quiescence which begins the fits of fever. If by any of the circumstances before described, or by two or more of them acting at the same time, a great degree of quiescence is induced on any considerable part of the circle of irritative motions, the whole class of them is more or less disturbed by their irritative associations. If this torpor be occasioned by a deficient supply of sensorial power, and happens to any of those parts of the system, which are accustomed to perpetual activity, as the vital motions, the torpor increases rapidly, because of the great expenditure of sensorial power by the incessant activity of those parts of the system, as shewn in No. 3. 2. of this Section. Hence a deficiency of all the secretions succeeds, and as animal heat is produced in proportion to the quantity of those secretions, the coldness of the skin is the first circumstance, which is attended to. Dr. Martin asserts, that some parts of his body were warmer than natural in the cold fit of fever; but it is certain, that those, which are uncovered, as the fingers, and nose, and ears, are much colder to the touch, and paler in appearance. It is possible, that his experiments were made at the beginning of the subsequent hot fits; which commence with partial distributions of heat, owing to some parts of the body regaining their natural irritability sooner than others.

From the quiescence of the anastomosing capillaries a paleness of the skin succeeds, and a less secretion of the perspirable matter; from the quiescence of the pulmonary capillaries a difficulty of respiration arises; and from the quiescence of the other glands less bile, less gastric and pancreatic juice, are secreted into the stomach and intestines, and less mucus and saliva are poured into the mouth; whence arises the dry tongue, costiveness, dry ulcers, and paucity of urine. From the quiescence of the absorbent system arises the great thirst, as less moisture is absorbed from the atmosphere. The absorption from the atmosphere was observed by Dr. Lyster to amount to eighteen ounces in one night, above what he had at the same time insensibly perspired. See Langrish. On the same account the urine is pale, though in small quantity, for the thinner part is not absorbed from it; and when repeated ague-fits continue long, the legs swell from the diminished absorption of the cellular absorbents.

From the quiescence of the intestinal canal a loss of appetite and flatulencies proceed. From the partial quiescence of the glandular viscera a swelling and tension about the præcordia becomes sensible to the touch; which is occasioned by the delay of the fluids from the defect of venous or lymphatic absorption. The pain of the forehead, and of the limbs, and of the small of the back, arises from the quiescence of the membranous fascia, or muscles of those parts, in the same manner as the skin becomes painful, when the vessels, of which it is composed, become quiescent from cold. The trembling in consequence of the pain of coldness, the restlessness, and the yawning, and stretching of the limbs, together with the shuddering, or rigours, are convulsive motions; and will be explained amongst the diseases of volition; Sect. XXXIV.

Sickness and vomiting is a frequent symptom in the beginnings of fever-fits, the muscular fibres of the stomach share the general torpor and debility of the system; their motions become first lessened, and then stop, and then become retrograde; for the act of vomiting, like the globus hystericus and the borborigmi of hypochondriasis, is always a symptom of debility, either from want of stimulus, as in hunger; or from want of sensorial power, as after intoxication; or from sympathy with some other torpid irritative motions, as in the cold fits of ague. See Sect. XII. 5. 5. XXIX. 11. and XXXV. 1. 3. where this act of vomiting is further explained.

The small pulse, which is said by some writers to be slow at the commencement of ague-fits, and which is frequently trembling and intermittent, is owing to the quiescence of the heart and arterial system, and to the resistance opposed to the circulating fluid from the inactivity of all the glands and capillaries. The great weakness and inability to voluntary motions, with the insensibility of the extremities, are owing to the general quiescence of the whole moving system; or, perhaps, simply to the deficient production of sensorial power.

If all these symptoms are further increased, the quiescence of all the muscles, including the heart and arteries, becomes complete, and death ensues. This is, most probably, the case of those who are starved to death with cold, and of those who are said to die in Holland from long skaiting on their frozen canals.

2. As soon as this general quiescence of the system ceases, either by the diminution of the cause, or by the accumulation of sensorial power, (as in syncope, Sect. XII. 7. 1.) which is the natural consequence of previous quiescence, the hot fit commences. Every gland of the body is now stimulated into stronger action than is natural, as its irritability is increased by accumulation of sensorial power during its late quiescence, a superabundance of all the secretions is produced, and an increase of heat in consequence of the increase of these secretions. The skin becomes red, and the perspiration great, owing to the increased action of the capillaries during the hot part of the paroxysm. The secretion of perspirable matter is perhaps greater during the hot fit than in the sweating fit which follows; but as the absorption of it also is greater, it does not stand on the skin in visible drops: add to this, that the evaporation of it also is greater, from the increased heat of the skin. But at the decline of the hot fit, as the mouths of the absorbents of the skin are exposed to the cooler air, or bed-clothes, these vessels sooner lose their increased activity, and cease to absorb more than their natural quantity: but the secerning vessels for some time longer, being kept warm by the circulating blood, continue to pour out an increased quantity of perspirable matter, which now stands on the skin in large visible drops; the exhalation of it also being lessened by the greater coolness of the skin, as well as its absorption by the diminished action of the lymphatics. See Class I. 1. 2. 3.

The increased secretion of bile and of other fluids poured into the intestines frequently induce a purging at the decline of the hot fit; for as the external absorbent vessels have their mouths exposed to the cold air, as above mentioned, they cease to be excited into unnatural activity sooner than the secretory vessels, whose mouths are exposed to the warmth of the blood: now, as the internal absorbents sympathize with the external ones, these also, which during the hot fit drank up the thinner part of the bile, or of other secreted fluids, lose their increased activity before the gland loses its increased activity, at the decline of the hot fit; and the loose dejections are produced from the same cause, that the increased perspiration stands on the surface of the skin, from the increased absorption ceasing sooner than the increased secretion.

The urine during the cold fit is in small quantity and pale, both from a deficiency of the secretion and a deficiency of the absorption.

During the hot fit it is in its usual quantity, but very high coloured and turbid, because a greater quantity had been secreted by the increased action of the kidnies, and also a greater quantity of its more aqueous part had been absorbed from it in the bladder by the increased action of the absorbents; and lastly, at the decline of the hot fit it is in large quantity and less coloured, or turbid, because the absorbent vessels of the bladder, as observed above, lose their increased action by sympathy with the cutaneous ones sooner than the secretory vessels of the kidnies lose their increased activity. Hence the quantity of the sediment, and the colour of the urine, in fevers, depend much on the quantity secreted by the kidnies, and the quantity absorbed from it again in the bladder: the kinds of sediment, as the lateritious, purulent, mucous, or bloody sediments, depend on other causes. It should be observed, that if the sweating be increased by the heat of the room, or of the bed-clothes, that a paucity of turbid urine will continue to be produced, as the absorbents of the bladder will have their activity increased by their sympathy with the vessels of the skin, for the purpose of supplying the fluid expended in perspiration.

The pulse becomes strong and full owing to the increased irritability of the heart and arteries, from the accumulation of sensorial power during their quiescence, and to the quickness of the return of the blood from the various glands and capillaries. This increased action of all the secretory vessels does not occur very suddenly, nor universally at the same time. The heat seems to begin about the center, and to be diffused from thence irregularly to the other parts of the system. This may be owing to the situation of the parts which first became quiescent and caused the fever-fit, especially when a hardness or tumour about the præcordia can be felt by the hand; and hence this part, in whatever viscus it is seated, might be the first to regain its natural or increased irritability.

3. It must be here noted, that, by the increased quantity of heat, and of the impulse of the blood at the commencement of the hot fit, a great increase of stimulus is induced, and is now added to the increased irritability of the system, which was occasioned by its previous quiescence. This additional stimulus of heat and momentum of the blood augments the violence of the movements of the arterial and glandular system in an increasing ratio. These violent exertions still producing more heat and greater momentum of the moving fluids, till at length the sensoral power becomes wasted by this great stimulus beneath its natural quantity, and predisposes the system to a second cold fit.

At length all these unnatural exertions spontaneously subside with the increased irritability that produced them; and which was itself produced by the preceding quiescence, in the same manner as the eye, on coming from darkness into day-light, in a little time ceases to be dazzled and pained, and gradually recovers its natural degree of irritability.

4. But if the increase of irritability, and the consequent increase of the stimulus of heat and momentum, produce more violent exertions than those above described; great pain arises in some part of the moving system, as in the membranes of the brain, pleura, or joints; and new motions of the vessels are produced in consequence of this pain, which are called inflammation; or delirium or stupor arises; as explained in Sect. XXI. and XXXIII.: for the immediate effect is the same, whether the great energy of the moving organs arises from an increase of stimulus or an increase of irritability; though in the former case the waste of sensorial power leads to debility, and in the latter to health.

Recapitulation.

X. Those muscles, which are less frequently exerted, and whose actions are interrupted by sleep, acquire less accumulation of sensorial power during their quiescent state, as the muscles of locomotion. In these muscles after great exertion, that is, after great exhaustion of sensorial power, the pain of fatigue ensues; and during rest there is a renovation of the natural quantity of sensorial power; but where the rest, or quiescence of the muscle, is long continued, a quantity of sensorial power becomes accumulated beyond what is necessary; as appears by the uneasiness occasioned by want of exercise; and which in young animals is one cause exciting them into action, as is seen in the play of puppies and kittens.

But when those muscles, which are habituated to perpetual actions, as those of the stomach by the stimulus of food, those of the vessels of the skin by the stimulus of heat, and those which constitute the arteries and glands by the stimulus of the blood, become for a time quiescent, from the want of their appropriated stimuli, or by their associations with other quiescent parts of the system; a greater accumulation of sensorial power is acquired during their quiescence, and a greater or quicker exhaustion of it is produced during their increased action.

This accumulation of sensorial power from deficient action, if it happens to the stomach from want of food, occasions the pain of hunger; if it happens to the vessels of the skin from want of heat, it occasions the pain of cold; and if to the arterial system from the want of its adapted stimuli, many disagreeable sensations are occasioned, such as are experienced in the cold fits of intermittent fevers, and are as various, as there are glands or membranes in the system, and are generally termed universal uneasiness.

When the quiescence of the arterial system is not owing to defect of stimulus as above, but to the defective quantity of sensorial power, as in the commencement of nervous fever, or irritative fever with weak pulse, a great torpor of this system is quickly induced; because both the irritation from the stimulus of the blood, and the association of the vascular motions with each other, continue to excite the arteries into action, and thence quickly exhaust the ill-supplied vascular muscles; for to rest is death; and therefore those vascular muscles continue to proceed, though with feebler action, to the extreme of weariness or faintness: while nothing similar to this affects the locomotive muscles, whose actions are generally caused by volition, and not much subject either to irritation or to other kinds of associations besides the voluntary ones, except indeed when they are excited by the lash of slavery.

In these vascular muscles, which are subject to perpetual action, and thence liable to great accumulation of sensorial power during their quiescence from want of stimulus, a great increase of activity occurs, either from the renewal of their accustomed stimulus, or even from much less quantities of stimulus than usual. This increase of action constitutes the hot fit of fever, which is attended with various increased secretions, with great concomitant heat, and general uneasiness. The uneasiness attending this hot paroxysm of fever, or fit of exertion, is very different from that, which attends the previous cold fit, or fit of quiescence, and is frequently the cause of inflammation, as in pleurisy, which is treated of in the next section.

A similar effect occurs after the quiescence of our organs of sense; those which are not subject to perpetual action, as the taste and smell, are less liable to an exuberant accumulation of sensorial power after their having for a time been inactive; but the eye, which is in perpetual action during the day, becomes dazzled, and liable to inflammation after a temporary quiescence.

Where the previous quiescence has been owing to a defect of sensorial power, and not to a defect of stimulus, as in the irritative fever with weak pulse, a similar increase of activity of the arterial system succeeds, either from the usual stimulus of the blood, or from a stimulus less than usual; but as there is in general in these cases of fever with weak pulse a deficiency of the quantity of the blood, the pulse in the hot fit is weaker than in health, though it is stronger than in the cold fit, as explained in No. 2. of this section. But at the same time in those fevers, where the defect of irritation is owing to the defect of the quantity of sensorial power, as well as to the defect of stimulus, another circumstance occurs; which consists in the partial distribution of it, as appears in partial flushings, as of the face or bosom, while the extremities are cold; and in the increase of particular secretions, as of bile, saliva, insensible perspiration, with great heat of the skin, or with partial sweats, or diarrhœa.

There are also many uneasy sensations attending these increased actions, which, like those belonging to the hot fit of fever with strong pulse, are frequently followed by inflammation, as in scarlet fever; which inflammation is nevertheless accompanied with a pulse weaker, though quicker, than the pulse during the remission or intermission of the paroxysms, though stronger than that of the previous cold fit.

From hence I conclude, that both the cold and hot fits of fever are necessary consequences of the perpetual and incessant action of the arterial and glandular system; since those muscular fibres and those organs of sense, which are most frequently exerted, become necessarily most affected both with defect and accumulation of sensorial power: and that hence *fever-fits are not an effort of nature to relieve herself,* and that therefore they should always be prevented or diminished as much as possible, by any means which decrease the general or partial vascular actions, when they are greater, or by increasing them when they are less than in health, as described in Sect. XII. 6. 1.

Thus have I endeavoured to explain, and I hope to the satisfaction of the candid and patient reader, the principal symptoms or circumstances of fever without the introduction of the supernatural power of spasm. To the arguments in favour of the doctrine of spasm it may be sufficient to reply, that in the evolution of medical as well as of dramatic catastrophe,

> Nec Deus intersit, nisi dignus vindice nodus inciderit.—
> HOR.

SECT. XXXIII
DISEASES OF SENSATION

I. 1. Motions excited by sensation. Digestion. Generation. Pleasure of existence. Hypochondriacism. 2. Pain introduced. Sensitive fevers of two kinds. 3. Two sensorial powers exerted in sensitive fevers. Size of the blood. Nervous fevers distinguished from putrid ones. The septic and antiseptic theory. 4. Two kinds of delirium. 5. Other animals are less liable to delirium, cannot receive our contagious diseases, and are less liable to madness. II. 1. Sensitive motions generated. 2. Inflammation explained. 3. Its remote causes from excess of irritation, or of irritability, not from those pains which are owing to defect of irritation. New vessels produced, and much heat. 4. Purulent matter secreted. 5. Contagion explained. 6. Received but once. 7. If common matter be contagious? 8. Why some contagions are received but once. 9. Why others may be received frequently. Contagions of small-pox and measles do not act at the same times. Two cases of such patients. 10. The blood from patients in the small-pox will not infect others. Cases of children thus inoculated. The variolous contagion is not received into the blood. It acts by sensitive association between the stomach and skin. III. 1. Absorption of solids and fluids. 2. Art of healing ulcers. 3. Mortification attended with less pain in weak people.

I. 1. As many motions of the body are excited and continued by irritations, so others require, either conjunctly with these, or separately, the pleasurable or painful sensations, for the purpose of producing them with due energy. Amongst these the business of digestion supplies us with an instance: if the food, which we swallow, is not attended with agreeable sensation, it digests less perfectly; and if very disagreeable sensation accompanies it, such as a nauseous idea, or very disgustful taste, the digestion becomes impeded; or retrograde motions of the stomach and œsophagus succeed, and the food is ejected.

The business of generation depends so much on agreeable sensation, that, where the object is disgustful, neither voluntary exertion nor irritation can effect the purpose; which is also liable to be interrupted by the pain of fear or bashfulness.

Besides the pleasure, which attends the irritations produced by the objects of lust and hunger, there seems to be a sum of pleasurable affection accompanying the various secretions of the numerous glands, which constitute the pleasure of life, in contradistinction to the tedium vitæ. This quantity or sum of pleasurable affection, seems to contribute to the due or energetic performance of the whole moveable system, as well that of the heart and arteries, as of digestion and of absorption; since without the due quantity of pleasurable sensation, flatulency and hypochondriacism affect the intestines, and a languor seizes the arterial pulsations and secretions; as occurs in great and continued anxiety of the mind.

2. Besides the febrile motions occasioned by irritation, described in Sect. XXXII. and termed irritative fever, it frequently happens that pain is excited by the violence of the fibrous contractions; and other new motions are then superadded, in consequence of sensation, which we shall term febris sensitiva, or sensitive fever. It must be observed, that most irritative fevers begin with a decreased exertion of irritation, owing to defect of stimulus; but that on the contrary the sensitive fevers, or inflammations, generally begin with the increased exertion of sensation, as mentioned in Sect. XXXI. on temperaments: for though the cold fit, which introduces inflammation, commences with decreased irritation, yet the inflammation itself commences in the hot fit during the increase of sensation. Thus a common pustule, or phlegmon, in a part of little sensibility does not excite an inflammatory fever; but if the stomach, intestines, or the tender substance beneath the nails, be injured, great sensation is produced, and the whole system is thrown into that kind of exertion, which constitutes inflammation.

These sensitive fevers, like the irritative ones, resolve themselves into those with arterial strength, and those with arterial debility, that is with excess or defect of sensorial power; these may be termed the febris sensitiva pulsu forti, sensitive fever with strong pulse, which is the synocha, or inflammatory fever; and the febris sensitiva pulsu debili, sensitive fever with weak pulse, which is the typhus gravior, or putrid fever of some writers.

3. The inflammatory fevers, which are here termed sensitive fevers with strong pulse, are generally attended with some topical inflammation, as pleurisy, peripneumony, or rheumatism, which distinguishes them from irritative fevers with strong pulse. The pulse is strong, quick, and full; for in this fever there is great irritation, as well as great sensation, employed in moving the arterial system. The size, or coagulable lymph, which appears on the blood, is probably an increased secretion from the inflamed internal

lining of the whole arterial system, the thinner part being taken away by the increased absorption of the inflamed lymphatics.

The sensitive fevers with weak pulse, which are termed putrid or malignant fevers, are distinguished from irritative fevers with weak pulse, called nervous fevers, described in the last section, as the former consist of inflammation joined with debility, and the latter of debility alone. Hence there is greater heat and more florid colour of the skin in the former, with petechiæ, or purple spots, and aphthæ, or sloughs in the throat, and generally with previous contagion.

When animal matter dies, as a slough in the throat, or the mortified part of a carbuncle, if it be kept moist and warm, as during its abhesion to a living body, it will soon putrify. This, and the origin of contagion from putrid animal substances, seem to have given rise to the septic and antiseptic theory of these fevers.

The matter in pustules and ulcers is thus liable to become putrid, and to produce microscopic animalcula; the urine, if too long retained, may also gain a putrescent smell, as well as the alvine feces; but some writers have gone so far as to believe, that the blood itself in these fevers has smelt putrid, when drawn from the arm of the patient: but this seems not well founded; since a single particle of putrid matter taken into the blood can produce fever, how can we conceive that the whole mass could continue a minute in a putrid state without destroying life? Add to this, that putrid animal substances give up air, as in gangrenes; and that hence if the blood was putrid, air should be given out, which in the blood-vessels is known to occasion immediate death.

In these sensitive fevers with strong pulse (or inflammations) there are two sensorial faculties concerned in producing the disease, viz. irritation and sensation; and hence, as their combined action is more violent, the general quantity of sensorial power becomes further exhausted during the exacerbation, and the system more rapidly weakened than in irritative fever with strong pulse; where the spirit of animation is weakened by but one mode of its exertion: so that this febris sensitiva pulsu forti (or inflammatory fever,) may be considered as the febris irritativa pulsu forti, with the addition of inflammation; and the febris sensitiva pulsu debili (or malignant fever) may be considered as the febris irritativa pulsu debili (or nervous fever), with the addition of inflammation.

4. In these putrid or malignant fevers a deficiency of irritability accompanies the increase of sensibility; and by this waste of sensorial power by the excess of sensation, which was already too small, arises the delirium and stupor which so perpetually attend these inflammatory fevers with

arterial debility. In these cases the voluntary power first ceases to act from deficiency of sensorial spirit; and the stimuli from external bodies have no effect on the exhausted sensorial power, and a delirium like a dream is the consequence. At length the internal stimuli cease to excite sufficient irritation, and the secretions are either not produced at all, or too parsimonious in quantity. Amongst these the secretion of the brain, or production of the sensorial power, becomes deficient, till at last all sensorial power ceases, except what is just necessary to perform the vital motions, and a stupor succeeds; which is thus owing to the same cause as the preceding delirium exerted in a greater degree.

This kind of delirium is owing to a suspension of volition, and to the disobedience of the senses to external stimuli, and is always occasioned by great debility, or paucity of sensorial power; it is therefore a bad sign at the end of inflammatory fevers, which had previous arterial strength, as rheumatism, or pleurisy, as it shews the presence of great exhaustion of sensorial power in a system, which having lately been exposed to great excitement, is not so liable to be stimulated into its healthy action, either by additional stimulus of food and medicines, or by the accumulation of sensorial power during its present torpor. In inflammatory fevers with debility, as those termed putrid fevers, delirium is sometimes, as well as stupor, rather a favourable sign; as less sensorial power is wasted during its continuance (see Class II. 1. 6. 8.), and the constitution not having been previously exposed to excess of stimulation, is more liable to be excited after previous quiescence.

When the sum of general pleasurable sensation becomes too great, another kind of delirium supervenes, and the ideas thus excited are mistaken for the irritations of external objects: such a delirium is produced for a time by intoxicating drugs, as fermented liquors, or opium: a permanent delirium of this kind is sometimes induced by the pleasures of inordinate vanity, or by the enthusiastic hopes of heaven. In these cases the power of volition is incapable of exertion, and in a great degree the external senses become incapable of perceiving their adapted stimuli, because the whole sensorial power is employed or expended on the ideas excited by pleasurable sensation.

This kind of delirium is distinguished from that which attends the fevers above mentioned from its not being accompanied with general debility, but simply with excess of pleasurable sensation; and is therefore in some measure allied to madness or to reverie; it differs from the delirium of dreams, as in this the power of volition is not totally suspended, nor are the senses precluded from external stimulation; there is therefore a degree of consistency, in this kind of delirium, and a degree of attention to external objects, neither of which exist in the delirium of fevers or in dreams.

5. It would appear, that the vascular system of other animals are less liable to be put into action by their general sum of pleasurable or painful sensation; and that the trains of their ideas, and the muscular motions usually associated with them, are less powerfully connected than in the human system. For other animals neither weep, nor smile, nor laugh; and are hence seldom subject to delirium, as treated of in Sect. XVI. on Instinct. Now as our epidemic and contagious diseases are probably produced by disagreeable sensation, and not simply by irritation; there appears a reason, why brute animals are less liable to epidemic or contagious diseases; and secondly, why none of our contagions, as the small-pox or measles, can be communicated to them, though one of theirs, viz. the hydrophobia, as well as many of their poisons, as those of snakes and of in insects, communicate their deleterious or painful effects to mankind.

Where the quantity of general painful sensation is too great in the system, inordinate voluntary exertions are produced either of our ideas, as in melancholy and madness, or of our muscles, as in convulsion. From these maladies also brute animals are much more exempt than mankind, owing to their greater inaptitude to voluntary exertion, as mentioned in Sect. XVI. on Instinct.

II. 1. When any moving organ is excited into such violent motions, that a quantity of pleasurable or painful sensation is produced, it frequently happens (but not always) that new motions of the affected organ are generated in consequence of the pain or pleasure, which are termed inflammation.

These new motions are of a peculiar kind, tending to distend the old, and to produce new fibres, and thence to elongate the straight muscles, which serve locomotion, and to form new vessels at the extremities or sides of the vascular muscles.

2. Thus the pleasurable sensations produce an enlargement of the nipples of nurses, of the papillæ of the tongue, of the penis, and probably produce the growth of the body from its embryon state to its maturity; whilst the new motions in consequence of painful sensation, with the growth of the fibres or vessels, which they occasion, are termed inflammation.

Hence when the straight muscles are inflamed, part of their tendons at each extremity gain new life and sensibility, and thus the muscle is for a time elongated; and inflamed bones become soft, vascular, and sensible. Thus new vessels shoot over the cornea of inflamed eyes, and into scirrhous tumours, when they become inflamed; and hence all inflamed parts grow together by intermixture, and inosculation of the new and old vessels.

The heat is occasioned from the increased secretions either of mucus, or of the fibres, which produce or elongate the vessels. The red colour is owing to the pellucidity of the newly formed vessels, and as the arterial parts of them are probably formed before their correspondent venous parts.

3. These new motions are excited either from the increased quantity of sensation in consequence of greater fibrous contractions, or from increased sensibility, that is, from the increased quantity of sensorial power in the moving organ. Hence they are induced by great external stimuli, as by wounds, broken bones; and by acrid or infectious materials; or by common stimuli on those organs, which have been some time quiescent; as the usual light of the day inflames the eyes of those, who have been confined in dungeons; and the warmth of a common fire inflames those, who have been previously exposed to much cold.

But these new motions are never generated by that pain, which arises from defect of stimulus, as from hunger, thirst, cold, or inanition, with all those pains, which are termed nervous. Where these pains exist, the motions of the affected part are lessened; and if inflammation succeeds, it is in some distant parts; as coughs are caused by coldness and moisture being long applied to the feet; or it is in consequence of the renewal of the stimulus, as of heat or food, which excites our organs into stronger action after their temporary quiescence; as kibed heels after walking in snow.

4. But when these new motions of the vascular muscles are exerted with greater violence, and these vessels are either elongated too much or too hastily, a new material is secreted from their extremities, which is of various kinds according to the peculiar animal motions of this new kind of gland, which secretes it; such is the pus laudabile or common matter, the variolous matter, venereal matter, catarrhous matter, and many others.

5. These matters are the product of an animal process; they are secreted or produced from the blood by certain diseased motions of the extremities of the blood-vessels, and are on that account all of them contagious; for if a portion of any of these matters is transmitted into the circulation, or perhaps only inserted into the skin, or beneath the cuticle of an healthy person, its stimulus in a certain time produces the same kind of morbid motions, by which itself was produced; and hence a similar kind is generated. See Sect. XXXIX. 6. 1.

6. It is remarkable, that many of these contagious matters are capable of producing a similar disease but once; as the small-pox and measles; and I suppose this is true of all those contagious diseases, which are spontaneously

cured by nature in a certain time; for if the body was capable of receiving the disease a second time, the patient must perpetually infect himself by the very matter, which he has himself produced, and is lodged about him; and hence he could never become free from the disease. Something similar to this is seen in the secondary fever of the confluent small-pox; there is a great absorption of variolous matter, a very minute part of which would give the genuine small-pox to another person; but here it only stimulates the system into common fever; like that which common puss, or any other acrid material might occasion.

7. In the pulmonary consumption, where common matter is daily absorbed, an irritative fever only, without new inflammation, is generally produced; which is terminated like other irritative fevers by sweats, or loose stools. Hence it does not appear, that this absorbed matter always acts as a contagious material producing fresh inflammation or new abscesses. Though there is reason to believe, that the first time any common matter is absorbed, it has this effect, but not the second time, like the variolous matter above mentioned.

This accounts for the opinion, that the pulmonary consumption is sometimes infectious, which opinion was held by the ancients, and continues in Italy at present; and I have myself seen three or four instances, where a husband and wife, who have slept together, and have thus much received each other's breath, who have infected each other, and both died in consequence of the original taint of only one of them. This also accounts for the abscesses in various parts of the body, that are sometimes produced after the inoculated small-pox is terminated; for this second absorption of variolous matter acts like common matter, and produces only irritative fever in those children, whose constitutions have already experienced the absorption of common matter; and inflammation with a tendency to produce new abscesses in those, whose constitutions have not experienced the absorptions of common matter.

It is probable, that more certain proofs might have been found to shew, that common matter is infectious the first time it is absorbed, tending to produce similar abscesses, but not the second time of its absorption, if this subject had been attended to.

8. These contagious diseases are very numerous, as the plague, small-pox, chicken-pox, measles, scarlet-fever, pemphigus, catarrh, chincough, venereal disease, itch, trichoma, tinea. The infectious material does not seem to be dissolved by the air, but only mixed with it perhaps in fine powder, which soon subsides; since many of these contagions can only be received by actual contact; and others of them only at small distances from the infected

person; as is evident from many persons having been near patients of the small-pox without acquiring the disease.

The reason, why many of these diseases are received but once, and others repeatedly, is not well understood; it appears to me, that the constitution becomes so accustomed to the stimuli of these infectious materials, by having once experienced them, that though irritative motions, as hectic fevers, may again be produced by them, yet no sensation, and in consequence no general inflammation succeeds; as disagreeable smells or tastes by habit cease to be perceived; they continue indeed to excite irritative ideas on the organs of sense, but these are not succeeded by sensation.

There are many irritative motions, which were at first succeeded by sensation, but which by frequent repetition cease to excite sensation, as explained in Sect. XX. on Vertigo. And, that this circumstance exists in respect to infectious matter appears from a known fact; that nurses, who have had the small-pox, are liable to experience small ulcers on their arms by the contact of variolous matter in lifting their patients; and that when patients, who have formerly had the small-pox have been inoculated in the arm, a phlegmon, or inflamed sore, has succeeded, but no subsequent fever. Which shews, that the contagious matter of the small-pox has not lost its power of stimulating the part it is applied to, but that the general system is not affected in consequence. See Section XII. 7. 6. XIX. 9.

9. From the accounts of the plague, virulent catarrh, and putrid dysentery, it seems uncertain, whether these diseases are experienced more than once; but the venereal disease and itch are doubtless repeatedly infectious; and as these diseases are never cured spontaneously, but require medicines, which act without apparent operation, some have suspected, that the contagious material produces similar matter rather by a chemical change of the fluids, than by an animal process; and that the specific medicines destroy their virus by chemically combining with it. This opinion is successfully combated by Mr. Hunter, in his Treatise on Venereal Disease, Part I. c. i.

But this opinion wants the support of analogy, as there is no known process in animal bodies, which is purely chemical, not even digestion; nor can any of these matters be produced by chemical processes. Add to this, that it is probable, that the insects, observed in the pustules of the itch, and in the stools of dysenteric patients, are the consequences, and not the causes of these diseases. And that the specific medicines, which cure the itch and lues venerea, as brimstone and mercury, act only by increasing the absorption of the matter in the ulcuscles of those diseases, and thence disposing them to heal; which would otherwise continue to spread.

Why the venereal disease, and itch, and tenia, or scald head, are repeatedly contagious, while those contagions attended with fever can be received but once, seems to depend on their being rather local diseases than universal ones, and are hence not attended with fever, except the purulent fever in their last stages, when the patient is destroyed by them. On this account the whole of the system does not become habituated to these morbid actions, so as to cease to be affected with sensation by a repetition of the contagion. Thus the contagious matter of the venereal disease, and of the tenia, affects the lymphatic glands, as the inquinal glands, and those about the roots of the hair and neck, where it is arrested, but does not seem to affect the blood-vessels, since no fever ensues.

Hence it would appear, that these kinds of contagion are propagated not by means of the circulation, but by sympathy of distant parts with each other; since if a distant part, as the palate, should be excited by sensitive association into the same kind of motions, as the parts originally affected by the contact of infectious matter; that distant part will produce the same kind of infectious matter; for every secretion from the blood is formed from it by the peculiar motions of the fine extremities of the gland, which secretes it; the various secreted fluids, as the bile, saliva, gastric juice, not previously existing, as such, in the blood-vessels.

And this peculiar sympathy between the genitals and the throat, owing to sensitive association, appears not only in the production of venereal ulcers in the throat, but in variety of other instances, as in the mumps, in the hydrophobia, some coughs, strangulation, the production of the beard, change of voice at puberty. Which are further described in Class IV. 1. 2. 7.

To evince that the production of such large quantities of contagious matter, as are seen in some variolous patients, so as to cover the whole skin almost with pustules, does not arise from any chemical fermentation in the blood, but that it is owing to morbid motions of the fine extremities of the capillaries, or glands, whether these be ruptured or not, appears from the quantity of this matter always corresponding with the quantity of the fever; that is, with the violent exertions of those glands and capillaries, which are the terminations of the arterial system.

The truth of this theory is evinced further by a circumstance observed by Mr. J. Hunter, in his Treatise on Venereal Disease; that in a patient, who was inoculated for the small-pox, and who appeared afterwards to have been previously infested with the measles, the progress of the small-pox was delayed till the measles had run their course, and that then the small-pox went through its usual periods.

Two similar cases fell under my care, which I shall here relate, as it confirms that of Mr. Hunter, and contributes to illustrate this part of the theory of contagious diseases. I have transcribed the particulars from a letter of Mr. Lightwood of Yoxal, the surgeon who daily attended them, and at my request, after I had seen them, kept a kind of journal of their cases.

Miss H. and Miss L. two sisters, the one about four and the other about three years old, were inoculated Feb. 7, 1791. On the 10th there was a redness on both arms discernible by a glass. On the 11th their arms were so much inflamed as to leave no doubt of the infection having taken place. On the 12th less appearance of inflammation on their arms. In the evening Miss L. had an eruption, which resembled the measles. On the 12th the eruption on Miss L. was very full on the face and breast, like the measles, with considerable fever. It was now known, that the measles were in a farm house in the neighbourhood. Miss H.'s arm less inflamed than yesterday. On the 14th Miss L.'s fever great, and the eruption universal. The arm appears to be healed. Miss H.'s arm somewhat redder. They were now put into separate rooms. On the 15th Miss L.'s arms as yesterday. Eruption continues. Miss H.'s arms have varied but little. 16th, the eruptions on Miss L. are dying away, her fever gone. Begins to have a little redness in one arm at the place of inoculation. Miss H.'s arms get redder, but she has no appearance of complaint. 20th, Miss L.'s arms have advanced slowly till this day, and now a few pustules appear. Miss H.'s arm has made little progress from the 16th to this day, and now she has some fever. 21st, Miss L. as yesterday. Miss H. has much inflammation, and an increase of the red circle on one arm to the size of half a crown, and had much fever at night, with fetid breath. 22d, Miss L.'s pustules continue advancing. Miss H.'s inflammation of her arm and red circle increases. A few red spots appear in different parts with some degree of fever this morning, 23d. Miss L. has a larger crop of pustules. Miss H. has small pustules and great inflammation of her arms, with but one pustule likely to suppurate. After this day they gradually got well, and the pustules disappeared.

In one of these cases the measles went through their common course with milder symptoms than usual, and in the other the measly contagion seemed just sufficient to stop the progress of variolous contagion, but without itself throwing the constitution into any disorder. At the same time both the measles and small-pox seem to have been rendered milder. Does not this give an idea, that if they were both inoculated at the same time, that neither of them might affect the patient?

From these cases I contend, that the contagious matter of these diseases does not affect the constitution by a fermentation, or chemical change of the blood, because then they must have proceeded together, and have produced a third something, not exactly similar to either of them: but that they produce new motions of the cutaneous terminations of the blood-vessels, which for a time proceed daily with increasing activity, like some paroxysms of fever, till they at length secrete or form a similar poison by these unnatural actions.

Now as in the measles one kind of unnatural motion takes place, and in the small-pox another kind, it is easy to conceive, that these different kinds of morbid motions cannot exist together; and therefore, that that which has first begun will continue till the system becomes habituated to the stimulus which occasions it, and has ceased to be thrown into action by it; and then the other kind of stimulus will in its turn produce fever, and new kinds of motions peculiar to itself.

10. On further considering the action of contagious matter, since the former part of this work was sent to the press; where I have asserted, in Sect. XXII. 3. 3. that it is probable, that the variolous matter is diffused through the blood; I prevailed on my friend Mr. Power, surgeon at Bosworth in Leicestershire to try, whether the small-pox could be inoculated by using the blood of a variolous patient instead of the matter from the pustules; as I thought such an experiment might throw some light at least on this interesting subject. The following is an extract from his letter:—

"March 11, 1793. I inoculated two children, who had not had the small-pox, with blood; which was taken from a patient on the second day after the eruption commenced, and before it was completed. And at the same time I inoculated myself with blood from the same person, in order to compare the appearances, which might arise in a person liable to receive the infection, and in one not liable to receive it. On the same day I inoculated four other children liable to receive the infection with blood taken from another person on the fourth day after the commencement of the eruption. The patients from whom the blood was taken had the disease mildly, but had the most pustules of any I could select from twenty inoculated patients; and as much of the blood was insinuated under the cuticle as I could introduce by elevating the skin without drawing blood; and three or four such punctures were made in each of their arms, and the blood was used in its fluid state.

"As the appearances in all these patients, as well as in myself, were similar, I shall only mention them in general terms. March 13. A slight subcuticular discoloration, with rather a livid appearance, without soreness or pain, was visible in them all, as well as in my own hand. 15. The discoloration somewhat less, without pain or soreness. Some patients inoculated on the same day with variolous matter have considerable inflammation. 17. The discoloration is quite gone in them all, and from my own hand, a dry mark only remaining. And they were all inoculated on the 18th, with variolous matter, which produced the disease in them all."

Mr. Power afterwards observes, that, as the patients from whom the blood was taken had the disease mildly, it may be supposed, that though the contagious matter might be mixed with the blood, it might still be in too dilute a state to convey the infection; but adds at the same time, that he has diluted recent matter with at least five times its quantity of water, and which has still given the infection; though he has sometimes diluted it so far as to fail.

The following experiments were instituted at my request by my friend Mr. Hadley, surgeon in Derby, to ascertain whether the blood of a person in the small-pox be capable of communicating the disease. "Experiment 1st. October 18th, 1793. I took some blood from a vein in the arm of a person who had the small-pox, on the second day of the eruption, and introduced a small quantity of it immediately with the point of a lancet between the scars and true skin of the right arm of a boy nine years old in two or three different places; the other arm was inoculated with variolous matter at the same time.

"19th. The punctured parts of the right arm were surrounded with some degree of subcuticular inflammation. 20th. The inflammation more considerable, with a slight degree of itching, but no pain upon pressure. 21st. Upon examining the arm this day with a lens I found the inflammation less extensive, and the redness changing to a deep yellow or orange-colour, 22d. Inflammation nearly gone. 23d. Nothing remained, except a slight discoloration and a little scurfy appearance on the punctures. At the same time the inflammation of the arm inoculated with variolous matter was increasing fast, and he had the disease mildly at the usual time.

"Experiment 2d. I inoculated another child at the same time and in the same manner, with blood taken on the first day of the eruption; but as the appearance and effects were similar to those in the preceding experiment, I shall not relate them minutely.

"Experiment 3d. October 20th. Blood was taken from a person who had the small-pox, on the third day of the eruption, and on the sixth from the commencement of the eruptive fever. I introduced some of it in its fluid state into both arms of a boy seven years old.

21st. There appeared to be some inflammation under the cuticle, where the punctures were made. 22d. Inflammation more considerable. 23d. On this day the inflammation was somewhat greater, and the cuticle rather elevated.

"24th. Inflammation much less, and only a brown or orange-colour remained. 25th. Scarcely any discoloration left. On this day he was inoculated with variolous matter, the progress of the infection went on in the usual way, and he had the small-pox very favourably.

"At this time I was requested to inoculate a young person, who was thought to have had the small-pox, but his parents were not quite certain; in one arm I introduced variolous matter, and in the other blood, taken as in experiment 3d. On the second day after the operation, the punctured parts were inflamed, though I think the arm in which I had inserted variolous matter was rather more so than the other. On the third the inflammation was increased, and looked much the same as in the preceding experiment. 4th. The inflammation was much diminished, and on the 5th almost gone. He was exposed at the same time to the natural infection, but has continued perfectly well.

"I have frequently observed (and believe most practitioners have done the same), that if variolous matter be inserted in the arm of a person who has previously had the small-pox, that the inflammation on the second or third days is much greater, than if they had not had the disease, but on the fourth or fifth it disappears.

"On the 23d I introduced blood into the arms of three more children, taken on the third and fourth days of the eruption. The appearances were much the same as mentioned in experiments first and third. They were afterwards inoculated with variolous matter, and had the disease in the regular way.

"The above experiments were made with blood taken from a small vein in the hand or foot of three or four different patients, whom I had at that time under inoculation. They were selected from 160, as having the greatest number of pustules. The part was washed with warm water before the blood was taken, to prevent the possibility of any matter being mixed with it from the surface."

Shall we conclude from hence, that the variolous matter never enters the blood-vessels? but that the morbid motions of the vessels of the skin around the insertion of it continue to increase in a larger and larger circle for six or seven days; that then their quantity of morbid action becomes great enough to produce a fever-fit, and to affect the stomach by association of motions? and finally, that a second association of motions is produced between the stomach and the other parts of the skin, inducing them into morbid actions similar to those of the circle round the insertion of the variolous matter? Many more experiments and observations are required before this important question can be satisfactorily answered.

It may be adduced, that as the matter inserted into the skin of the arm frequently swells the lymphatic in the axilla, that in that circumstance it seems to be there arrested in its progress, and cannot be imagined to enter the blood by that lymphatic gland till the swelling of it subsides. Some other phænomena of the disease are more easily reconcileable to this theory of sympathetic motions than to that of absorption; as the time taken up between the insertion of the matter, and the operation of it on the system, as mentioned above. For the circle around the insertion is seen to increase, and to inflame; and I believe, undergoes a kind of diurnal paroxysm of torpor and paleness with a succeeding increase of action and colour, like a topical fever-fit. Whereas if the matter is conceived to circulate for six or seven days with the blood, without producing disorder, it ought to be rendered milder, or the blood-vessels more familiarized to its acrimony.

It is much easier to conceive from this doctrine of associated or sympathetic motions of distant parts of the system, how it happens, that the variolous infection can be received but once, as before explained; than by supposing, that a change is effected in the mass of blood by any kind of fermentative process.

The curious circumstance of the two contagions of small-pox and measles not acting at the same time, but one of them resting or suspending its action till that of the other ceases, may be much easier explained from sympathetic or associated actions of the infected part with other parts of the system, than it can from supposing the two contagions to enter the circulation.

The skin of the face is subject to more frequent vicissitudes of heat and cold, from its exposure to the open air, and is in consequence more liable to sensitive association with the stomach than any other part of the surface of the body, because their actions have been more frequently thus associated. Thus in a surfeit from drinking cold water, when a person is very hot and fatigued, an eruption is liable to appear on the face in consequence of this sympathy. In the same manner the rosy eruption on the faces of drunkards more probably arises from the sympathy of the face with the stomach, rather than between the face and the liver, as is generally supposed.

This sympathy between the stomach and the skin of the face is apparent in the eruption of the small-pox; since, where the disease is in considerable quantity, the eruption on the face first succeeds the sickness of the stomach. In the natural disease the stomach seems to be frequently primarily affected, either alone or along with the tonsils, as the matter seems to be only diffused in the air, and by being mixed with the saliva, or mucus of the tonsils, to be swallowed into the stomach.

After some days the irritative circles of motions become disordered by this new stimulus, which acts upon the mucus lining of the stomach; and sickness, vertigo, and a diurnal fever succeed. These disordered irritative motions become daily increased for two or three days, and then by their increased action certain sensitive motions, or inflammation, is produced, and at the next cold fit of fever, when the stomach recovers from its torpor, an inflammation of the external skin is formed in points (which afterwards suppurate), by sensitive association, in the same manner as a cough is produced in consequence of exposing the feet to cold, as described in Sect. XXV. 17. and Class IV. 2. I. 7. If the inoculated skin of the arm, as far as it appears inflamed, was to be cut out, or destroyed by caustic, before the fever commenced, as suppose on the fourth day after inoculation, would this prevent the disease? as it is supposed to prevent the hydrophobia.

III. 1. Where the new vessels, and enlarged old ones, which constitute inflammation, are not so hastily distended as to burst, and form a new kind of gland for the secretion of matter, as above mentioned; if such circumstances happen as diminish the painful sensation, the tendency to growth ceases, and by and by an absorption commences, not only of the superabundant quantity of fluids deposited in the inflamed part, but of the solids likewise, and this even of the hardest kind.

Thus during the growth of the second set of teeth in children, the roots of the first set are totally absorbed, till at length nothing of them remains but the crown; though a few weeks before, if they are drawn immaturely, their roots are found complete. Similar to this Mr. Hunter has observed, that where a dead piece of bone is to exfoliate, or to separate from a living one, that the dead part does not putrify, but remains perfectly sound, while the surface of the living part of the bone, which is in contact with the dead part, becomes absorbed, and thus effects its separation. Med. Comment. Edinb. V. 1. 425. In the same manner the calcareous matter of gouty concretions, the coagulable lymph deposited on inflamed membranes in rheumatism and extravasated blood become absorbed; which are all as solid and as indissoluble materials as the new vessels produced in inflammation.

This absorption of the new vessels and deposited fluids of inflamed parts is called resolution: it is produced by first using such internal means as decrease the pain of the part, and in consequence its new motions, as repeated bleeding, cathartics, diluent potations, and warm bath.

After the vessels are thus emptied, and the absorption of the new vessels and deposited fluids is evidently begun, it is much promoted by stimulating the part externally by solutions of lead, or other metals, and internally by the bark, and small doses of opium. Hence when an ophthalmy begins to become paler, any acrid eye-water, as a solution of six grains of white vitriol in an ounce of water, hastens the absorption, and clears the eye in a very short time. But the same application used a few days sooner would have increased the inflammation. Hence after evacuation opium in small doses may contribute to promote the absorption of fluids deposited on the brain, as observed by Mr. Bromfield in his treatise of surgery.

2. Where an abscess is formed by the rupture of these new vessels, the violence of inflammation ceases, and a new gland separates a material called pus: at the same time a less degree of inflammation produces new vessels called vulgarly proud flesh; which, if no bandage confines its growth, nor any other circumstance promotes absorption in the wound, would rise to a great height above the usual size of the part.

Hence the art of healing ulcers consists in producing a tendency to absorption in the wound greater than the deposition. Thus when an ill-conditioned ulcer separates a copious and thin discharge, by the use of any stimulus, as of salts of lead, or mercury, or copper externally applied, the discharge becomes diminished in quantity, and becomes thicker, as the thinner parts are first absorbed.

But nothing so much contributes to increase the absorption in a wound as covering the whole limb above the sore with a bandage, which should be spread with some plaster, as with emplastrum de minio, to prevent it from slipping. By this artificial tightness of the skin, the arterial pulsations act with double their usual power in promoting the ascending current of the fluid in the valvular lymphatics.

Internally the absorption from ulcers should be promoted first by evacuation, then by opium, bark, mercury, steel.

3. Where the inflammation proceeds with greater violence or rapidity, that is, when by the painful sensation a more inordinate activity of the organ is produced, and by this great activity an additional quantity of painful sensation follows in an increasing ratio, till the whole of the sensorial power, or spirit of animation, in the part becomes exhausted, a mortification ensues, as in a carbuncle, in inflammations of the bowels, in the extremities of old people, or in the limbs of those who are brought near a fire after having been much benumbed with cold. And from hence it appears, why weak people are more subject to mortification than strong ones, and why in weak persons less pain will produce mortification, namely, because the sensorial power is sooner exhausted by any excess of activity. I remember seeing a gentleman who had the preceding day travelled two stages in a chaise with what he termed a bearable pain in his bowels; which when I saw him had ceased rather suddenly, and without a passage through him;

his pulse was then weak, though not very quick; but as nothing which he swallowed would continue in his stomach many minutes, I concluded that the bowel was mortified; he died on the next day. It is usual for patients sinking under the small-pox with mortified pustules, and with purple spots intermixed, to complain of no pain, but to say they are pretty well to the last moment.

Recapitulation.

IV. When the motions of any part of the system, in consequence of previous torpor, are performed with more energy than in the irritative fevers, a disagreeable sensation is produced, and new actions of some part of the system commence in consequence of this sensation conjointly with the irritation: which motions constitute inflammation. If the fever be attended with a strong pulse, as in pleurisy, or rheumatism, it is termed synocha sensitiva, or sensitive fever with strong pulse; which is usually termed inflammatory fever. If it be attended with weak pulse, it is termed typhus sensitivus, or sensitive fever with weak pulse, or typhus gravior, or putrid malignant fever.

The synocha sensitiva, or sensitive fever with strong pulse, is generally attended with some topical inflammation, as in peripneumony, hepatitis, and is accompanied with much coagulable lymph, or size; which rises to the surface of the blood, when taken into a bason, as it cools; and which is believed to be the increased mucous secretion from the coats of the arteries, inspissated by a greater absorption of its aqueous and saline part, and perhaps changed by its delay in the circulation.

The typhus sensitivus, or sensitive fever with weak pulse, is frequently attended with delirium, which is caused by the deficiency of the quantity of sensorial power, and with variety of cutaneous eruptions.

Inflammation is caused by the pains occasioned by excess of action, and not by those pains which are occasioned by defect of action. These morbid actions, which are thus produced by two sensorial powers, viz. by irritation and sensation, secrete new living fibres, which elongate the old vessels, or form new ones, and at the same time much heat is evolved from these combinations. By the rupture of these vessels, or by a new construction of their apertures, purulent matters are secreted of various kinds; which are infectious the first time they are applied to the skin beneath the cuticle, or swallowed with the saliva into the stomach. This contagion acts not by its being absorbed into the circulation, but by the sympathies, or associated actions, between the part first stimulated by the contagious matter and the other parts of the system. Thus in the natural small-pox the contagion is swallowed with the saliva, and by its stimulus inflames the stomach; this variolous inflammation of the stomach increases every day, like the circle round the puncture of an inoculated arm, till it becomes great enough to disorder the circles of irritative and sensitive motions, and thus produces fever-fits, with sickness and vomiting. Lastly, after the cold paroxysm, or fit of torpor, of the stomach has increased for two or three successive days, an inflammation of the skin commences in points; which generally first appear upon the face, as the associated actions between the skin of the face and that of the stomach have been more frequently exerted together than those of any other parts of the external surface.

Contagious matters, as those of the measles and small-pox, do not act upon the system at the same time; but the progress of that which was last received is delayed, till the action of the former infection ceases. All kinds of matter, even that from common ulcers, are probably contagious the first time they are inserted beneath the cuticle or swallowed into the stomach; that is, as they were formed by certain morbid actions of the extremities of the vessels, they have the power to excite similar morbid actions in the extremities of other vessels, to which they are applied; and these by sympathy, or associations of motion, excite similar morbid actions in distant parts of the system, without entering the circulation; and hence the blood of a patient in the small-pox will not give that disease by inoculation to others.

When the new fibres or vessels become again absorbed into the circulation, the inflammation ceases; which is promoted, after sufficient evacuations, by external stimulants and bandages: but where the action of the vessels is very great, a mortification of the part is liable to ensue, owing to the exhaustion of sensorial power; which however occurs in weak people without much pain, and without very violent previous inflammation; and, like partial paralysis, may be esteemed one mode of natural death of old people, a part dying before the whole.

SECT. XXXIV
DISEASES OF VOLITION

I. 1. *Volition defined. Motions termed involuntary are caused by volition. Desires opposed to each other. Deliberation. Ass between two hay-cocks. Saliva swallowed against one's desire. Voluntary motions distinguished from those associated with sensitive motions. 2. Pains from excess, and from defect of motion. No pain is felt during vehement voluntary exertion; as in cold fits of ague, labour-pains, strangury, tenesmus, vomiting, restlessness in fevers, convulsion of a wounded muscle. 3. Of holding the breath and screaming in pain; why swine and dogs cry out in pain, and not sheep and horses. Of grinning and biting in pain; why mad animals bite others. 4. Epileptic convulsions explained, why the fits begin with quivering of the under jaw, biting the tongue, and setting the teeth; why the convulsive motions are alternately relaxed. The phenomenon of laughter explained. Why children cannot tickle themselves. How some have died from immoderate laughter. 5. Of cataleptic spasms, of the locked jaw, of painful cramps. 6. Syncope explained. Why no external objects are perceived in syncope. 7. Of palsy and apoplexy from violent exertions. Case of Mrs. Scot. From dancing, scating, swimming. Case of Mr. Nairn. Why palsies are not always immediately preceded by violent exertions. Palsy and epilepsy from diseased livers. Why the right arm more frequently paralytic than the left. How paralytic limbs regain their motions. II. Diseases of the sensual motions from excess or defect of voluntary exertion. 1. Madness. 2. Distinguished from delirium. 3. Why mankind more liable to insanity than brutes. 4. Suspicion. Want of shame, and of cleanliness. 5. They bear cold, hunger, and fatigue. Charles XII. of Sweden. 6. Pleasureable delirium, and insanity. Child riding on a stick. Pains of martyrdom not felt. 7. Dropsy. 8. Inflammation cured by insanity. III. 1. Pain relieved by reverie. Reverie is an exertion of voluntary and sensitive motions. 2. Case of reverie. 3. Lady supposed to have two souls. 4. Methods of relieving pain.*

I. 1. Before we commence this Section on Diseased Voluntary Motions, it may be necessary to premise, that the word volition is not used in this work exactly in its common acceptation. Volition is said in Section V. to bear the same analogy to desire and aversion, which sensation does to pleasure and pain. And hence that, when desire or aversion produces any action of the muscular fibres, or of the organs of sense, they are termed volition; and the actions produced in consequence are termed voluntary actions. Whence it appears, that motions of our muscles or ideas may be produced in consequence of desire or aversion without our having the power to prevent them, and yet these motions may be termed voluntary, according to our definition of the word; though in common language they would be called involuntary.

The objects of desire and aversion are generally at a distance, whereas those of pleasure and pain are immediately acting upon our organs. Hence, before desire or aversion are exerted, so as to cause any actions, there is generally time for deliberation; which consists in discovering the means to obtain the object of desire, or to avoid the object of aversion; or in examining the good or bad consequences, which may result from them. In this case it is evident, that we have a power to delay the proposed action, or to perform it; and this power of choosing, whether we shall act or not, is in common language expressed by the word volition, or will. Whereas in this work the word volition means simply the active state of the sensorial faculty in producing motion in consequence of desire or aversion: whether we have the power of restraining that action, or not; that is, whether we exert any actions in consequence of opposite desires or aversions, or not.

For if the objects of desire or aversion are present, there is no necessity to investigate or compare the *means* of obtaining them, nor do we always deliberate about their consequences; that is, no deliberation necessarily intervenes, and in consequence the power of choosing to act or not is not exerted. It is probable, that this twofold use of the word volition in all languages has confounded the metaphysicians, who have disputed about free will and necessity. Whereas from the above analysis it would appear, that during our sleep, we use no voluntary exertions at all; and in our waking hours, that they are the consequence of desire or aversion.

To will is to act in consequence of desire; but to desire means to desire something, even if that something be only to become free from the pain, which causes the desire; for to desire nothing is not to desire; the word desire, therefore, includes both the action and the object or motive; for the object and motive of desire are the same thing. Hence to desire without an object, that is, without a motive, is a solecism in language. As if one should ask, if you could eat without food, or breathe without air.

From this account of volition it appears, that convulsions of the muscles, as in epileptic fits, may in the common sense of that word be termed involuntary; because no deliberation is interposed between the desire or aversion and the consequent action; but in the sense of the word, as above defined, they belong to the class of voluntary motions, as delivered in Vol. II. Class III. If this use of the word be discordant to the ear of the reader, the term morbid voluntary motions, or motions in consequence of aversion, may be substituted in its stead.

If a person has a desire to be cured of the ague, and has at the same time an aversion (or contrary desire) to swallowing an ounce of Peruvian bark; he balances desire against desire, or aversion against aversion; and thus he acquires the power of choosing, which is the common acceptation of the word *willing*. But in the cold fit of ague, after having discovered that the act of shuddering, or exerting the subcutaneous muscles, relieves the pain of cold; he immediately exerts this act of volition, and shudders, as soon as the pain and consequent aversion return, without any deliberation intervening; yet is this act, as well as that of swallowing an ounce of the bark, caused by volition; and that even though he endeavours in vain to prevent it by a weaker contrary volition. This recalls to our minds the story of the hungry ass between two hay-stacks, where the two desires are supposed so exactly to counteract each other, that he goes to neither of the stacks, but perishes by want. Now as two equal and opposite desires are thus supposed to balance each other, and prevent all action, it follows, that if one of these hay-stacks was suddenly removed, that the ass would irresistibly be hurried to the other, which in the common use of the word might be called an involuntary act; but which, in our acceptation of it, would be classed amongst voluntary actions, as above explained.

Hence to deliberate is to compare opposing desires or aversions, and that which is the most interesting at length prevails, and produces action. Similar to this, where two pains oppose each other, the stronger or more interesting one produces action; as in pleurisy the pain from suffocation would produce expansion of the lungs, but the pain occasioned by extending the inflamed membrane, which lines the chest, opposes this expansion, and one or the other alternately prevails.

When any one moves his hand quickly near another person's eyes, the eye-lids instantly close; this act in common language is termed involuntary, as we have not time to deliberate or to exert any contrary desire or aversion, but in this work it would be termed a voluntary act, because it is caused by the faculty of volition, and after a few trials the nictitation can be prevented by a contrary or opposing volition.

The power of opposing volitions is best exemplified in the story of Mutius Scævola, who is said to have thrust his hand into the fire before Porcenna, and to have suffered it to be consumed for having failed him in his attempt on the life of that general. Here the aversion for the loss of same, or the unsatisfied desire to serve his country, the two prevalent enthusiasms at that time, were more powerful than the desire of withdrawing his hand, which must be occasioned by the pain of combustion; of these opposing volitions

Vincit amor patriæ, laudumque immensa cupido.

If any one is told not to swallow his saliva for a minute, he soon swallows it contrary to his will, in the common sense of that word; but this also is a voluntary action, as it is performed by the faculty of volition, and is thus to be understood. When the power of volition is exerted on any of our senses, they become more acute, as in our attempts to hear small noises in the night. As explained in Section XIX. 6. Hence by our attention to the fauces from our desire not to swallow our saliva; the fauces become more sensible; and the stimulus of the saliva is followed by greater sensation, and consequent desire of swallowing it. So that the desire or volition in consequence of the increased sensation of the saliva is more powerful, than the previous desire not to swallow it. See Vol. II. Deglutitio invita. In the same manner if a modest man wishes not to want to make water, when he is confined with ladies in a coach or an assembly-room; that very act of volition induces the circumstance, which he wishes to avoid, as above explained; insomuch that I once saw a partial insanity, which might be called a voluntary diabetes, which was occasioned by the fear (and consequent aversion) of not being able to make water at all.

It is further necessary to observe here, to prevent any confusion of voluntary, with sensitive, or associate motions, that in all the instances of violent efforts to relieve pain, those efforts are at first voluntary exertions; but after they have been frequently repeated for the purpose of relieving certain pains, they become associated with those pains, and cease at those times to be subservient to the will; as in coughing, sneezing, and strangury. Of these motions those which contribute to remove or dislodge the offending cause, as the actions of the abdominal muscles in parturition or in vomiting, though they were originally excited by volition, are in this work termed sensitive motions; but those actions of the muscles or organs of sense, which do not contribute to remove the offending cause, as in general convulsions or in madness, are in this work termed voluntary motions, or motions in consequence of aversion, though in common language they are called involuntary ones. Those sensitive unrestrainable actions, which contribute to remove the cause of pain are uniformly and invariably exerted, as in coughing or sneezing; but those motions which are exerted in consequence of aversion without contributing to remove the painful cause, but only to prevent the sensation of it, as in epileptic, or cataleptic fits, are not uniformly and invariably exerted, but change from one set of muscles to another, as will be further explained; and may by this criterion also be distinguished from the former.

At the same time those motions, which are excited by perpetual stimulus, or by association with each other, or immediately by pleasureable or painful sensation, may properly be termed involuntary motions, as those of the heart and arteries; as the faculty of volition seldom affects those, except when it exists in unnatural quantity, as in maniacal people.

2. It was observed in Section XIV. on the Production of Ideas, that those parts of the system, which are usually termed the organs of sense, are liable to be excited into pain by the excess of the stimulus of those objects, which are by nature adapted to affect them; as of too great light, sound, or pressure. But that these organs receive no pain from the defect or absence of these stimuli, as in darkness or silence. But that our other organs of perception, which have generally been called appetites, as of hunger, thirst, want of heat, want of fresh air, are liable to be affected with pain by the defect, as well as by the excess of their appropriated stimuli.

This excess or defect of stimulus is however to be considered only as the remote cause of the pain, the immediate cause being the excess or defect of the natural action of the affected part, according to Sect. IV. 5. Hence all the pains of the body may be divided into those from excess of motion, and those from defect of motion; which distinction is of great importance in the knowledge and the cure of many diseases. For as the pains from excess of motion either gradually subside, or are in general succeeded by inflammation; so those from defect of motion either gradually subside, or are in general succeeded by convulsion, or madness. These pains are easily distinguishable from each other by this circumstance, that the former are attended with heat of the pained part, or of the whole body; whereas the latter exists without increase of heat in the pained part, and is generally attended with coldness of the extremities of the body; which is the true criterion of what have been called nervous pains.

Thus when any acrid material, as snuff or lime, falls into the eye, pain and inflammation and heat are produced from the excess of stimulus; but violent hunger, hemicrania, or the clavus hystericus, are attended with coldness of the extremities, and defect of circulation. When we are exposed to great cold, the pain we experience from the deficiency of heat is attended with a quiescence of the motions of the vascular system; so that no inflammation is produced, but a great desire of heat, and a tremulous motion of the subcutaneous muscles, which is properly a convulsion in consequence of this pain from defect of the stimulus of heat.

It was before mentioned, that as sensation consists in certain movements of the sensorium, beginning at some of the extremities of it, and propagated to the central parts of it; so volition consists of certain other movements of the sensorium, commencing in the central parts of it, and propagated to some of its extremities. This idea of these two great powers of motion in the animal machine is confirmed from observing, that they never exist in a great degree or universally at the same time; for while we strongly exert our voluntary motions, we cease to feel the pains or uneasinesses, which occasioned us to exert them.

Hence during the time of fighting with fists or swords no pain is felt by the combatants, till they cease to exert themselves. Thus in the beginning of ague-fits the painful sensation of cold is diminished, while the patient exerts himself in the shivering and gnashing of his teeth. He then ceases to exert himself, and the pain of cold returns; and he is thus perpetually induced to reiterate these exertions, from which he experiences a temporary relief. The same occurs in labour-pains, the exertion of the parturient woman relieves the violence of the pains for a time, which recur again soon after she has ceased to use those exertions. The same is true in many other painful diseases, as in the strangury, tenesmus, and the efforts of vomiting; all these disagreeable sensations are diminished or removed for a time by the various exertions they occasion, and recur alternately with those exertions.

The restlessness in some fevers is an almost perpetual exertion of this kind, excited to relieve some disagreeable sensations; the reciprocal opposite exertions of a wounded worm, the alternate emprosthotonos and opisthotonos of some spasmodic diseases, and the intervals of all convulsions, from whatever cause, seem to be owing to this circumstance of the laws of animation; that great or universal exertion cannot exist at the same time with great or universal sensation, though they can exist reciprocally; which is probably resolvable into the more general law, that the whole sensorial power being expended in one mode of exertion, there is none to spare for any other. Whence syncope, or temporary apoplexy, succeeds to epileptic convulsions.

3. Hence when any violent pain afflicts us, of which we can neither avoid nor remove the cause, we soon learn to endeavour to alleviate it, by exerting some violent voluntary effort, as mentioned above; and are naturally induced to use those muscles for this purpose, which have been in the early periods of our lives most frequently or most powerfully exerted.

Now the first muscles, which infants use most frequently, are those of respiration; and on this account we gain a habit of holding our breath, at the same time that we use great efforts to exclude it, for this purpose of alleviating unavoidable pain; or we press out our breath through a small aperture of the larynx, and scream violently, when the pain is greater than is relievable by the former mode of exertion. Thus children scream to relieve any pain either of body or mind, as from anger, or fear of being beaten.

Hence it is curious to observe, that those animals, who have more frequently exerted their muscles of respiration violently, as in talking, barking, or grunting, as children, dogs, hogs, scream much more, when they are in pain, than those other animals, who use little or no language in their common modes of life; as horses, sheep, and cows.

The next most frequent or most powerful efforts, which infants are first tempted to produce, are those with the muscles in biting hard substances; indeed the exertion of these muscles is very powerful in common mastication, as appears from the pain we receive, if a bit of bone is unexpectedly found amongst our softer food; and further appears from their acting to so great mechanical disadvantage, particularly when we bite with the incisores, or canine teeth; which are first formed, and thence are first used to violent exertion.

Hence when a person is in great pain, the cause of which he cannot remove, he sets his teeth firmly together, or bites some substance between them with great vehemence, as another mode of violent exertion to produce a temporary relief. Thus we have a proverb where no help can be had in pain, "to grin and abide;" and the tortures of hell are said to be attended with "gnashing of teeth."

Hence in violent spasmodic pains I have seen people bite not only their tongues, but their arms or fingers, or those of the attendants, or any object which was near them; and also strike, pinch, or tear, others or themselves, particularly the part of their own body, which is painful at the time. Soldiers, who die of painful wounds in battle, are said in Homer to bite the ground. Thus also in the bellon, or colica saturnina, the patients are said to bite their own flesh, and dogs in this disease to bite up the ground they lie upon. It is probable that the great endeavours to bite in mad dogs, and the violence of other mad animals, is owing to the same cause.

4. If the efforts of our voluntary motions are exerted with still greater energy for the relief of some disagreeable sensation, convulsions are produced; as the various kinds of epilepsy, and in some hysteric paroxysms. In all these diseases a pain, or disagreeable sensation is produced, frequently

by worms, or acidity in the bowels, or by a diseased nerve in the side, or head, or by the pain of a diseased liver.

In some constitutions a more intolerable degree of pain is produced in some part at a distance from the cause by sensitive association, as before explained; these pains in such constitutions arise to so great a degree, that I verily believe no artificial tortures could equal some, which I have witnessed; and am confident life would not have long been preserved, unless they had been soon diminished or removed by the universal convulsion of the voluntary motions, or by temporary madness.

In some of the unfortunate patients I have observed, the pain has risen to an inexpressible degree, as above described, before the convulsions have supervened; and which were preceded by screaming, and grinning; in others, as in the common epilepsy, the convulsion has immediately succeeded the commencement of the disagreeable sensations; and as a stupor frequently succeeds the convulsions, they only seemed to remember that a pain at the stomach preceded the fit, or some other uneasy feel; or more frequently retained no memory at all of the immediate cause of the paroxysm. But even in this kind of epilepsy, where the patient does not recollect any preceding pain, the paroxysms generally are preceded by a quivering motion of the under jaw, with a biting of the tongue; the teeth afterwards become pressed together with vehemence, and the eyes are then convulsed, before the commencement of the universal convulsion; which are all efforts to relieve pain.

The reason why these convulsive motions are alternately exerted and remitted was mentioned above, and in Sect. XII. 1. 3. when the exertions are such as give a temporary relief to the pain, which excites them, they cease for a time, till the pain is again perceived; and then new exertions are produced for its relief. We see daily examples of this in the loud reiterated laughter of some people; the pleasureable sensation, which excites this laughter, arises for a time so high as to change its name and become painful: the convulsive motions of the respiratory muscles relieve the pain for a time; we are, however, unwilling to lose the pleasure, and presently put a stop to this exertion, and immediately the pleasure recurs, and again as instantly rises into pain. All of us have felt the pain of immoderate laughter; children have been tickled into convulsions of the whole body; and others have died in the act of laughing; probably from a paralysis succeeding the long continued actions of the muscles of respiration.

Hence we learn the reason, why children, who are so easily excited to laugh by the tickling of other people's fingers, cannot tickle themselves into laughter. The exertion of their hands in the endeavour to tickle themselves prevents the necessity of any exertion of the respiratory muscles to relieve the excess of pleasurable affection. See Sect. XVII. 3. 5.

Chrysippus is recorded to have died laughing, when an ass was invited to sup with him. The same is related of one of the popes, who, when he was ill, saw a tame monkey at his bedside put on the holy thiara. Hall. Phys. T. III. p. 306.

There are instances of epilepsy being produced by laughing recorded by Van Swieten, T. III. 402 and 308. And it is well known, that many people have died instantaneously from the painful excess of joy, which probably might have been prevented by the exertions of laughter.

Every combination of ideas, which we attend to, occasions pain or pleasure; those which occasion pleasure, furnish either social or selfish pleasure, either malicious or friendly, or lascivious, or sublime pleasure; that is, they give us pleasure mixed with other emotions, or they give us unmixed pleasure, without occasioning any other emotions or exertions at the same time. This unmixed pleasure, if it be great, becomes painful, like all other animal motions from stimuli of every kind; and if no other exertions are occasioned at the same time, we use the exertion of laughter to relieve this pain. Hence laughter is occasioned by such wit as excites simple pleasure without any other emotion, such as pity, love, reverence. For sublime ideas are mixed with admiration, beautiful ones with love, new ones with surprise; and these exertions of our ideas prevent the action of laughter from being necessary to relieve the painful pleasure above described. Whence laughable wit consists of frivolous ideas, without connections of any consequence, such as puns on words, or on phrases, incongruous junctions of ideas; on which account laughter is so frequent in children.

Unmixed pleasure less than that, which causes laughter, causes sleep, as in singing children to sleep, or in slight intoxication from wine or food. See Sect. XVIII. 12.

5. If the pains, or disagreeable sensations, above described do not obtain a temporary relief from these convulsive exertions of the muscles, those convulsive exertions continue without remission, and one kind of catalepsy is produced. Thus when a nerve or tendon produces great pain by its being inflamed or wounded, the patient sets his teeth firmly together, and grins violently, to diminish the pain; and if the pain is not relieved by this exertion, no relaxation of the maxillary muscles takes place, as in the convulsions above described, but the jaws remain firmly fixed together. This locked jaw is the most frequent instance of cataleptic spasm, because we are more inclined to exert the muscles subservient to mastication from their early obedience to violent efforts of volition.

But in the case related in Sect. XIX. on Reverie, the cataleptic lady had pain in her upper teeth; and pressing one of her hands vehemently against her cheek-bone to diminish this pain, it remained in that attitude for about half an hour twice a day, till the painful paroxysm was over.

I have this very day seen a young lady in this disease, (with which she has frequently been afflicted,) she began to-day with violent pain shooting from one side of the forehead to the occiput, and after various struggles lay on the bed with her fingers and wrists bent and stiff for about two hours; in other respects she seemed in a syncope with a natural pulse. She then had intervals of pain and of spasm, and took three grains of opium every hour till she had taken nine grains, before the pains and spasm ceased.

There is, however, another species of fixed spasm, which differs from the former, as the pain exists in the contracted muscle, and would seem rather to be the consequence than the cause of the contraction, as in the cramp in the calf of the leg, and in many other parts of the body.

In these spasms it should seem, that the muscle itself is first thrown into contraction by some disagreeable sensation, as of cold; and that then the violent pain is produced by the great contraction of the muscular fibres extending its own tendons, which are said to be sensible to extension only; and is further explained in Sect. XVIII. 15.

6. Many instances have been given in this work, where after violent motions excited by irritation, the organ has become quiescent to less, and even to the great irritation, which induced it into violent motion; as after looking long at the sun or any bright colour, they cease to be seen; and after removing from bright day-light into a gloomy room, the eye cannot at first perceive the objects, which stimulate it less. Similar to this is the syncope, which succeeds after the violent exertions of our voluntary motions, as after epileptic fits, for the power of volition acts in this case as the stimulus in the other. This syncope is a temporary palsy, or apoplexy, which ceases after a time, the muscles recovering their power of being excited into action by the efforts of volition; as the eye in the circumstance above mentioned recovers in a little time its power of seeing objects in a gloomy room; which were invisible immediately after coming out of a stronger light. This is owing to an accumulation of sensorial power during the inaction of those fibres, which were before accustomed to perpetual exertions, as explained in Sect. XII. 7. 1. A slighter degree of this disease is experienced by every one after great fatigue, when the muscles gain such inability to further action, that we are obliged to rest them for a while, or to summon a greater power of volition to continue their motions.

In all the syncopes, which I have seen induced after convulsive fits, the pulse has continued natural, though the organs of sense, as well as the locomotive muscles, have ceased to perform their functions; for it is necessary for the perception of objects, that the external organs of sense should be properly excited by the voluntary power, as the eye-lids must be open, and perhaps the muscles of the eye put into action to distend, and thence give greater pellucidity to the cornea, which in syncope, as in death, appears flat and less transparent.

The tympanum of the ear also seems to require a voluntary exertion of its muscles, to gain its due tension, and it is probable the other external organs of sense require a similar voluntary exertion to adapt them to the distinct perception of objects. Hence in syncope as in sleep, as the power of volition is suspended, no external objects are perceived. See Sect. XVIII. 5. During the time which the patient lies in a fainting fit, the spirit of animation becomes accumulated; and hence the muscles in a while become irritable by their usual stimulation, and the fainting fit ceases. See Sect. XII. 7. 1.

7. If the exertion of the voluntary motions has been still more energetic, the quiescence, which succeeds, is so complete, that they cannot again be excited into action by the efforts of the will. In this manner the palsy, and apoplexy (which is an universal palsy) are frequently produced after convulsions, or other violent exertions; of this I shall add a few instances.

Platernus mentions some, who have died apoplectic from violent exertions in dancing; and Dr. Mead, in his Essay on Poisons, records a patient in the hydrophobia, who at one effort broke the cords which bound him, and at the same instant expired. And it is probable, that those, who have expired from immoderate laughter, have died from this paralysis consequent to violent exertion. Mrs. Scott of Stafford was walking in her garden in perfect health with her neighbour Mrs. — —; the latter accidentally fell into a muddy rivulet, and tried in vain to disengage herself by the assistance of Mrs. Scott's hand. Mrs. Scott exerted her utmost power for many minutes, first to assist her friend, and next to prevent herself from being pulled into the morass, as her distressed companion would not disengage her hand. After other assistance was procured by their united screams, Mrs. Scott walked to a chair about twenty yards from the brook, and was seized with an apoplectic stroke: which continued many days, and terminated in a total loss of her right arm, and her speech; neither of which she ever after perfectly recovered.

It is said, that many people in Holland have died after skating too long or too violently on their frozen canals; it is probable the death of these, and of others, who have died suddenly in swimming, has been owing to this great quiescence or paralysis; which has succeeded very violent exertions, added to the concomitant cold, which has had greater effect after the sufferers had been heated and exhausted by previous exercise.

I remember a young man of the name of Nairne at Cambridge, who walking on the edge of a barge fell into the river. His cousin and fellow-student of the same name, knowing the other could not swim, plunged into the water after him, caught him by his clothes, and approaching the bank by a vehement exertion propelled him safe to the land, but that instant, seized, as was supposed, by the cramp, or paralysis, sunk to rise no more. The reason why the cramp of the muscles, which compose the calf of the leg, is so liable to affect swimmers, is, because these muscles have very weak antagonists, and are in walking generally elongated again after their contraction by the weight of the body on the ball of the toe, which is very much greater than the resistance of the water in swimming. See Section XVIII. 15.

It does not follow that every apoplectic or paralytic attack is immediately preceded by vehement exertion; the quiescence, which succeeds exertion, and which is not so great as to be termed paralysis, frequently recurs afterwards at certain periods; and by other causes of quiescence, occurring with those periods, as was explained in treating of the paroxysms of intermitting fevers; the quiescence at length, becomes so great as to be incapable of again being removed by the efforts of volition, and complete paralysis is formed. See Section XXXII. 3. 2.

Many of the paralytic patients, whom I have seen, have evidently had diseased livers from the too frequent potation of spirituous liquors; some of them have had the gutta rosea on their faces and breasts; which has in some degree receded either spontaneously, or by the use of external remedies, and the paralytic stroke has succeeded; and as in several persons, who have drank much vinous spirits, I have observed epileptic fits to commence at about forty or fifty years of age, without any hereditary taint, from the stimulus, as I believed, of a diseased liver; I was induced to ascribe many paralytic cases to the same source; which were not evidently the effect of age, or of unacquired debility. And the account given before of dropsies, which very frequently are owing to a paralysis of the absorbent system, and are generally attendant on free drinkers of spirituous liquors, confirmed me in this opinion.

The disagreeable irritation of a diseased liver produces exertions and consequent quiescence; these by the accidental concurrence of other causes of quiescence, as cold, solar or lunar periods, inanition, the want of their usual portion of spirit of wine, at length produces paralysis.

This is further confirmed by observing, that the muscles, we most frequently, or most powerfully exert, are most liable to palsy; as those of the voice and of articulation, and of those paralytics which I have seen, a much greater proportion have lost the use of their right arm; which is so much more generally exerted than the left.

I cannot dismiss this subject without observing, that after a paralytic stroke, if the vital powers are not much injured, that the patient has all the movements of the affected limb to learn over again, just as in early infancy; the limb is first moved by the irritation of its muscles, as in stretching, (of which a case was related in Section VII. 1. 3.) or by the electric concussion; afterwards it becomes obedient to sensation, as in violent danger or fear; and lastly, the muscles become again associated with volition, and gradually acquire their usual habits of acting together.

Another phænomenon in palsies is, that when the limbs of one side are disabled, those of the other are in perpetual motion. This can only be explained from conceiving that the power of motion, whatever it is, or wherever it resides, and which is capable of being exhausted by fatigue, and accumulated in rest, is now less expended, whilst one half of the body is capable of receiving its usual proportion of it, and is hence derived with greater ease or in greater abundance into the limbs, which remain unaffected.

II. 1. The excess or defect of voluntary exertion produces similar effects upon the sensual motions, or ideas of the mind, as those already mentioned upon the muscular fibres. Thus when any violent pain, arising from the defect of some peculiar stimulus, exists either in the muscular or sensual systems of fibres, and which cannot be removed by acquiring the defective stimulus; as in some constitutions convulsions of the muscles are produced to procure a temporary relief, so in other constitutions vehement voluntary exertions of the ideas of the mind are produced for the same purpose; for during this exertion, like that of the muscles, the pain either vanishes or is diminished: this violent exertion constitutes madness; and in many cases I have seen the madness take place, and the convulsions cease, and reciprocally the madness cease, and the convulsions supervene. See Section III. 5. 8.

2. Madness is distinguishable from delirium, as in the latter the patient knows not the place where he resides, nor the persons of his friends or attendants, nor is conscious of any external objects, except when spoken to with a louder voice, or stimulated with unusual force, and even then he soon relapses into a state of inattention to every thing about him. Whilst in the former he is perfectly sensible to every thing external, but has the voluntary powers of his mind intensely exerted on some particular object of his desire

or aversion, he harbours in his thoughts a suspicion of all mankind, lest they should counteract his designs; and while he keeps his intentions, and the motives of his actions profoundly secret; he is perpetually studying the means of acquiring the object of his wish, or of preventing or revenging the injuries he suspects.

3. A late French philosopher, Mr. Helvetius, has deduced almost all our actions from this principle of their relieving us from the ennui or tædium vitæ; and true it is, that our desires or aversions are the motives of all our voluntary actions; and human nature seems to excel other animals in the more facil use of this voluntary power, and on that account is more liable to insanity than other animals. But in mania this violent exertion of volition is expended on mistaken objects, and would not be relieved, though we were to gain or escape the objects, that excite it. Thus I have seen two instances of madmen, who conceived that they had the itch, and several have believed they had the venereal infection, without in reality having a symptom of either of them. They have been perpetually thinking upon this subject, and some of them were in vain salivated with design of convincing them to the contrary.

4. In the minds of mad people those volitions alone exist, which are unmixed with sensation; immoderate suspicion is generally the first symptom, and want of shame, and want of delicacy about cleanliness. Suspicion is a voluntary exertion of the mind arising from the pain of fear, which it is exerted to relieve: shame is the name of a peculiar disagreeable sensation, see Fable of the Bees, and delicacy about cleanliness arises from another disagreeable sensation. And therefore are not found in the minds of maniacs, which are employed solely in voluntary exertions. Hence the most modest women in this disease walk naked amongst men without any kind of concern, use obscene discourse, and have no delicacy about their natural evacuations.

5. Nor are maniacal people more attentive to their natural appetites, or to the irritations which surround them, except as far as may respect their suspicions or designs; for the violent and perpetual exertions of their voluntary powers of mind prevents their perception of almost every other object, either of irritation or of sensation. Hence it is that they bear cold, hunger, and fatigue, with much greater pertinacity than in their sober hours, and are less injured by them in respect to their general health. Thus it is asserted by historians, that Charles the Twelfth of Sweden slept on the snow, wrapped only in his cloak, at the siege of Frederickstad, and bore extremes of cold and hunger, and fatigue, under which numbers of his soldiers perished; because the king was insane with ambition, but the

soldier had no such powerful stimulus to preserve his system from debility and death.

6. Besides the insanities arising from exertions in consequence of pain, there is also a pleasurable insanity, as well as a pleasurable delirium; as the insanity of personal vanity, and that of religious fanaticism. When agreeable ideas excite into motion the sensorial power of sensation, and this again causes other trains of agreeable ideas, a constant stream of pleasurable ideas succeeds, and produces pleasurable delirium. So when the sensorial power of volition excites agreeable ideas, and the pleasure thus produced excites more volition in its turn, a constant flow of agreeable voluntary ideas succeeds; which when thus exerted in the extreme constitutes insanity.

Thus when our muscular actions are excited by our sensations of pleasure, it is termed play; when they are excited by our volition, it is termed work; and the former of these is attended with less fatigue, because the muscular actions in play produce in their turn more pleasurable sensation; which again has the property of producing more muscular action. An agreeable instance of this I saw this morning. A little boy, who was tired with walking, begged of his papa to carry him. "Here," says the reverend doctor, "ride upon my gold-headed cane;" and the pleased child, putting it between his legs, gallopped away with delight, and complained no more of his fatigue. Here the aid of another sensorial power, that of pleasurable sensation, superadded vigour to the exertion of exhausted volition. Which could otherwise only have been excited by additional pain, as by the lash of slavery. On this account where the whole sensorial power has been exerted on the contemplation of the promised joys of heaven, the saints of all persecuted religions have borne the tortures of martyrdom with otherwise unaccountable firmness.

7. There are some diseases, which obtain at least a temporary relief from the exertions of insanity; many instances of dropsies being thus for a time cured are recorded. An elderly woman labouring with ascites I twice saw relieved for some weeks by insanity, the dropsy ceased for several weeks, and recurred again alternating with the insanity. A man afflicted with difficult respiration on lying down, with very irregular pulse, and œdematous legs, whom I saw this day, has for above a week been much relieved in respect to all those symptoms by the accession of insanity, which is shewn by inordinate suspicion, and great anger.

In cases of common temporary anger the increased action of the arterial system is seen by the red skin, and increased pulse, with the immediate increase of muscular activity. A friend of mine, when he was painfully fatigued by riding on horseback, was accustomed to call up ideas into his mind, which used to excite his anger or indignation, and thus for a time at least relieved the pain of fatigue. By this temporary insanity, the effect of the voluntary power upon the whole of his system was increased; as in the cases of dropsy above mentioned, it would appear, that the increased action of the voluntary faculty of the sensorium affected the absorbent system, as well as the secerning one.

8. In respect to relieving inflammatory pains, and removing fever, I have seen many instances, as mentioned in Sect. XII. 2. 4. One lady, whom I attended, had twice at some years interval a locked jaw, which relieved a pain on her sternum with peripneumony. Two other ladies I saw, who towards the end of violent peripneumony, in which they frequently lost blood, were at length cured by insanity supervening. In the former the increased voluntary exertion of the muscles of the jaw, in the latter that of the organs of sense, removed the disease; that is, the disagreeable sensation, which had produced the inflammation, now excited the voluntary power, and these new voluntary exertions employed or expended the superabundant sensorial power, which had previously been exerted on the arterial system, and caused inflammation.

Another case, which I think worth relating, was of a young man about twenty; he had laboured under an irritative fever with debility for three or four weeks, with very quick and very feeble pulse, and other usual symptoms of that species of typhus, but at this time complained much and frequently of pain of his legs and feet. When those who attended him were nearly in despair of his recovery, I observed with pleasure an insanity of mind supervene: which was totally different from delirium, as he knew his friends, calling them by their names, and the room in which he lay, but became violently suspicious of his attendants, and calumniated with vehement oaths his tender mother, who sat weeping by his bed. On this his pulse became slower and firmer, but the quickness did not for some time intirely cease, and he gradually recovered. In this case the introduction of an increased quantity of the power of volition gave vigour to those movements of the system, which are generally only actuated by the power of irritation, and of association.

Another case I recollect of a young man, about twenty-five, who had the scarlet-fever, with very quick pulse, and an universal eruption on his skin, and was not without reason esteemed to be in great danger of his life. After a few days an insanity supervened, which his friends mistook for delirium, and he gradually recovered, and the cuticle peeled off. From these and a few other cases I have always esteemed insanity to be a favourable sign in fevers, and have cautiously distinguished it from delirium.

III. Another mode of mental exertion to relieve pain, is by producing a train of ideas not only by the efforts of volition, as in insanity; but by those of sensation likewise, as in delirium and sleep. This mental effort is termed reverie, or somnambulation, and is described more at large in Sect. XIX. on that subject. But I shall here relate another case of that wonderful disease, which fell yesterday under my eye, and to which I have seen many analogous alienations of mind, though not exactly similar in all circumstances. But as all of them either began or terminated with pain or convulsion, there can be no doubt but that they are of epileptic origin, and constitute another mode of mental exertion to relieve some painful sensation.

1. Master A. about nine years old, had been seized at seven every morning for ten days with uncommon fits, and had had slight returns in the afternoon. They were supposed to originate from worms, and had been in vain attempted to be removed by vermifuge purges. As his fit was expected at seven yesterday morning, I saw him before that hour; he was asleep, seemed free from pain, and his pulse natural. About seven he began to complain of pain about his navel, or more to the left side, and in a few minutes had exertions of his arms and legs like swimming. He then for half an hour hunted a pack of hounds; as appeared by his hallooing, and calling the dogs by their names, and discoursing with the attendants of the chase, describing exactly a day of hunting, which (I was informed) he had witnessed a year before, going through all the most minute circumstances of it; calling to people, who were then present, and lamenting the absence of others, who were then also absent. After this scene he imitated, as he lay in bed, some of the plays of boys, as swimming and jumping. He then sung an English and then an Italian song; part of which with his eyes open, and part with them closed, but could not be awakened or excited by any violence, which it was proper to use.

After about an hour he came suddenly to himself with apparent surprise, and seemed quite ignorant of any part of what had passed, and after being apparently well for half an hour, he suddenly fell into a great stupor, with slower pulse than natural, and a slow moaning respiration, in which he continued about another half hour, and then recovered.

The sequel of this disease was favourable; he was directed one grain of opium at six every morning, and then to rise out of bed; at half past six he was directed fifteen drops of laudanum in a glass of wine and water. The first day the paroxysm became shorter, and less violent. The dose of opium was increased to one-half more, and in three or four days the fits left him. The bark and filings of iron were also exhibited twice a day; and I believe the complaint returned no more.

2. In this paroxysm it must be observed, that he began with pain, and ended with stupor, in both circumstances resembling a fit of epilepsy. And that therefore the exertions both of mind and body, both the voluntary ones, and those immediately excited by pleasurable sensation, were exertions to relieve pain.

The hunting scene appeared to be rather an act of memory than of imagination, and was therefore rather a voluntary exertion, though attended with the pleasurable eagerness, which was the consequence of those ideas recalled by recollection, and not the cause of them.

These ideas thus voluntarily recollected were succeeded by sensations of pleasure, though his senses were unaffected by the stimuli of visible or audible objects; or so weakly excited by them as not to produce sensation or attention. And the pleasure thus excited by volition produced other ideas and other motions in consequence of the sensorial power of sensation. Whence the mixed catenations of voluntary and sensitive ideas and muscular motions in reverie; which, like every other kind of vehement exertion, contribute to relieve pain, by expending a large quantity of sensorial power.

Those fits generally commence during sleep, from whence I suppose they have been thought to have some connection with sleep, and have thence been termed Somnambulism; but their commencement during sleep is owing to our increased excitability by internal sensations at that time, as explained in Sect. XVIII. 14. and 15., and not to any similitude between reverie and sleep.

3. I was once concerned for a very elegant and ingenious young lady, who had a reverie on alternate days, which continued nearly the whole day; and as in her days of disease she took up the same kind of ideas, which she had conversed about on the alternate day before, and could recollect nothing of them on her well-day; she appeared to her friends to possess two minds. This case also was of epileptic kind, and was cured, with some relapses, by opium administered before the commencement of the paroxysm.

4. Whence it appears, that the methods of relieving inflammatory pains, is by removing all stimulus, as by venesection, cool air, mucilaginous diet, aqueous potation, silence, darkness.

The methods of relieving pains from defect of stimulus is by supplying the peculiar stimulus required, as of food, or warmth.

And the general method of relieving pain is by exciting into action some great part of the system for the purpose of expending a part of the sensorial power. This is done either by exertion of the voluntary ideas and muscles, as in insanity and convulsion; or by exerting both voluntary and sensitive motions, as in reverie; or by exciting the irritative motions by wine or opium internally, and by the warm bath or blisters externally; or lastly, by exciting the sensitive ideas by good news, affecting stories, or agreeable passions.

SECT. XXXV
DISEASES OF ASSOCIATION

I. 1. *Sympathy or consent of parts. Primary and secondary parts of an associated train of motions reciprocally affect each other. Parts of irritative trains of motion affect each other in four ways. Sympathies of the skin and stomach. Flushing of the face after a meal. Eruption of the small-pox on the face. Chilness after a meal. 2. Vertigo from intoxication. 3. Absorption from the lungs and pericardium by emetics. In vomiting the actions of the stomach are decreased, not increased. Digestion strengthened after an emetic. Vomiting from deficiency of sensorial power. 4. Dyspnœa from cold bathing. Slow pulse from digitalis. Death from gout in the stomach. II. 1. Primary and secondary parts of sensitive associations affect each other. Pain from gall-stone, from urinary stone, Hemicrania. Painful epilepsy. 2. Gout and red face from inflamed liver. Shingles from inflamed kidney. 3. Coryza from cold applied to the feet. Pleurisy. Hepatitis. 4. Pain of shoulders from inflamed liver. III. Diseases from the associations of ideas.*

I. 1. Many synchronous and successive motions of our muscular fibres, and of our organs of sense, or ideas, become associated so as to form indissoluble tribes or trains of action, as shewn in Section X. on Associate Motions. Some constitutions more easily establish these associations, whether by voluntary, sensitive, or irritative repetitions, and some more easily lose them again, as shewn in Section XXXI. on Temperaments.

When the beginning of such a train of actions becomes by any means disordered, the succeeding part is liable to become disturbed in consequence, and this is commonly termed sympathy or consent of parts by the writers of medicine. For the more clear understanding of these sympathies we must consider a tribe or train of actions as divided into two parts, and call one of them the primary or original motions, and the other the secondary or sympathetic ones.

The primary and secondary parts of a train of irritative actions may reciprocally affect each other in four different manners. 1. They may both be exerted with greater energy than natural. 2. The former may act with greater, and the latter with less energy. 3. The former may act with less, and the latter with greater energy. 4. They may both act with less energy than natural. I shall now give an example of each kind of these modes of action, and endeavour to shew, that though the primary and secondary parts of these trains or tribes of motion are connected by irritative association, or their previous habits of acting together, as described in Sect. XX. on Vertigo. Yet that their acting with similar or dissimilar degrees of energy, depends on the greater or less quantity of sensorial power, which the primary part of the train expends in its exertions.

The actions of the stomach constitute so important a part of the associations of both irritative and sensitive motions, that it is said to sympathize with almost every part of the body; the first example, which I shall adduce to shew that both the primary and secondary parts of a train of irritative associations of motion act with increased energy, is taken from the consent of the skin with this organ. When the action of the fibres of the stomach is increased, as by the stimulus of a full meal, the exertions of the cutaneous arteries of the face become increased by their irritative associations with those of the stomach, and a glow or flushing of the face succeeds. For the small vessels of the skin of the face having been more accustomed to the varieties of action, from their frequent exposure to various degrees of cold and heat become more easily excited into increased action, than those of the covered parts of our bodies, and thus act with more energy from their irritative or sensitive associations with the stomach. On this account in small-pox the eruption in consequence of the previous affection of the stomach breaks out a day sooner on the face than on the hands, and two days sooner than on the trunk, and recedes in similar times after maturation.

But secondly, in weaker constitutions, that is, in those who possess less sensorial power, so much of it is expended in the increased actions of the fibres of the stomach excited by the stimulus of a meal, that a sense of chilness succeeds instead of the universal glow above mentioned; and thus the secondary part of the associated train of motions is diminished in energy, in consequence of the increased activity of the primary part of it.

2. Another instance of a similar kind, where the secondary part of the train acts with less energy in consequence of the greater exertions of the primary part, is the vertigo attending intoxication; in this circumstance so much sensorial power is expended on the stomach, and on its nearest or more strongly associated motions, as those of the subcutaneous vessels, and probably of the membranes of some internal viscera, that the irritative motions of the retina become imperfectly exerted from deficiency of

sensorial power, as explained in Sect. XX. and XXI. 3. on Vertigo and on Drunkenness, and hence the staggering inebriate cannot completely balance himself by such indistinct vision.

3. An instance of the third circumstance, where the primary part of a train of irritative motions acts with less, and the secondary part with greater energy, may be observed by making the following experiment. If a person lies with his arms and shoulders out of bed, till they become cold, a temporary coryza or catarrh is produced; so that the passage of the nostrils becomes totally obstructed; at least this happens to many people; and then on covering the arms and shoulders, till they become warm, the passage of the nostrils ceases again to be obstructed, and a quantity of mucus is discharged from them. In this case the quiescence of the vessels of the skin of the arms and shoulders, occasioned by exposure to cold air, produces by irritative association an increased action of the vessels of the membrane of the nostrils; and the accumulation of sensorial power during the torpor of the arms and shoulders is thus expended in producing a temporary coryza or catarrh.

Another instance may be adduced from the sympathy or consent of the motions of the stomach with other more distant links of the very extensive tribes or trains of irritative motions associated with them, described in Sect. XX. on Vertigo. When the actions of the fibres of the stomach are diminished or inverted, the actions of the absorbent vessels, which take up the mucus from the lungs, pericardium, and other cells of the body, become increased, and absorb the fluids accumulated in them with greater avidity, as appears from the exhibition of foxglove, antimony, or other emetics in cases of anasarca, attended with unequal pulse and difficult respiration.

That the act of nausea and vomiting is a decreased exertion of the fibres of the stomach may be thus deduced; when an emetic medicine is administered, it produces the pain of sickness, as a disagreeable taste in the mouth produces the pain of nausea; these pains, like that of hunger, or of cold, or like those, which are usually termed nervous, as the head-ach or hemicrania, do not excite the organ into greater action; but in this case I imagine the pains of sickness or of nausea counteract or destroy the pleasurable sensation, which seems necessary to digestion, as shewn in Sect. XXXIII. 1. 1. The peristaltic motions of the fibres of the stomach become enfeebled by the want of this stimulus of pleasurable sensation, and in consequence stop for a time, and then become inverted; for they cannot become inverted without being previously stopped. Now that this inversion of the trains of motion of the fibres of the stomach is owing to the deficiency of pleasurable sensation is evinced from this circumstance, that a nauseous idea excited by words will produce vomiting as effectually us a nauseous drug.

Hence it appears, that the act of nausea or vomiting expends less sensorial power than the usual peristaltic motions of the stomach in the digestion of our aliment; and that hence there is a greater quantity of sensorial power becomes accumulated in the fibres of the stomach, and more of it in consequence to spare for the action of those parts of the system, which are thus associated with the stomach, as of the whole absorbent series of vessels, and which are at the same time excited by their usual stimuli.

From this we can understand, how after the operation of an emetic the stomach becomes more irritable and sensible to the stimulus, and the pleasure of food; since as the sensorial power becomes accumulated during the nausea and vomiting, the digestive power is afterwards exerted more forceably for a time. It should, however, be here remarked, that though vomiting is in general produced by the defect of this stimulus of pleasurable sensation, as when a nauseous drug is administered; yet in long continued vomiting, as in sea-sickness, or from habitual dram-drinking, it arises from deficiency of sensorial power, which in the former case is exhausted by the increased exertion of the irritative ideas of vision, and in the latter by the frequent application of an unnatural stimulus.

4. An example of the fourth circumstance above mentioned, where both the primary and secondary parts of a train of motions proceed with energy less than natural, may be observed in the dyspnœa, which occurs in going into a very cold bath, and which has been described and explained in Sect. XXXII. 3. 2.

And by the increased debility of the pulsations of the heart and arteries during the operation of an emetic. Secondly, from the slowness and intermission of the pulsations of the heart from the incessant efforts to vomit occasioned by an overdose of digitalis. And thirdly, from the total stoppage of the motions of the heart, or death, in consequence of the torpor of the stomach, when affected with the commencement or cold paroxysm of the gout. See Sect. XXV. 17.

II. 1. The primary and secondary parts of the trains of sensitive association reciprocally affect each other in different manners. 1. The increased sensation of the primary part may cease, when that of the secondary part commences. 2. The increased action of the primary part may cease, when that of the secondary part commences. 3. The primary part may have increased sensation, and the secondary part increased action. 4. The primary part may have increased action, and the secondary part increased sensation.

Examples of the first mode, where the increased sensation of the primary part of a train of sensitive association ceases, when that of the secondary part commences, are not unfrequent; as this is the general origin of those pains, which continue some time without being attended with inflammation, such as the pain at the pit of the stomach from a stone at the neck of the gall-bladder, and the pain of strangury in the glans penis from a stone at the neck of the urinary bladder. In both these cases the part, which is affected secondarily, is believed to be much more sensible than the part primarily affected, as described in the catalogue of diseases, Class II. 1. 1. 11. and IV. 2. 2. 2. and IV. 2. 2. 4.

The hemicrania, or nervous headach, as it is called, when it originates from a decaying tooth, is another disease of this kind; as the pain of the carious tooth always ceases, when the pain over one eye and temple commences. And it is probable, that the violent pains, which induce convulsions in painful epilepsies, are produced in the same manner, from a more sensible part sympathizing with a diseased one of less sensibility. See Catalogue of Diseases, Class IV. 2. 2. 8. and III. 1. 1. 6.

The last tooth, or dens sapientiæ, of the upper jaw most frequently decays first, and is liable to produce pain over the eye and temple of that side. The last tooth of the under-jaw is also liable to produce a similar hemicrania, when it begins to decay. When a tooth in the upper-jaw is the cause of the headach, a slighter pain is sometimes perceived on the cheek-bone. And when a tooth in the lower-jaw is the cause of headach, a pain sometimes affects the tendons of the muscles of the neck, which are attached near the jaws. But the clavus hystericus, or pain about the middle of the parietal bone on one side of the head, I have seen produced by the second of the molares, or grinders, of the under-jaw; of which I shall relate the following case. See Class IV. 2. 2. 8.

Mrs. — —, about 30 years of age, was seized with great pain about the middle of the right parietal bone, which had continued a whole day before I saw her, and was so violent as to threaten to occasion convulsions. Not being able to detect a decaying tooth, or a tender one, by examination with my eye, or by striking them with a tea-spoon, and fearing bad consequences from her tendency to convulsion, I advised her to extract the last tooth of the under-jaw on the affected side; which was done without any good effect. She was then directed to lose blood, and to take a brisk cathartic; and after that had operated, about 60 drops of laudanum were given her, with large doses of bark; by which the pain was removed. In about a fortnight she took a cathartic medicine by ill advice, and the pain returned with greater violence in the same place; and, before I could arrive, as she lived 30 miles from me, she suffered a paralytic stroke; which affected her limbs and her face on one side, and relieved the pain of her head.

About a year afterwards I was again called to her on account of a pain as violent as before exactly on the same part of the other parietal bone. On examining her mouth I found the second molaris of the under-jaw on the side before affected was now decayed, and concluded, that this tooth had occasioned the stroke of the palsy by the pain and consequent exertion it had caused. On this account I earnestly entreated her to allow the sound molaris of the same jaw opposite to the decayed one to be extracted; which was forthwith done, and the pain of her head immediately ceased, to the astonishment of her attendants.

In the cases above related of the pain existing in a part distant from the seat of the disease, the pain is owing to defect of the usual motions of the painful part. This appears from the coldness, paleness, and emptiness of the affected vessels, or of the extremities of the body in general, and from there being no tendency to inflammation. The increased action of the primary part of these associated motions, as of the hepatic termination of the bile-duct; from the stimulus of a gall-stone, or of the interior termination of the urethra from the stimulus of a stone in the bladder, or lastly, of a decaying tooth in hemicrania, deprives the secondary part of these associated motions, namely, the exterior terminations of the bile-duct or urethra, or the pained membranes of the head in hemicrania, of their natural share of sensorial power: and hence the secondary parts of these sensitive trains of association become pained from the deficiency of their usual motions, which is accompanied with deficiency of secretions and of heat. See Sect. IV. 5. XII. 5. 3. XXXIV. 1.

Why does the pain of the primary part of the association cease, when that of the secondary part commences? This is a question of intricacy, but perhaps not inexplicable. The pain of the primary part of these associated trains of motion was owing to too great stimulus, as of the stone at the neck of the bladder, and was consequently caused by too great action of the pained part. This greater action than natural of the primary part of these associated motions, by employing or expending the sensorial power of irritation belonging to the whole associated train of motions, occasioned torpor, and consequent pain in the secondary part of the associated train; which was possessed of greater sensibility than the primary part of it. Now the great pain of the secondary part of the train, as soon as it commences, employs or expends the sensorial power of sensation belonging to the whole associated train of motions; and in consequence the motions of the primary part, though increased by the stimulus of an extraneous body, cease to be accompanied with pain or sensation.

If this mode of reasoning be just it explains a curious fact, why when two parts of the body are strongly stimulated, the pain is felt only in one of them, though it is possible by voluntary attention it may be alternately perceived in them both. In the same manner, when two new ideas are presented to us from the stimulus of external bodies, we attend to but one of them at a time. In other words, when one set of fibres, whether of the muscles or organs of sense, contract so strongly as to excite much sensation; another set of fibres contracting more weakly do not excite sensation at all, because the sensorial power of sensation is pre-occupied by the first set of fibres. So we cannot will more than one effect at once, though by associations previously formed we can move many fibres in combination.

Thus in the instances above related, the termination of the bile duct in the duodenum, and the exterior extremity of the urethra, are more sensible than their other terminations. When these parts are deprived of their usual motions by deficiency of sensorial power, as above explained, they become painful according to law the fifth in Section IV. and the less pain originally excited by the stimulus of concreted bile, or of a stone at their other extremities ceases to be perceived. Afterwards, however, when the concretions of bile, or the stone on the urinary bladder, become more numerous or larger, the pain from their increased stimulus becomes greater than the associated pain; and is then felt at the neck of the gall bladder or urinary bladder; and the pain of the glans penis, or at the pit of the stomach, ceases to be perceived.

2. Examples of the second mode, where the increased action of the primary part of a train of sensitive association ceases, when that of the secondary part commences, are also not unfrequent; as this is the usual manner of the translation of inflammations from internal to external parts of the system, such as when an inflammation of the liver or stomach is translated to the membranes of the foot, and forms the gout; or to the skin of the face, and forms the rosy drop; or when an inflammation of the membranes of the kidneys is translated to the skin of the loins, and forms one kind of herpes, called shingles; in these cases by whatever cause the original inflammation may have been produced, as the secondary part of the train of sensitive association is more sensible, it becomes exerted with greater violence than the first part of it; and by both its increased pain, and the increased motion of its fibres, so far diminishes or exhausts the sensorial power of sensation; that the primary part of the train being less sensible ceases both to feel pain, and to act with unnatural energy.

3. Examples of the third mode, where the primary part of a train of sensitive association of motions may experience increased sensation, and the secondary part increased action, are likewise not unfrequent; as it is in

this manner that most inflammations commence. Thus, after standing some time in snow, the feet become affected with the pain of cold, and a common coryza, or inflammation of the membrane of the nostrils, succeeds. It is probable that the internal inflammations, as pleurisies, or hepatitis, which are produced after the cold paroxysm of fever, originate in the same manner from the sympathy of those parts with some others, which were previously pained from quiescence; as happens to various parts of the system during the cold fits of fevers. In these cases it would seem, that the sensorial power of sensation becomes accumulated during the pain of cold, as the torpor of the vessels occasioned by the defect of heat contributes to the increase or accumulation of the sensorial power of irritation, and that both these become exerted on some internal part, which was not rendered torpid by the cold which affected the external parts, nor by its association with them; or which sooner recovered its sensibility. This requires further consideration.

4. An example of the fourth mode, or where the primary part of a sensitive association of motions may have increased action, and the secondary part increased sensation, may be taken from the pain of the shoulder, which attends inflammation of the membranes of the liver, see Class IV. 2. 2. 9.; in this circumstance so much sensorial power seems to be expended in the violent actions and sensations of the inflamed membranes of the liver, that the membranes associated with them become quiescent to their usual stimuli, and painful in consequence.

There may be other modes in which the primary and secondary parts of the trains of associated sensitive motions may reciprocally affect each other, as may be seen by looking over Class IV. in the catalogue of diseases; all which may probably be resolved into the plus and minus of sensorial power, but we have not yet had sufficient observations made upon them with a view to this doctrine.

III. The associated trains of our ideas may have sympathies, and their primary and secondary parts affect each other in some manner similar to those above described; and may thus occasion various curious phenomena not yet adverted to, besides those explained in the Sections on Dreams, Reveries, Vertigo, and Drunkenness; and may thus disturb the deductions of our reasonings, as well as the streams of our imaginations; present us with false degrees of fear, attach unfounded value to trivial circumstances; give occasion to our early prejudices and antipathies; and thus embarrass the happiness of our lives. A copious and curious harvest might be reaped from this province of science, in which, however, I shall not at present wield my sickle.

SECT. XXXVI
OF THE PERIODS OF DISEASES

I. Muscles excited by volition soon cease to contract, or by sensation, or by irritation, owing to the exhaustion of sensorial power. Muscles subjected to less stimulus have their sensorial power accumulated. Hence the periods of some fevers. Want of irritability after intoxication. II. 1. Natural actions catenated with daily habits of life. 2. With solar periods. Periods of sleep. Of evacuating the bowels. 3. Natural actions catenated with lunar periods. Menstruation. Venereal orgasm of animals. Barrenness. III. Periods of diseased animal actions from stated returns of nocturnal cold, from solar and lunar influence. Periods of diurnal fever, hectic fever, quotidian, tertian, quartan fever. Periods of gout, pleurisy, of fevers with arterial debility, and with arterial strength, Periods of rhaphania, of nervous cough, hemicrania, arterial hæmorrhages, hæmorrhoids, hæmoptoe, epilepsy, palsy, apoplexy, madness. IV. Critical days depend on lunar periods. Lunar periods in the small pox.

I. If any of our muscles be made to contract violently by the power of volition, as those of the fingers, when any one hangs by his hands on a swing, fatigue soon ensues; and the muscles cease to act owing to the temporary exhaustion of the spirit of animation; as soon as this is again accumulated in the muscles, they are ready to contract again by the efforts of volition.

Those violent muscular actions induced by pain become in the same manner intermitted and recurrent; as in labour-pains, vomiting, tenesmus, strangury; owing likewise to the temporary exhaustion of the spirit of animation, as above mentioned.

When any stimulus continues long to act with unnatural violence, so as to produce too energetic action of any of our moving organs, those motions soon cease, though the stimulus continues to act; as in looking long on a bright object, as on an inch-square of red silk laid on white paper in the sunshine. See Plate I. in Sect. III. 1.

On the contrary, where less of the stimulus of volition, sensation, or irritation, have been applied to a muscle than usual; there appears to be an accumulation of the spirit of animation in the moving organ; by which it is liable to act with greater energy from less quantity of stimulus, than was previously necessary to excite it into so great action; as after having been immersed in snow the cutaneous vessels of our hands are excited into stronger action by the stimulus of a less degree of heat, than would previously have produced that effect.

From hence the periods of some fever-fits may take their origin, either simply, or by their accidental coincidence with lunar and solar periods, or with the diurnal periods of heat and cold, to be treated of below; for during the cold fit at the commencement of a fever, from whatever cause that cold fit may have been induced, it follows, 1. That the spirit of animation must become accumulated in the parts, which exert during this cold fit less than their natural quantity of action. 2. If the cause producing the cold fit does not increase, or becomes diminished; the parts before benumbed or inactive become now excitable by smaller stimulus, and are thence thrown into more violent action than is natural; that is a hot fit succeeds the cold one. 3. By the energetic action of the system during the hot fit, if it continues long, an exhaustion of the spirit of animation takes place; and another cold fit is liable to succeed, from the moving system not being excitable into action from its usual stimulus. This inirritability of the system from a too great previous stimulus, and consequent exhaustion of sensorial power, is the cause of the general debility, and sickness, and head-ach, some hours after intoxication. And hence we see one of the causes of the periods of fever-fits; which however are frequently combined with the periods of our diurnal habits, or of heat and cold, or of solar or lunar periods.

When besides the tendency to quiescence occasioned by the expenditure of sensorial power during the hot fit of fever, some other cause of torpor, as the solar or lunar periods, is necessary to the introduction of a second cold fit; the fever becomes of the intermittent kind; that is, there is a space of time intervenes between the end of the hot fit, and the commencement of the next cold one. But where no exteriour cause is necessary to the introduction of the second cold fit; no such interval of health intervenes; but the second cold fit commences, as soon as the sensorial power is sufficiently exhausted by the hot fit; and the fever becomes continual.

II. 1. The following are natural animal actions, which are frequently catenated with our daily habits of life, as well as excited by their natural irritations. The periods of hunger and thirst become catenated with certain portions of time, or degrees of exhaustion, or other diurnal habits of life. And if the pain of hunger be not relieved by taking food at the usual time, it is liable to cease till the next period of time or other habits recur; this is not only true in respect to our general desire of food, but the kinds of it also are governed by this periodical habit; insomuch that beer taken to breakfast will disturb the digestion of those, who have been accustomed to tea; and tea taken at dinner will disagree with those, who have been accustomed to beer. Whence it happens, that those, who have weak stomachs, will be able to digest more food, if they take their meals at regular hours; because they have both the stimulus of the aliment they take, and the periodical habit, to assist their digestion.

The periods of emptying the bladder are not only dependent on the acrimony or distention of the water in it, but are frequently catenated with external cold applied to the skin, as in cold bathing, or washing the hands; or with other habits of life, as many are accustomed to empty the bladder before going to bed, or into the house after a journey, and this whether it be full or not.

Our times of respiration are not only governed by the stimulus of the blood in the lungs, or our desire of fresh air, but also by our attention to the hourly objects before us. Hence when a person is earnestly contemplating an idea of grief, he forgets to breathe, till the sensation in his lungs becomes very urgent; and then a sigh succeeds for the purpose of more forceably pushing forwards the blood, which is accumulated in the lungs.

Our times of respiration are also frequently governed in part by our want of a steady support for the actions of our arms, and hands, as in threading a needle, or hewing wood, or in swimming; when we are intent upon these objects, we breathe at the intervals of the exertion of the pectoral muscles.

2. The following natural animal actions are influenced by solar periods. The periods of sleep and of waking depend much on the solar period, for we are inclined to sleep at a certain hour, and to awake at a certain hour, whether we have had more or less fatigue during the day, if within certain limits; and are liable to wake at a certain hour, whether we went to bed earlier or later, within certain limits. Hence it appears, that those who complain of want of sleep, will be liable to sleep better or longer, if they accustom themselves to go to rest, and to rise, at certain hours.

The periods of evacuating the bowels are generally connected with some part of the solar day, as well as with the acrimony or distention occasioned by the feces. Hence one method of correcting costiveness is by endeavouring to establish a habit of evacuation at a certain hour of the day, as recommended by Mr. Locke, which may be accomplished by using daily voluntary efforts at those times, joined with the usual stimulus of the material to be evacuated.

3. The following natural animal actions are connected with lunar periods. 1. The periods of female menstruation are connected with lunar periods to great exactness, in some instances even to a few hours. These do not commence or terminate at the full or change, or at any other particular part of the lunation, but after they have commenced at any part of it, they continue to recur at that part with great regularity, unless disturbed by some violent circumstance, as explained in Sect. XXXII. No. 6. their return is immediately caused by deficient venous absorption, which is owing to the want of the stimulus, designed by nature, of amatorial copulation, or of the growing fetus. When the catamenia returns sooner than the period of lunation, it shows a tendency of the constitution to inirritability; that is to debility, or deficiency of sensorial power, and is to be relieved by small doses of steel and opium.

The venereal orgasm of birds and quadrupeds seems to commence, or return about the most powerful lunations at the vernal or autumnal equinoxes; but if it be disappointed of its object, it is said to recur at monthly periods; in this respect resembling the female catamenia. Whence it is believed, that women are more liable to become pregnant at or about the time of their catamenia, than at the intermediate times; and on this account they are seldom much mistaken in their reckoning of nine lunar periods from the last menstruation; the inattention to this may sometimes have been the cause of supposed barrenness, and is therefore worth the observation of those, who wish to have children.

III. We now come to the periods of diseased animal actions. The periods of fever-fits, which depend on the stated returns of nocturnal cold, are discussed in Sect. XXXII. 3. Those, which originate or recur at solar or lunar periods, are also explained in Section XXXII. 6. These we shall here enumerate; observing, however, that it is not more surprising, that the influence of the varying attractions of the sun and moon, should raise the ocean into mountains, than that it should affect the nice sensibilities of animal bodies; though the manner of its operation on them is difficult to be understood. It is probable however, that as this influence gradually lessens during the course of the day, or of the lunation, or of the year, some actions of our system become less and less; till at length a total quiescence of some part is induced; which is the commencement of the paroxysms of fever, of menstruation, of pain with decreased action of the affected organ, and of consequent convulsion.

1. A diurnal fever in some weak people is distinctly observed to come on towards evening, and to cease with a moist skin early in the morning, obeying the solar periods. Persons of weak constitutions are liable to get into better spirits at the access of the hot fit of this evening fever; and are thence inclined to sit up late; which by further enfeebling them increases the disease; whence they lose their strength and their colour.

2. The periods of hectic fever, supposed to arise from absorption of matter, obeys the diurnal periods like the above, having the exacerbescence towards evening, and its remission early in the morning, with sweats, or diarrhœa, or urine with white sediment.

3. The periods of quotidian fever are either catenated with solar time, and return at the intervals of twenty-four hours; or with lunar time, recurring at the intervals of about twenty-five hours. There is great use in knowing with what circumstances the periodical return or new morbid motions are conjoined, as the most effectual times of exhibiting the proper medicines are thus determined. So if the torpor, which ushers in an ague fit, is catenated with the lunar day: it is known, when the bark or opium must be given, so as to exert its principal effect about the time of the expected return. Solid opium should be given about an hour before the expected cold fit; liquid opium and wine about half an hour; the bark repeatedly for six or eight hours previous to the expected return.

4. The periods of tertian fevers, reckoned from the commencement of one cold fit to the commencement of the next cold fit, recur with solar intervals of forty-eight hours, or with lunar ones of about fifty hours. When these of recurrence begin one or two hours earlier than the solar period, it shews, that the torpor or cold fit is produced by less external influence; and therefore that it is more liable to degenerate into a fever with only remissions; so when menstruation recurs sooner than the period of lunation, it shews a tendency of the habit to torpor of inirritability.

5. The periods of quartan fevers return at solar intervals of seventy-two hours, or at lunar ones of about seventy-four hours and an half. This kind of ague appears most in moist cold autumns, and in cold countries replete with marshes. It is attended with greater debility, and its cold access more difficult to prevent. For where there is previously a deficiency of sensorial power, the constitution is liable to run into greater torpor from any further diminution of it; two ounces of bark and some steel should be given on the

day before the return of the cold paroxysm, and a pint of wine by degrees a few hours before its return, and thirty drops of laudanum one hour before the expected cold fit.

6. The periods of the gout generally commence about an hour before sun-rise, which is usually the coldest part of the twenty-four hours. The greater periods of the gout seem also to observe the solar influence, returning about the same season of the year.

7. The periods of the pleurisy recur with exacerbation of the pain and fever about sun-set, at which time venesection is of most service. The same may be observed of the inflammatory rheumatism, and other fevers with arterial strength, which seem to obey solar periods; and those with debility seem to obey lunar ones.

8. The periods of fevers with arterial debility seem to obey the lunar day, having their access daily nearly an hour later; and have sometimes two accesses in a day, resembling the lunar effects upon the tides.

9. The periods of rhaphania, or convulsions of the limbs from rheumatic pains, seem to be connected with solar influence, returning at nearly the same hour for weeks together, unless disturbed by the exhibition of powerful doses of opium.

So the periods of Tussis ferina, or violent cough with slow pulse, called nervous cough, recurs by solar periods. Five grains of opium, given at the time the cough commenced disturbed the period, from seven in the evening to eleven, at which time it regularly returned for some days, during which time the opium was gradually omitted. Then 120 drops of laudanum were given an hour before the access of the cough, and it totally ceased. The laudanum was continued a fortnight, and then gradually discontinued.

10. The periods of hemicrania, and of painful epilepsy, are liable to obey lunar periods, both in their diurnal returns, and in their greater periods of weeks, but are also induced by other exciting causes.

11. The periods of arterial hæmorrhages seem to return at solar periods about the same hour of the evening or morning. Perhaps the venous hæmorrhages obey the lunar periods, as the catamenia, and hæmorrhoids.

12. The periods of the hæmorrhoids, or piles, in some recur monthly, in others only at the greater lunar influence about the equinoxes.

13. The periods of hæmoptoe sometimes obey solar influence, recurring early in the morning for several days; and sometimes lunar periods, recurring monthly; and sometimes depend on our hours of sleep. See Class I. 2. 1. 9.

14. Many of the first periods of epileptic fits obey the monthly lunation with some degree of accuracy; others recur only at the most powerful lunations before the vernal equinox, and after the autumnal one; but when the constitution has gained a habit of relieving disagreeable sensations by this kind of exertion, the fit recurs from any slight cause.

15. The attack of palsy and apoplexy are known to recur with great frequency about the equinoxes.

16. There are numerous instances of the effect of the lunations upon the periods of insanity, whence the name of lunatic has been given to those afflicted with this disease.

IV. The critical days, in which fevers are supposed to terminate, have employed the attention of medical philosophers from the days of Hippocrates to the present time. In whatever part of a lunation a fever commences, which owes either its whole cause to solar and lunar influence, or to this in conjunction with other causes; it would seem, that the effect would be the greatest at the full and new moon, as the tides rise highest at those times, and would be the least at the quadratures; thus if a fever-fit should commence at the new or full moon, occasioned by the solar and lunar attraction diminishing some chemical affinity of the particles of blood, and thence decreasing their stimulus on our sanguiferous system, as mentioned in Sect. XXXII. 6. this effect will daily decrease for the first seven days, and will then increase till about the fourteenth day, and will again decrease till about the twenty-first day, and increase again till the end of the lunation. If a fever-fit from the above cause should commence on the seventh day after either lunation, the reverse of the above circumstances would happen. Now it is probable, that those fevers, whose crisis or terminations are influenced by lunations, may begin at one or other of the above times, namely at the changes or quadratures; though sufficient observations have not been made to ascertain this circumstance. Hence I conclude, that the small-pox and measles have their critical days, not governed by the times required for certain chemical changes in the blood, which affect or alter the stimulus of the contagious matter, but from the daily increasing or decreasing effect of this lunar link of catenation, as explained in Section XVII. 3. 3. And as other fevers terminate most frequently about the seventh, fourteenth, twenty-first, or about the end of four weeks, when no medical assistance has disturbed their periods, I conclude, that these crises, or terminations, are governed by periods of the lunations; though we are still ignorant of their manner of operation.

In the distinct small-pox the vestiges of lunation are very apparent, after inoculation a quarter of a lunation precedes the commencement of the fever, another quarter terminates with the complete eruption, another quarter with the complete maturation, and another quarter terminates the complete absorption of a material now rendered inoffensive to the constitution.

SECT. XXXVII
OF DIGESTION, SECRETION, NUTRITION

I. Crystals increase by the greater attraction of their sides. Accretion by chemical precipitations, by welding, by pressure, by agglutination. II. Hunger, digestion, why it cannot be imitated out of the body. Lacteals absorb by animal selection or appetency. III. The glands and pores absorb nutritious particles by animal selection. Organic particles of Buffon. Nutrition applied at the time of elongation of fibres. Like inflammation. IV. It seems easier to have preserved animals than to reproduce them. Old age and death from inirritability. Three causes of this. Original fibres of the organs of sense and muscles unchanged. V. Art of producing long life.

I. The larger crystals of saline bodies may be conceived to arise from the combination of smaller crystals of the same form, owing to the greater attractions of their sides than of their angles. Thus if four cubes were floating in a fluid, whose friction or resistance is nothing, it is certain the sides of these cubes would attract each other stronger than their angles; and hence that these four smaller cubes would so arrange themselves as to produce one larger one.

There are other means of chemical accretion, such as the depositions of dissolved calcareous or siliceous particles, as are seen in the formation of the stalactites of limestone in Derbyshire, or of calcedone in Cornwall. Other means of adhesion are produced by heat and pressure, as in the welding of iron-bars; and other means by simple pressure, as in forcing two pieces of caoutchou, or elastic gum, to adhere; and lastly, by the agglutination of a third substance penetrating the pores of the other two, as in the agglutination of wood by means of animal gluten. Though the ultimate particles of animal bodies are held together during life, as well as after death, by their specific attraction of cohesion, like all other matter; yet it does not appear, that their original organization was produced by chemical laws, and their production and increase must therefore only be looked for from the laws of animation.

II. When the pain of hunger requires relief, certain parts of the material world, which surround us, when applied to our palates, excite into action the muscles of deglutition; and the material is swallowed into the stomach. Here the new aliment becomes mixed with certain animal fluids, and undergoes a chemical process, termed digestion; which however chemistry has not yet learnt to imitate out of the bodies of living animals or vegetables. This process seems very similar to the saccharine process in the lobes of farinaceous seeds, as of barley, when it begins to germinate; except that, along with the sugar, oil and mucilage are also produced; which form the chyle of animals, which is very similar to their milk.

The reason, I imagine, why this chyle-making, or saccharine process, has not yet been imitated by chemical operations, is owing to the materials being in such a situation in respect to warmth, moisture, and motion; that they will immediately change into the vinous or acetous fermentation; except the new sugar be absorbed by the numerous lacteal or lymphatic vessels, as soon as it is produced; which is not easy to imitate in the laboratory.

These lacteal vessels have mouths, which are irritated into action by the stimulus of the fluid, which surrounds them; and by animal selection, or appetency, they absorb such part of the fluid as is agreeable to their palate; those parts, for instance, which are already converted into chyle, before they have time to undergo another change by a vinous or acetous fermentation. This animal absorption of fluid is almost visible to the naked eye in the action of the puncta lacrymalia; which imbibe the tears from the eye, and discharge them again into the nostrils.

III. The arteries constitute another reservoir of a changeful fluid; from which, after its recent oxygenation in the lungs, a further animal selection of various fluids is absorbed by the numerous glands; these select their respective fluids from the blood, which is perpetually undergoing a chemical change; but the selection by these glands, like that of the lacteals, which open their mouths into the digesting aliment in the stomach, is from animal appetency, not from chemical affinity; secretion cannot therefore be imitated in the laboratory, as it consists in a selection of part of a fluid during the chemical change of that fluid.

The mouths of the lacteals, and lymphatics, and the ultimate terminations of the glands, are finer than can easily be conceived; yet it is probable, that the pores, or interstices of the parts, or coats, which constitute these ultimate vessels, may still have greater tenuity; and that these pores from the above analogy must posses a similar power of irritability, and absorb by their living energy the particles of fluid adapted to their purposes, whether to replace the parts abraded or dissolved, or to elongate and enlarge themselves. Not only every kind of gland is thus endued with its peculiar appetency, and selects the material agreeable to its taste from the blood, but every individual pore acquires by animal selection the material, which it wants; and thus nutrition seems to be performed in a manner so similar to secretion; that they only differ in the one retaining, and the other parting again with the particles, which they have selected from the blood.

This way of accounting for nutrition from stimulus, and the consequent animal selection of particles, is much more analogous to other phenomena of the animal microcosm, than by having recourse to the microscopic animalcula, or organic particles of Buffon, and Needham; which being already compounded must themselves require nutritive particles to continue their own existence. And must be liable to undergo a change by our digestive or secretory organs; otherwise mankind would soon resemble by their theory the animals, which they feed upon. He, who is nourished by beef or venison, would in time become horned; and he, who feeds on pork or bacon, would gain a nose proper for rooting into the earth, as well as for the perception of odours.

The whole animal system may be considered as consisting of the extremities of the nerves, or of having been produced from them; if we except perhaps the medullary part of the brain residing in the head and spine, and in the trunks of the nerves. These extremities of the nerves are either of those of locomotion, which are termed muscular fibres; or of those of sensation, which constitute the immediate organs of sense, and which have also their peculiar motions. Now as the fibres, which constitute the bones and membranes, possessed originally sensation and motion; and are liable again to possess them, when they become inflamed; it follows, that those were, when first formed, appendages to the nerves of sensation or locomotion, or were formed from them. And that hence all these solid parts of the body, as they have originally consisted of extremities of nerves, require an apposition of nutritive particles of a similar kind, contrary to the opinion of Buffon and Needham above recited.

Lastly, as all these filaments have possessed, or do possess, the power of contraction, and of consequent inertion or elongation; it seems probable, that the nutritive particles are applied during their times of elongation; when their original constituent particles are removed to a greater distance from each other. For each muscular or sensual fibre may be considered as a row or string of beads; which approach, when in contraction, and recede during its rest or elongation; and our daily experience shews us, that great action emaciates the system, and that it is repaired during rest.

Something like this is seen out of the body; for if a hair, or a single untwisted fibre of flax or silk, be soaked in water; it becomes longer and thicker by the water, which is absorbed into its pores. Now if a hair could be supposed to be thus immersed in a solution of particles similar to those, which compose it; one may imagine, that it might be thus increased in weight and magnitude; as the particles of oak-bark increase the substance of the hides of beasts in the process of making leather. I mention these not as philosophic analogies, but as similes to facilitate our ideas, how an accretion of parts may be effected by animal appetences, or selections, in a manner somewhat similar to mechanical or chemical attractions.

If those new particles of matter, previously prepared by digestion and sanguification, only supply the places of those, which have been abraded by the actions of the system, it is properly termed nutrition. If they are applied to the extremities of the nervous fibrils, or in such quantity as to increase the length or crassitude of them, the body becomes at the same time enlarged, and its growth is increased, as well as its deficiences repaired.

In this last case something more than a simple apposition or selection of particles seems to be necessary; as many parts of the system during its growth are caused to recede from those, with which they were before in contact; as the ends of the bones, or cartilages, recede from each other, as their growth advances: this process resembles inflammation, as appears in ophthalmy, or in the production of new flesh in ulcers, where old vessels are enlarged, and new ones produced; and like that is attended with sensation. In this situation the vessels become distended with blood, and acquire greater sensibility, and may thus be compared to the erection of the penis, or of the nipples of the breasts of women; while new particles become added at the same time; as in the process of nutrition above described.

When only the natural growth of the various parts of the body are produced, a pleasurable sensation attends it, as in youth, and perhaps in those, who are in the progress of becoming fat. When an unnatural growth is the consequence, as in inflammatory diseases, a painful sensation attends the enlargement of the system.

IV. This apposition of new parts, as the old ones disappear, selected from the aliment we take, first enlarges and strengthens our bodies for twenty years, for another twenty years it keeps us in health and vigour, and adds strength and solidity to the system; and then gradually ceases to nourish us properly, and for another twenty years we gradually sink into decay, and finally cease to act, and to exist.

On considering this subject one should have imagined at first view, that it might have been easier for nature to have supported her progeny for ever in health and life, than to have perpetually reproduced them by the wonderful and mysterious process of generation. But it seems our bodies by long habit cease to obey the stimulus of the aliment, which should support us. After we have acquired our height and solidity we make no more new parts, and the system obeys the irritations, sensations, volitions; and associations, with, less and less energy, till the whole sinks into inaction.

Three causes may conspire to render our nerves less excitable, which have been already mentioned, 1. If a stimulus be greater than natural, it produces too great an exertion of the stimulated organ, and in consequence exhausts the spirit of animation; and the moving organ ceases to act, even though the stimulus be continued. And though rest will recruit this exhaustion, yet some degree of permanent injury remains, as is evident after exposing the eyes long to too strong a light. 2. If excitations weaker than natural be applied, so as not to excite the organ into action, (as when small doses of aloe or rhubarb are exhibited,) they may be gradually increased, without exciting the organ into action; which will thus acquire a habit of disobedience to the stimulus; thus by increasing the dose by degrees, great quantities of opium or wine may be taken without intoxication. See Sect. XII. 3. 1.

3. Another mode, by which life is gradually undermined, is when irritative motions continue to be produced in consequence of stimulus, but are not succeeded by sensation; hence the stimulus of contagious matter is not capable of producing fever a second time, because it is not succeeded by sensation. See Sect. XII. 3. 6. And hence, owing to the want of the general pleasurable sensation, which ought to attend digestion and glandular secretion, an irksomeness of life ensues; and, where this is in greater excess, the melancholy of old age occurs, with torpor or debility.

From hence I conclude, that it is probable that the fibrillæ, or moving filaments at the extremities of the nerves of sense, and the fibres which constitute the muscles (which are perhaps the only parts of the system that are endued with contractile life) are not changed, as we advance in years, like the other parts of the body; but only enlarged or elongated with our growth; and in consequence they become less and less excitable into action. Whence, instead of gradually changing the old animal, the generation of a totally new one becomes necessary with undiminished excitability; which many years will continue to acquire new parts, or new solidity, and then losing its excitability in time, perish like its parent.

V. From this idea the art of preserving long health and life may be deduced; which must consist in using no greater stimulus, whether of the quantity or kind of our food and drink, or of external circumstances, such as heat, and exercise, and wakefulness, than is sufficient to preserve us in vigour; and gradually, as we grow old to increase the stimulus of our aliment, as the irritability of our system increases.

The debilitating effects ascribed by the poet MARTIAL to the excessive use of warm bathing in Italy, may with equal propriety be applied to the warm rooms of England; which, with the general excessive stimulus of spirituous or fermented liquors, and in some instances of immoderate venery, contribute to shorten our lives.

> *Balnea, vina, venus, corrumpunt corpora nostra,*
>
> *At faciunt vitam balnea, vina, venus!*
>
> Wine, women, warmth, against our lives combine;
>
> But what is life without warmth, women, wine!

SECT. XXXVIII
OF THE OXYGENATION OF THE BLOOD IN THE LUNGS, AND IN THE PLACENTA

I. Blood absorbs oxygene from the air, whence phosphoric acid changes its colour, gives out heat, and some phlogistic material, and acquires an ethereal spirit, which is dissipated in fibrous motion. II. The placenta is a pulmonary organ like the gills of fish. Oxygenation of the blood from air, from water, by lungs, by gills, by the placenta; necessity of this oxygenation to quadrupeds, to fish, to the fœtus in utero. Placental vessels inserted into the arteries of the mother. Use of cotyledons in cows. Why quadrupeds have not sanguiferous lochia. Oxygenation of the chick in the egg, of feeds. III. The liquor amnii is not excrementitious. It is nutritious. It is found in the esophagus and stomach, and forms the meconium. Monstrous births without heads. Question of Dr. Harvey.

I. From the recent discoveries of many ingenious philosophers it appears, that during respiration the blood imbibes the vital part of the air, called oxygene, through the membranes of the lungs; and that hence respiration may be aptly compared to a slow combustion. As in combustion the oxygene of the atmosphere unites with some phlogistic or inflammable body, and forms an acid (as in the production of vitriolic acid from sulphur, or carbonic acid from charcoal,) giving out at the same time a quantity of the matter of heat; so in respiration the oxygene of the air unites with the phlogistic part of the blood, and probably produces phosphoric or animal acid, changing the colour of the blood from a dark to a bright red; and probably some of the matter of heat is at the same time given out according to the theory of Dr. Crawford. But as the evolution of heat attends almost all chemical combinations, it is probable, that it also attends the secretions of the various fluids from the blood; and that the constant combinations or productions of new fluids by means of the glands constitute the more general source of animal heat; this seems evinced by the universal evolution of the matter of heat in the blush of shame or of anger; in which at the same time an increased secretion of the perspirable matter occurs; and the partial evolution of it from topical inflammations, as in gout or rheumatism, in which there is a secretion of new blood-vessels.

Some medical philosophers have ascribed the heat of animal bodies to the friction of the particles of the blood against the sides of the vessels. But no perceptible heat has ever been produced by the agitation of water, or oil, or quicksilver, or other fluids; except those fluids have undergone at the same time some chemical change, as in agitating milk or wine, till they become sour.

Besides the supposed production of phosphoric acid, and change of colour of the blood, and the production of carbonic acid, there would appear to be something of a more subtile nature perpetually acquired from the atmosphere; which is too fine to be long contained in animal vessels, and therefore requires perpetual renovation; and without which life cannot continue longer than a minute or two; this ethereal fluid is probably secreted from the blood by the brain, and perpetually dissipated in the actions of the muscles and organs of sense.

That the blood acquires something from the air, which is immediately necessary to life, appears from an experiment of Dr. Hare (Philos. Transact. abridged, Vol. III. p. 239.) who found, "that birds, mice, &c. would live as long again in a vessel, where he had crowded in double the quantity of air by a condensing engine, than they did when confined in air of the common density." Whereas if some kind of deleterious vapour only was exhaled from the blood in respiration; the air, when condensed into half its compass, could not be supposed to receive so much of it.

II. Sir Edward Hulse, a physician of reputation at the beginning of the present century, was of opinion, that the placenta was a respiratory organ, like the gills of fish; and not an organ to supply nutriment to the fœtus; as mentioned in Derham's Physico-theology. Many other physicians seem to have espoused the same opinion, as noticed by Haller. Elem. Physiologiæ, T. 1. Dr. Gipson published a defence of this theory in the Medical Essays of Edinburgh, Vol. I. and II. which doctrine is there controverted at large by the late Alexander Monro; and since that time the general opinion has been, that the placenta is an organ of nutrition only, owing perhaps rather to the authority of so great a name, than to the validity of the arguments adduced in its support. The subject has lately been resumed by Dr. James Jeffray, and by Dr. Forester French, in their inaugural dissertations at Edinburgh and at Cambridge; who have defended the contrary opinion in an able and ingenious manner; and from whose Theses I have extracted many of the following remarks.

First, by the late discoveries of Dr. Priestley, M. Lavoisier, and other philosophers, it appears, that the basis of atmospherical air, called oxygene, is received by the blood through the membranes of the lungs; and that by this addition the colour of the blood is changed from a dark to a light red. Secondly, that water possesses oxygene also as a part of its composition, and contains air likewise in its pores; whence the blood of fish receives oxygenc from the water, or from the air it contains, by means of their gills, in the same manner as the blood is oxygenated in the lungs of air-breathing animals; it changes its colour at the same time from a dark to a light red in the vessels of their gills, which constitute a pulmonary organ adapted to the medium in which they live. Thirdly, that the placenta consists of arteries carrying the blood to its extremities, and a vein bringing it back, resembling exactly in structure the lungs and gills above mentioned; and that the blood changes its colour from a dark to a light red in passing through these vessels.

This analogy between the lungs and gills of animals, and the placenta of the fetus, extends through a great variety of other circumstances; thus air-breathing creatures and fish can live but a few minutes without air or water; or when they are confined in such air or water, as has been spoiled by their own respiration; the same happens to the fetus, which, as soon as the placenta is separated from the uterus, must either expand its lungs, and receive air, or die. Hence from the structure, as well as the use of the placenta, it appears to be a respiratory organ, like the gills of fish, by which the blood in the fetus becomes oxygenated.

From the terminations of the placental vessels not being observed to bleed after being torn from the uterus, while those of the uterus effuse a great quantity of florid arterial blood, the terminations of the placental vessels would seem to be inserted into the arterial ones of the mother; and to receive oxygenation from the passing currents of her blood through their coats or membranes; which oxygenation is proved by the change of the colour of the blood from dark to light red in its passage from the placental arteries to the placental vein.

The curious structure of the cavities or lacunæ of the placenta, demonstrated by Mr. J. Hunter, explain this circumstance. That ingenious philosopher has shewn, that there are numerous cavities of lacunæ formed on that side of the placenta, which is in contact with the uterus; those cavities or cells are filled with blood from the maternal arteries, which open into them; which blood is again taken up by the maternal veins, and is thus perpetually changed. While the terminations of the placental arteries and veins are spread in fine reticulation on the sides of these cells. And thus, as the growing fetus requires greater oxygenation, an apparatus is produced resembling exactly the air-cells of the lungs.

In cows, and other ruminating animals, the internal surface of the uterus is unequal like hollow cups, which have been called cotyledons; and into these cavities the prominencies of the numerous placentas, with which the fetus of those animals is furnished, are inserted, and strictly adhere; though they may be extracted without effusion of blood. These inequalities of the uterus, and the numerous placentas in consequence, seem to be designed for the purpose of expanding a greater surface for the terminations of the placental vessels for the purpose of receiving oxygenation from the uterine ones; as the progeny of this class of animals are more completely formed before their nativity, than that of the carnivorous classes, and must thence in the latter weeks of pregnancy require greater oxygenation. Thus calves and lambs can walk about in a few minutes after their birth; while puppies and kittens remain many days without opening their eyes. And though on the separation of the cotyledons of ruminating animals no blood is effused, yet this is owing clearly to the greater power of contraction of their uterine lacunæ or alveoli. See Medical Essays, Vol. V. page 144. And from the same cause they are not liable to a sanguiferous menstruation.

The necessity of the oxygenation of the blood in the fetus is farther illustrated by the analogy of the chick in the egg; which appears to have its blood oxygenated at the extremities of the vessels surrounding the yolk; which are spread on the air-bag at the broad end of the egg, and may absorb oxygene through that moist membrane from the air confined behind it; and which is shewn by experiments in the exhausted receiver to be changeable though the shell.

This analogy may even be extended to the growing seeds of vegetables; which were shewn by Mr. Scheele to require a renovation of the air over the water, in which they were confined. Many vegetable seeds are surrounded with air in their pods or receptacles, as peas, the fruit of staphylea, and lichnis vesicaria; but it is probable, that those seeds, after they are shed, as well as the spawn of fish, by the situation of the former on or near the moist and aerated surface of the earth, and of the latter in the ever-changing and ventilated water, may not be in need of an apparatus for the oxygenation of their first blood, before the leaves of one, and the gills of the other, are produced for this purpose.

III. 1. There are many arguments, besides the strict analogy between the liquor amnii and the albumen ovi, which shew the former to be a nutritive fluid; and that the fetus in the latter months of pregnancy takes it into its stomach; and that in consequence the placenta is produced for some other important purpose.

First, that the liquor amnii is not an excrementitious fluid is evinced, because it is found in greater quantity, when the fetus is young, decreasing after a certain period till birth. Haller asserts, "that in some animals but a small quantity of this fluid remains at the birth. In the eggs of hens it is consumed on the eighteenth day, so that at the exclusion of the chick scarcely any remains. In rabbits before birth there is none." Elem. Physiol. Had this been an excrementitious fluid, the contrary would probably have occurred. Secondly, the skin of the fetus is covered with a whitish crust or pellicle, which would seem to preclude any idea of the liquor amnii being produced by any exsudation of perspirable matter. And it cannot consist of urine, because in brute animals the urachus passes from the bladder to the alantois for the express purpose of carrying off that fluid; which however in the human fetus seems to be retained in the distended bladder, as the feces are accumulated in the bowels of all animals.

2. The nutritious quality of the liquid, which surrounds the fetus, appears from the following considerations. 1. It is coagulable by heat, by nitrous acid, and by spirit of wine, like milk, serum of blood, and other fluids, which daily experience evinces to be nutritious. 2. It has a saltish taste according to the accurate Baron Haller, not unlike the whey of milk, which it even resembles in smell. 3. The white of the egg which constitutes the food of the chick, is shewn to be nutritious by our daily experience; besides the experiment of its nutritious effects mentioned by Dr. Fordyce in his late Treatise on Digestion, p. 178; who adds, that it much resembles the essential parts of the serum of blood.

3. A fluid similar to the fluid, with which the fetus is surrounded, except what little change may be produced by a beginning digestion, is found in the stomach of the fetus; and the white of the egg is found, in the same manner in the stomach of the chick.

Numerous hairs, similar to those of its skin, are perpetually found among the contents of the stomach in new-born calves; which must therefore have licked themselves before their nativity. Blasii Anatom. See Sect. XVI. 2. on Instinct.

The chick in the egg is seen gently to move in its surrounding fluid, and to open and shut its mouth alternately. The same has been observed in puppies. Haller's El. Phys. I. 8. p. 201.

A column of ice has been seen to reach down the œsophagus from the mouth to the stomach in a frozen fetus; and this ice was the liquor amnii frozen.

The meconium, or first fæces, in the bowels of new-born infants evince, that something has been digested; and what could this be but the liquor amnii together with the recrements of the gastric juice and gall, which were necessary for its digestion?

There have been recorded some monstrous births of animals without heads, and consequently without mouths, which seem to have been delivered on doubtful authority, or from inaccurate observation. There are two of such monstrous productions however better attested; one of a human fetus, mentioned by Gipson in the Scots Medical Essays; which having the gula impervious was furnished with an aperture into the wind-pipe, which communicated below into the gullet; by means of which the liquor amnii might be taken into the stomach before nativity without danger of suffocation, while the fetus had no occasion to breathe. The other monstrous fetus is described by Vander Wiel, who asserts, that he saw a monstrous lamb, which had no mouth; but instead of it was furnished with an opening in the lower part of the neck into the stomach. Both these instances evidently favour the doctrine of the fetus being nourished by the mouth; as otherwise there had been no necessity for new or unnatural apertures into the stomach, when the natural ones were deficient?

From these facts and observations we may safely infer, that the fetus in the womb is nourished by the fluid which surrounds it; which during the first period of gestation is absorbed by the naked lacteals; and is afterwards swallowed into the stomach and bowels, when these organs are perfected; and lastly that the placenta is an organ for the purpose of giving due oxygenation to the blood of the fetus; which is more necessary, or at least more frequently necessary, than even the supply of food.

The question of the great Harvey becomes thus easily answered. "Why is not the fetus in the womb suffocated for want of air, when it remains there even to the tenth month without respiration: yet if it be born in the seventh or eighth month, and has once respired, it becomes immediately suffocated for want of air, if its respiration be obstructed?"

For further information on this subject, the reader is referred to the Tentamen Medicum of Dr. Jeffray, printed at Edinburgh in 1786. And it is hoped that Dr. French will some time give his theses on this subject to the public.

SECT. XXXIX
OF GENERATION

Felix, qui causas altà caligine mersas

Pandit, et evolvit tenuissima vincula rerum.

I. Habits of acting and feeling of individuals attend the soul into a future life, and attend the new embryon at the time of its production. The new speck of entity absorbs nutriment, and receives oxygene. Spreads the terminations of its vessels on cells, which communicate with the arteries of the uterus; sometimes with those of the peritoneum. Afterwards it swallows the liquor amnii, which it produces by its irritation from the uterus, or peritoneum. Like insects in the heads of calves and sheep. Why the white of egg is of two consistencies. Why nothing is found in quadrupeds similar to the yolk, nor in most vegetable seeds. II. 1. Eggs of frogs and fish impregnated out of their bodies. Eggs of fowls which are not fecundated, contain only the nutriment for the embryon. The embryon is produced by the male, and the nutriment by the female. Animalcula in semine. Profusion of nature's births. 2. Vegetables viviparous. Buds and bulbs have each a father but no mother. Vessels of the leaf and bud inosculate. The paternal offspring exactly resembles the parent. 3. Insects impregnated for six generations. Polypus branches like buds. Creeping roots. Viviparous flowers. Tænia, volvox. Eve from Adam's rib. Semen not a stimulus to the egg. III. 1. Embryons not originally created within other embryons. Organized matter is not so minute. 2. All the parts of the embryon are not formed in the male parent. Crabs produce their legs, worms produce their heads and tails. In wens, cancers, and inflammations, new vessels are formed. Mules partake of the forms of both parents. Hair and nails grow by elongation, not by distention. 3. Organic particles of Buffon. IV. 1. Rudiment of the embryon a simple living filament, becomes a living ring, and then a living tube. 2. It acquires irritabilities, and sensibilities with new organizations, as in wounded snails, polypi, moths, gnats, tadpoles. Hence new parts are acquired by addition not by distention. 3. All parts of the body grow if not confined. 4. Fetuses deficient at their extremities, or have a duplicature of parts. Monstrous births. Double parts of vegetables. 5. Mules cannot be formed by distention of the seminal ens. 6. Families of animals from a mixture of their orders. Mules imperfect. 7. Animal appetency like chemical affinity. Vis fabricatrix and medicatrix of nature. 8. The changes of animals before and after nativity. Similarity of their structure. Changes in them from lust, hunger, and danger. All warm-blooded animals derived from one living filament. Cold-blooded animals, insects, worms, vegetables, derived also from one living filament. Male animals have teats. Male pigeon gives milk. The world itself generated. The cause of causes. A state of probation and responsibility. V. 1. Efficient cause of the colours of birds eggs, and of hair and feathers, which become white in snowy countries. Imagination of the female colours the egg. Ideas or motions of the retina imitated by the extremities of the nerves of touch, or rete mucosum. 2. Nutriment supplied by the female of three kinds. Her imagination can only affect the first kind. Mules how produced, and mulattoes. Organs of reproduction why deficient in mules. Eggs with double yolks. VI. 1. Various secretions produced by the extremities of the vessels, as in the glands. Contagious matter. Many glands affected by pleasurable ideas, as those which secrete the semen. 2. Snails and worms are hermaphrodite, yet cannot impregnate themselves. Final cause of this. 3. The imagination of the male forms the sex. Ideas, or motions of the nerves of vision or of touch, are imitated by the ultimate extremities of the glands of the testes, which mark the sex. This effect of the imagination belongs only to the male. The sex of the embryon is not owing to accident. 4. Causes of the changes in animals from imagination as in monsters. From the male. From the female. 5. Miscarriages from fear. 6. Power of the imagination of the male over the colour, form, and sex of the progeny. An instance of. 7. Act of generation accompanied with ideas of the male or female form. Art of begetting beautiful children of either sex. VII. Recapitulation. VIII. Conclusion. Of cause and effect. The atomic philosophy leads to a first cause.

I. The ingenious Dr. Hartley in his work on man, and some other philosophers, have been of opinion, that our immortal part acquires during this life certain habits of action or of sentiment, which become for ever indissoluble, continuing after death in a future state of existence; and add, that if these habits are of the malevolent kind, they must render the possessor miserable even in heaven. I would apply this ingenious idea to the generation or production of the embryon, or new animal, which partakes so much of the form and propensities of the parent.

Owing to the imperfection of language the offspring is termed a *new* animal, but is in truth a branch or elongation of the parent; since a part of the embryon-animal is, or was, a part of the parent; and therefore in strict language it cannot be said to be entirely *new* at the time of its production; and therefore it may retain some of the habits of the parent-system.

At the earliest period of its existence the embryon, as secreted from the blood of the male, would seem to consist of a living filament with certain capabilities of irritation, sensation, volition, and association; and also with some acquired habits or propensities peculiar to the parent: the former of these are in common with other animals; the latter seem to distinguish or produce the kind of animal, whether man or quadruped, with the similarity of feature or form to the parent. It is difficult to be conceived, that a living entity can be separated or produced from the blood by the action of a gland; and which shall afterwards become an animal similar to that in whose vessels it is formed; even though we should suppose with some modern theorists, that the blood is alive; yet every other hypothesis concerning generation rests on principles still more difficult to our comprehension.

At the time of procreation this speck of entity is received into an appropriated nidus, in which it must acquire two circumstances necessary to its life and growth; one of these is food or sustenance, which is to be received by the absorbent mouths of its vessels; and the other is that part of atmospherical air, or of water, which by the new chemistry is termed oxygene, and which affects the blood by passing through the coats of the vessels which contain it. The fluid surrounding the embryon in its new habitation, which is called liquor amnii, supplies it with nourishment; and as some air cannot but be introduced into the uterus along with a new embryon, it would seem that this same fluid would for a short time, suppose for a few hours, supply likewise a sufficient quantity of the oxygene for its immediate existence.

On this account the vegetable impregnation of aquatic plants is performed in the air; and it is probable that the honey-cup or nectary of vegetables requires to be open to the air, that the anthers and stigmas of the flower may have food of a more oxygenated kind than the common vegetable sap-juice.

On the introduction of this primordium of entity into the uterus the irritation of the liquor amnii, which surrounds it, excites the absorbent mouths of the new vessels into action; they drink up a part of it, and a pleasurable sensation accompanies this new action; at the same time the chemical affinity of the oxygene acts through the vessels of the rubescent blood; and a previous want, or disagreeable sensation, is relieved by this process.

As the want of this oxygenation of the blood is perpetual, (as appears from the incessant necessity of breathing by lungs or gills,) the vessels become extended by the efforts of pain or desire to seek this necessary object of oxygenation, and to remove the disagreeable sensation, which that want occasions. At the same time new particles of matter are absorbed, or applied to these extended vessels, and they become permanently elongated, as the fluid in contact with them soon loses the oxygenous part, which it at first possessed, which was owing to the introduction of air along with the embryon. These new blood-vessels approach the sides of the uterus, and penetrate with their fine terminations into the vessels of the mother; or adhere to them, acquiring oxygene through their coats from the passing currents of the arterial blood of the mother. See Sect. XXXVIII. 2.

This attachment of the placental vessels to the internal side of the uterus by their own proper efforts appears further illustrated by the many instances of extra-uterine fetuses, which have thus attached or inserted their vessels into the peritoneum; or on the viscera, exactly in the same manner as they naturally insert or attach them to the uterus.

The absorbent vessels of the embryon continue to drink up nourishment from the fluid in which they swim, or liquor amnii; and which at first needs no previous digestive preparation; but which, when the whole apparatus of digestion becomes complete, is swallowed by the mouth into the stomach, and being mixed with saliva, gastric juice, bile, pancreatic juice, and mucus of the intestines, becomes digested, and leaves a recrement, which produces the first feces of the infant, called meconium.

The liquor amnii is secreted into the uterus, as the fetus requires it, and may probably be produced by the irritation of the fetus as an extraneous body; since a similar fluid is acquired from the peritoneum in cases of extra-uterine gestation. The young caterpillars of the gadfly placed in the skins of cows, and the young of the ichneumon-fly placed in the backs of the caterpillars on cabbages, seem to produce their nourishment by their irritating the sides of their nidus. A vegetable secretion and concretion is thus produced on oak-leaves by the gall-insect, and by the cynips in the bedeguar of the rose; and by the young grasshopper on many plants, by which the animal surrounds itself with froth. But in no circumstance is extra-uterine gestation so exactly resembled as by the eggs of a fly, which are deposited in the frontal sinus of sheep and calves. These eggs float in some ounces of fluid collected in a thin pellicle or hydatide. This bag of fluid compresses the optic nerve on one side, by which the vision being less distinct in that eye, the animal turns in perpetual circles towards the side affected, in order to get a more accurate view of objects; for the same reason as in squinting the affected eye is turned away from the object contemplated. Sheep in the warm months keep their noses close to the ground to prevent this fly from so readily getting into their nostrils.

The liquor amnii is secreted into the womb as it is required, not only in respect to quantity, but, as the digestive powers of the fetus become formed, this fluid becomes of a different consistence and quality, till it is exchanged for milk after nativity. Haller. Physiol. V. 1. In the egg the white part, which is analogous to the liquor amnii of quadrupeds, consists of two distinct parts; one of which is more viscid, and probably more difficult of digestion, and more nutritive than the other; and this latter is used in the last week of incubation. The yolk of the egg is a still stronger or more nutritive fluid, which is drawn up into the bowels of the chick just at its exclusion from the shell, and serves it for nourishment for a day or two, till it is able to digest, and has learnt to choose the harder seeds or grains, which are to afford it sustenance. Nothing analogous to this yolk is found in the fetus of lactiferous animals, as the milk is another nutritive fluid ready prepared for the young progeny.

The yolk therefore is not necessary to the spawn of fish, the eggs of insects, or for the seeds of vegetables; as their embryons have probably their food presented to them as soon as they are excluded from their shells, or have extended their roots. Whence it happens that some insects produce a living progeny in the spring and summer, and eggs in the autumn; and some vegetables have living roots or buds produced in the place of seeds, as the polygonum viviparum, and magical onions. See Botanic Garden, p. 11. art. anthoxanthum.

There seems however to be a reservoir of nutriment prepared for some seeds besides their cotyledons or seed-leaves, which may be supposed in some measure analogous to the yolk of the egg. Such are the saccharine juices of apples, grapes and other fruits, which supply nutrition to the seeds after they fall on the ground. And such is the milky juice in the centre of the cocoa-nut, and part of the kernel of it; the same I suppose of all other monocotyledon seeds, as of the palms, grasses, and lilies.

II. 1. The process of generation is still involved in impenetrable obscurity, conjectures may nevertheless be formed concerning some of its circumstances. First, the eggs of fish and frogs are impregnated, after they leave the body of the female; because they are deposited in a fluid, and are not therefore covered with a hard shell. It is however remarkable, that neither frogs nor fish will part with their spawn without the presence of the male; on which account female carp and gold-fish in small ponds, where there are no males, frequently die from the distention of their growing spawn. 2. The eggs of fowls, which are laid without being impregnated, are seen to contain only the yolk and white, which are evidently the food or sustenance for the future chick. 3. As the cicatricula of these eggs is given by the cock, and is evidently the rudiment of the new animal; we may conclude, that the embryon is produced by the male, and the proper food and nidus by the female. For if the female be supposed to form an equal part of the embryon, why should she form the whole of the apparatus for nutriment and for oxygenation? the male in many animals is larger, stronger, and digests more food than the female, and therefore should contribute as much or more towards the reproduction of the species; but if he contributes only half the embryon and none of the apparatus for sustenance and oxygenation, the division is unequal; the strength of the male, and his consumption of food are too great for the effect, compared with that of the female, which is contrary to the usual course of nature.

In objection to this theory of generation it may be said, if the animalcula in femine, as seen by the microscope, be all of them rudiments of homunculi, when but one of them can find a nidus, what a waste nature has made of her productions? I do not assert that these moving particles, visible by the microscope, are homunciones; perhaps they may be the creatures of stagnation or putridity, or perhaps no creatures at all; but if they are supposed to be rudiments of homunculi, or embryons, such a profusion of them corresponds with the general efforts of nature to provide for the continuance of her species of animals. Every individual tree produces innumerable seeds, and every individual fish innumerable spawn, in such inconceivable abundance as would in a short space of time crowd the earth and ocean with inhabitants; and these are much more perfect animals than the animalcula in femine can be supposed to be, and perish in uncounted millions. This argument only shews, that the productions of nature are governed by general laws; and that by a wise superfluity of provision she has ensured their continuance.

2. That the embryon is secreted or produced by the male, and not by the conjunction of fluids from both male and female, appears from the analogy of vegetable seeds. In the large flowers, as the tulip, there is no similarity of apparatus between the anthers and the stigma: the seed is produced according to the observations of Spallanzani long before the flowers open, and in consequence long before it can be impregnated, like the egg in the

pullet. And after the prolific dust is shed on the stigma, the seed becomes coagulated in one point first, like the cicatricula of the impregnated egg. See Botanic Garden, Part I. additional note 38. Now in these simple products of nature, if the female contributed to produce the new embryon equally with the male, there would probably have been some visible similarity of parts for this purpose, besides those necessary for the nidus and sustenance of the new progeny. Besides in many flowers the males are more numerous than the females, or than the separate uterine cells in their germs, which would shew, that the office of the male was at least as important as that of the female; whereas if the female, besides producing the egg or seed, was to produce an equal part of the embryon, the office of reproduction would be unequally divided between them.

Add to this, that in the most simple kind of vegetable reproduction, I mean the buds of trees, which are their viviparous offspring, the leaf is evidently the parent of the bud, which rises in its bosom, according to the observation of Linnaeus. This leaf consists of absorbent vessels, and pulmonary ones, to obtain its nutriment, and to impregnate it with oxygene. This simple piece of living organization is also furnished with a power of reproduction; and as the new offspring is thus supported adhering to its father, it needs no mother to supply it with a nidus, and nutriment, and oxygenation; and hence no female leaf has existence.

I conceive that the vessels between the bud and the leaf communicate or inosculate; and that the bud is thus served with vegetable blood, that is, with both nutriment and oxygenation, till the death of the parent-leaf in autumn. And in this respect it differs from the fetus of viviparous animals. Secondly, that then the bark-vessels belonging to the dead-leaf, and in which I suppose a kind of manna to have been deposited, become now the placental vessels, if they may be so called, of the new bud. From the vernal sap thus produced of one sugar-maple-tree in New-York and in Pennsylvania, five or six pounds of good sugar may be made annually without destroying the tree. Account of maple-sugar by B. Rushes. London, Phillips. (See Botanic Garden, Part I. additional note on vegetable placentation.)

These vessels, when the warmth of the vernal sun hatches the young bud, serve it with a saccharine nutriment, till it acquires leaves of its own, and shoots a new system of absorbents down the bark and root of the tree, just as the farinaceous or oily matter in seeds, and the saccharine matter in fruits, serve their embryons with nutriment, till they acquire leaves and roots. This analogy is as forceable in so obscure a subject, as it is curious, and may in large buds, as of the horse-chesnut, be almost seen by the naked eye; if with a penknife the remaining rudiment of the last year's leaf, and of the new bud in its bosom, be cut away slice by slice. The seven ribs of the last year's leaf will be seen to have arisen from the pith in seven distinct points making a curve; and the new bud to have been produced in their centre, and to have pierced the alburnum and cortex, and grown without the assistance of a mother. A similar process may be seen on dissecting a tulip-root in winter; the leaves, which inclosed the last year's flower-stalk, were not necessary for the flower; but each of these was the father of a new bud, which may be now found at its base; and which, as it adheres to the parent, required no mother.

This paternal offspring of vegetables, I mean their buds and bulbs, is attended with a very curious circumstance; and that is, that they exactly resemble their parents, as is observable in grafting fruit-trees, and in propagating flower-roots; whereas the seminal offspring of plants, being supplied with nutriment by the mother, is liable to perpetual variation. Thus also in the vegetable class dioicia, where the male flowers are produced on one tree, and the female ones on another; the buds of the male trees uniformly produce either male flowers, or other buds similar to themselves; and the buds of the female trees produce either female flowers, or other buds similar to themselves; whereas the seeds of these trees produce either male or female plants. From this analogy of the production of vegetable buds without a mother, I contend that the mother does not contribute to the formation of the living ens in animal generation, but is necessary only for supplying its nutriment and oxygenation.

There is another vegetable fact published by M. Koelreuter, which he calls "a complete metamorphosis of one natural species of plants into another," which shews, that in seeds as well as in buds, the embryon proceeds from the male parent, though the form of the subsequent mature plant is in part dependant on the female. M. Koelreuter impregnated a stigma of the nicotiana rustica with the farina of the nicotiana paniculata, and obtained prolific seeds from it. With the plants which sprung from these seeds, he repeated the experiment, impregnating them with the farina of the nicotiana paniculata. As the mule plants which he thus produced were prolific, he continued to impregnate them for many generations with the farina of the nicotiana paniculata, and they became more and more like the male parent, till he at length obtained six plants in every respect perfectly similar to the nicotiana paniculata; and in no respect resembling their female parent the nicotiana rustica. *Blumenbach* on Generation.

3. It is probable that the insects, which are said to require but one impregnation for six generations, as the aphis (see Amenit. Academ.) produce their progeny in the manner above described, that is, without a mother, and not without a father; and thus experience a lucina sine concubitu. Those who have attended to the habits of the polypus, which

is found in the stagnant water of our ditches in July, affirm, that the young ones branch out from the side of the parent like the buds of trees, and after a time separate themselves from them. This is so analogous to the manner in which the buds of trees appear to be produced, that these polypi may be considered as all male animals, producing embryons, which require no mother to supply them with a nidus, or with nutriment, and oxygenation.

This lateral or lineal generation of plants, not only obtains in the buds of trees, which continue to adhere to them, but is beautifully seen in the wires of knot-grass, polygonum aviculare, and in those of strawberries, fragaria vesca. In these an elongated creeping bud is protruded, and, where it touches the ground, takes root, and produces a new plant derived from its father, from which it acquires both nutriment and oxygenation; and in consequence needs no maternal apparatus for these purposes. In viviparous flowers, as those of allium magicum, and polygonum viviparum, the anthers and the stigmas become effete and perish; and the lateral or paternal offspring succeeds instead of seeds, which adhere till they are sufficiently mature, and then fall upon the ground, and take root like other bulbs.

The lateral production of plants by wires, while each new plant is thus chained to its parent, and continues to put forth another and another, as the wire creeps onward on the ground, is exactly resembled by the tape-worm, or tænia, so often found in the bowels, stretching itself in a chain quite from the stomach to the rectum. Linnæus asserts, "that it grows old at one extremity, while it continues to generate young ones at the other, proceeding ad infinitum, like a root of grass. The separate joints are called gourd-worms, and propagate new joints like the parent without end, each joint being furnished with its proper mouth, and organs of digestion." Systema naturæ. Vermes tenia. In this animal there evidently appears a power of reproduction without any maternal apparatus for the purpose of supplying nutriment and oxygenation to the embryon, as it remains attached to its father till its maturity. The volvox globator, which is a transparent animal, is said by Linnæus to bear within it sons and grand-sons to the fifth generation. These are probably living fetuses, produced by the father, of different degrees of maturity, to be detruded at different periods of time, like the unimpregnated eggs of various sizes, which are found in poultry; and as they are produced without any known copulation, contribute to evince, that the living embryon in other orders of animals is formed by the male-parent, and not by the mother, as one parent has the power to produce it.

This idea of the reproduction of animals from a single living filament of their fathers, appears to have been shadowed or allegorized in the curious account in sacred writ of the formation of Eve from a rib of Adam.

From all these analogies I conclude, that the embryon is produced solely by the male, and that the female supplies it with a proper nidus, with sustenance, and with oxygenation; and that the idea of the semen of the male constituting only a stimulus to the egg of the female, exciting it into life, (as held by some philosophers) has no support from experiment or analogy.

III. 1. Many ingenious philosophers have found so great difficulty in conceiving the manner of the reproduction of animals, that they have supposed all the numerous progeny, to have existed in miniature in the animal originally created; and that these infinitely minute forms are only evolved or distended, as the embryon increases in the womb. This idea, besides its being unsupported by any analogy we are acquainted with, ascribes a greater tenuity to organized matter, than we can readily admit; as these included embryons are supposed each of them to consist of the various and complicate parts of animal bodies: they must possess a much greater degree of minuteness, than that which was ascribed to the devils that tempted St. Anthony; of whom 20,000 were said to have been able to dance a saraband on the point of the finest needle without incommoding each other.

2. Others have supposed, that all the parts of the embryon are formed in the male, previous to its being deposited in the egg or uterus; and that it is then only to have its parts evolved or distended as mentioned above; but this is only to get rid of one difficulty by proposing another equally incomprehensible: they found it difficult to conceive, how the embryon could be formed in the uterus or egg, and therefore wished it to be formed before it came thither. In answer to both these doctrines it may be observed, 1st, that some animals, as the crab-fish, can reproduce a whole limb, as a leg which has been broken off; others, as worms and snails, can reproduce a head, or a tail, when either of them has been cut away; and that hence in these animals at least a part can be formed anew, which cannot be supposed to have existed previously in miniature.

Secondly, there are new parts or new vessels produced in many diseases, as on the cornea of the eye in ophthalmy, in wens and cancers, which cannot be supposed to have had a prototype or original miniature in the embryon.

Thirdly, how could mule-animals be produced, which partake of the forms of both the parents, if the original embryon was a miniature existing in the semen of the male parent? if an embryon of the male ass was only expanded, no resemblance to the mare could exist in the mule.

This mistaken idea of the extension of parts seems to have had its rise from the mature man resembling the general form of the fetus; and from thence it was believed, that the parts of the fetus were distended into the man; whereas they have increased 100 times in weight, as well as 100 times in size; now no one will call the additional 99 parts a distention of the original one part in respect to weight. Thus the uterus during pregnancy is greatly enlarged in thickness and solidity as well as in capacity, and hence must have acquired this additional size by accretion of new parts, not by an extension of the old ones; the familiar act of blowing up the bladder of an animal recently slaughtered has led our imaginations to apply this idea of distention to the increase of size from natural growth; which however must be owing to the apposition of new parts; as it is evinced from the increase of weight along with the increase of dimension; and is even visible to our eyes in the elongation of our hair from the colour of its ends; or when it has been dyed on the head; and in the growth of our nails from the specks sometimes observable on them; and in the increase of the white crescent at their roots, and in the growth of new flesh in wounds, which consists of new nerves as well as of new blood-vessels.

3. Lastly, Mr. Buffon has with great ingenuity imagined the existence of certain organic particles, which are supposed to be partly alive, and partly mechanic springs. The latter of these were discovered by Mr. Needham in the milt or male organ of a species of cuttle fish, called calmar; the former, or living animalcula, are found in both male and female secretions, in the infusions of seeds, as of pepper, in the jelly of roasted veal, and in all other animal and vegetable substances. These organic particles he supposes to exist in the spermatic fluids of both sexes, and that they are derived thither from every part of the body, and must therefore resemble, as he supposes, the parts from whence they are derived. These organic particles he believes to be in constant activity, till they become mixed in the womb, and then they instantly join and produce an embryon or fetus similar to the two parents.

Many objections might be adduced to this fanciful theory, I shall only mention two. First, that it is analogous to no known animal laws. And secondly, that as these fluids, replete with organic particles derived both from the male and female organs, are supposed to be similar; there is no reason why the mother should not produce a female embryon without the assistance of the male, and realize the lucina sine concubitu.

IV. 1. I conceive the primordium, or rudiment of the embryon, as secreted from the blood of the parent, to consist of a simple living filament as a muscular fibre; which I suppose to be an extremity of a nerve of loco-motion, as a fibre of the retina is an extremity of a nerve of sensation; as for instance one of the fibrils, which compose the mouth of an absorbent vessel; I suppose this living filament, of whatever form it may be, whether sphere, cube, or cylinder, to be endued with the capability of being excited into action by certain kinds of stimulus. By the stimulus of the surrounding fluid, in which it is received from the male, it may bend into a ring; and thus form the beginning of a tube. Such moving filaments, and such rings, are described by those, who have attended to microscopic animalcula. This living ring may now embrace or absorb a nutritive particle of the fluid, in which it swims; and by drawing it into its pores, or joining it by compression to its extremities, may increase its own length or crassitude; and by degrees the living ring may become a living tube.

2. With this new organization, or accretion of parts, new kinds of irritability may commence; for so long as there was but one living organ, it could only be supposed to possess irritability; since sensibility may be conceived to be an extension of the effect of irritability over the rest of the system. These new kinds of irritability and of sensibility in consequence of new organization, appear from variety of facts in the more mature animal; thus the formation of the testes, and consequent secretion of the semen, occasion the passion of lust; the lungs must be previously formed before their exertions to obtain fresh air can exist; the throat or œsophagus must be formed previous to the sensation or appetites of hunger and thirst; one of which seems to reside at the upper end, and the other at the lower end of that canal.

Thus also the glans penis, when it is distended with blood, acquires a new sensibility, and a new appetency. The same occurs to the nipples of the breasts of female animals, when they are distended with blood, they acquire the new appetency of giving milk. So inflamed tendons and membranes, and even bones, acquire new sensations; and the parts of mutilated animals, as of wounded snails, and polypi, and crabs, are reproduced; and at the same time acquire sensations adapted to their situations. Thus when the head of a snail is reproduced after decollation with a sharp rasor, those curious telescopic eyes are also reproduced, and acquire their sensibility to light, as well as their adapted muscles for retraction on the approach of injury.

With every new change, therefore, of organic form, or addition of organic parts, I suppose a new kind of irritability or of sensibility to be produced; such varieties of irritability or of sensibility exist in our adult state in the glands; every one of which is furnished with an irritability, or a taste, or appetency, and a consequent mode of action peculiar to itself.

In this manner I conceive the vessels of the jaws to produce those of the teeth, those of the fingers to produce the nails, those of the skin to produce the hair; in the same manner as afterwards about the age of puberty the beard and other great changes in the form of the body, and disposition of the mind, are produced in consequence of the new secretion of semen; for if the animal is deprived of this secretion those changes do not take place.

These changes I conceive to be formed not by elongation or distention of primeval stamina, but by apposition of parts; as the mature crab-fish, when deprived of a limb, in a certain space of time has power to regenerate it; and the tadpole puts forth its feet long after its exclusion from the spawn; and the caterpillar in changing into a butterfly acquires a new form, with new powers, new sensations, and new desires.

The natural history of butterflies, and moths, and beetles, and gnats, is full of curiosity; some of them pass many months, and others even years, in their caterpillar or grub state; they then rest many weeks without food, suspended in the air, buried in the earth, or submersed in water; and change themselves during this time into an animal apparently of a different nature; the stomachs of some of them, which before digested vegetable leaves or roots, now only digest honey; they have acquired wings for the purpose of seeking this new food, and a long proboscis to collect it from flowers, and I suppose a sense of smell to detect the secret places in flowers, where it is formed. The moths, which fly by night, have a much longer proboscis rolled up under their chins like a watch spring; which they extend to collect the honey from flowers in their sleeping state; when they are closed, and the nectaries in consequence more difficult to be plundered. The beetle kind are furnished with an external covering of a hard material to their wings, that they may occasionally again make holes in the earth, in which they passed the former state of their existence.

But what most of all distinguishes these new animals is, that they are new furnished with the powers of reproduction; and that they now differ from each other in sex, which does not appear in their caterpillar or grub state. In some of them the change from a caterpillar into a butterfly or moth seems to be accomplished for the sole purpose of their propagation; since they immediately die after this is finished, and take no food in the interim, as the silk-worm in this climate; though it is possible, it might take honey as food, if it was presented to it. For in general it would seem, that food of a more stimulating kind, the honey of vegetables instead of their leaves, was necessary for the purpose of the seminal reproduction of these animals, exactly similar to what happens in vegetables; in these the juices of the earth are sufficient for their purpose of reproduction by buds or bulbs; in which the new plant seems to be formed by irritative motions, like the growth of their other parts, as their leaves or roots; but for the purpose of seminal or amatorial reproduction, where sensation is required, a more stimulating food becomes necessary for the anther, and stigma; and this food is honey; as explained in Sect. XIII. on Vegetable Animation.

The gnat and the tadpole resemble each other in their change from natant animals with gills into aerial animals with lungs; and in their change of the element in which they live; and probably of the food, with which they are supported; and lastly, with their acquiring in their new state the difference of sex, and the organs of seminal or amatorial reproduction. While the polypus, who is their companion in their former state of life, not being allowed to change his form and element, can only propagate like vegetable buds by the same kind of irritative motions, which produces the growth of his own body, without the seminal or amatorial propagation, which requires sensation; and which in gnats and tadpoles seems to require a change both of food and of respiration.

From hence I conclude, that with the acquisition of new parts, new sensations, and new desires, as well as new powers, are produced; and this by accretion to the old ones, and not by distention of them. And finally, that the most essential parts of the system, as the brain for the purpose of distributing the power of life, and the placenta for the purpose of oxygenating the blood, and the additional absorbent vessels for the purpose of acquiring aliment, are first formed by the irritations above mentioned, and by the pleasurable sensations attending those irritations, and by the exertions in consequence of painful sensations, similar to those of hunger and suffocation. After these an apparatus of limbs for future uses, or for the purpose of moving the body in its present natant state, and of lungs for future respiration, and of testes for future reproduction, are formed by the irritations and sensations, and consequent exertions of the parts previously existing, and to which the new parts are to be attached.

3. In confirmation of these ideas it may be observed, that all the parts of the body endeavour to grow, or to make additional parts to themselves throughout our lives; but are restrained by the parts immediately containing them; thus, if the skin be taken away, the fleshy parts beneath soon shoot out new granulations, called by the vulgar proud flesh. If the periosteum be removed, a similar growth commences from the bone. Now in the case of the imperfect embryon, the containing or confining parts are not yet supposed to be formed, and hence there is nothing to restrain its growth.

4. By the parts of the embryon being thus produced by new apportions, many phenomena both of animal and vegetable productions receive an easier explanation; such as that many fetuses are deficient at the extremities, as in a finger or a toe, or in the end of the tongue, or in what is called a hare-lip with deficiency of the palate. For if there should be a deficiency in the quantity of the first nutritive particles laid up in the egg for the reception of the first living filament, the extreme parts, as being last formed, must shew this deficiency by their being imperfect.

This idea of the growth of the embryon accords also with the production of some monstrous births, which consist of a duplicature of the limbs, as chickens with four legs; which could not occur, if the fetus was formed by the distention of an original stamen, or miniature. For if there should be a superfluity of the first nutritive particles laid up in the egg for the first living filament; it is easy to conceive, that a duplicature of some parts may be formed. And that such superfluous nourishment sometimes exists, is evinced by the double yolks in some eggs, which I suppose were thus formed previous to their impregnation by the exuberant nutriment of the hen.

This idea is confirmed by the analogy of the monsters in the vegetable world also; in which a duplicate or triplicate production of various parts of the flower is observable, as a triple nectary in some columbines, and a triple petal in some primroses; and which are supposed to be produced by abundant nourishment.

5. If the embryon be received into a fluid, whose stimulus is different in some degree from the natural, as in the production of mule-animals, the new irritabilities or sensibilities acquired by the increasing or growing organized parts may differ, and thence produce parts not similar to the father, but of a kind belonging in part to the mother; and thus, though the original stamen or living ens was derived totally from the father, yet new irritabilities or sensibilities being excited, a change of form corresponding with them will be produced. Nor could the production of mules exist, if the stamen or miniature of all the parts of the embryon is previously formed in the male semen, and is only distended by nourishment in the female uterus. Whereas this difficulty ceases, if the embryon be supposed to consist of a living filament, which acquires or makes new parts with new irritabilities, as it advances in its growth.

The form, solidity, and colour, of the particles of nutriment laid up for the reception of the first living filament, as well as their peculiar kind of stimulus, may contribute to produce a difference in the form, solidity, and colour of the fetus, so as to resemble the mother, as it advances in life. This also may especially happen during the first state of the existence of the embryon, before it has acquired organs, which can change these first nutritive particles, as explained in No. 5. 2. of this Section. And as these nutritive particles are supposed to be similar to those, which are formed for her own nutrition, it follows that the fetus should so far resemble the mother.

This explains, why hereditary diseases may be derived either from the male or female parent, as well as the peculiar form of either of their bodies. Some of these hereditary diseases are simply owing to a deficient activity of a part of the system, as of the absorbent vessels, which open into the cells or cavities of the body, and thus occasion dropsies. Others are at the same time owing to an increase of sensation, as in scrophula and consumption; in these the obstruction of the fluids is first caused by the inirritability of the vessels, and the inflammation and ulcers which succeed, are caused by the consequent increase of sensation in the obstructed part. Other hereditary diseases, as the epilepsy, and other convulsions, consist in too great voluntary exertions in consequence of disagreeable sensation in some particular diseased part. Now as the pains, which occasion these convulsions, are owing to defect of the action of the diseased part, as shewn in Sect. XXXIV. it is plain, that all these hereditary diseases may have their origin either from defective irritability derived from the father, or from deficiency of the stimulus of the nutriment derived from the mother. In either case the effect would be similar; as a scrophulous race is frequently produced among the poor from the deficient stimulus of bad diet, or of hunger; and among the rich, by a deficient irritability from their having been long accustomed to too great stimulus, as of vinous spirit.

6. From this account of reproduction it appears, that all animals have a similar origin, viz. from a single living filament; and that the difference of their forms and qualities has arisen only from the different irritabilities and sensibilities, or voluntarities, or associabilities, of this original living filament; and perhaps in some degree from the different forms of the particles of the fluids, by which it has been at first stimulated into activity. And that from hence, as Linnæus has conjectured in respect to the vegetable world, it is not impossible, but the great variety of species of animals, which now tenant the earth, may have had their origin from the mixture of a few natural orders. And that those animal and vegetable mules, which could continue their species, have done so, and constitute the numerous families of animals and vegetables which now exist; and that those mules, which were produced with imperfect organs of generation, perished without reproduction, according to the observation of Aristotle; and are the animals, which we now call mules. See Botanic Garden, Part II. Note on Dianthus.

Such a promiscuous intercourse of animals is said to exist at this day in New South Wales by Captain Hunter. And that not only amongst the quadrupeds and birds of different kinds, but even amongst the fish, and, as he believes, amongst the vegetables. He speaks of an animal between the opossum and the kangaroo, from the size of a sheep to that of a rat. Many fish seemed to partake of the shark; some with a shark's head and shoulders, and the hind part of a shark; others with a shark's head and the body of a mullet; and some with a shark's head and the flat body of a stingray. Many birds partake of the parrot; some have the head, neck, and bill of a parrot, with long straight feet and legs; others with legs and feet of a parrot, with head and neck of a sea gull. Voyage to South Wales by Captain John Hunter, p. 68.

7. All animals therefore, I contend, have a similar cause of their organization, originating from a single living filament, endued indeed with different kinds of irritabilities and sensibilities, or of animal appetencies; which exist in every gland, and in every moving organ of the body, and are as essential to living organization as chemical affinities are to certain combinations of inanimate matter.

If I might be indulged to make a simile in a philosophical work, I should say, that the animal appetencies are not only perhaps less numerous originally than the chemical affinities; but that like these latter, they change with every new combination; thus vital air and azote, when combined, produce nitrous acid; which now acquires the property of dissolving silver; so with every new additional part to the embryon, as of the throat or lungs, I suppose a new animal appetency to be produced.

In this early formation of the embryon from the irritabilities, sensibilities, and associabilities, and consequent appetencies, the faculty of volition can scarcely be supposed to have had its birth. For about what can the fetus deliberate, when it has no choice of objects? But in the more advanced state of the fetus, it evidently possesses volition; as it frequently changes its attitude, though it seems to sleep the greatest part of its time; and afterwards the power of volition contributes to change or alter many parts of the body during its growth to manhood, by our early modes of exertion in the various departments of life. All these faculties then constitute the vis fabricatrix, and the vis conservatrix, as well as the vis medicatrix of nature, so much spoken of, but so little understood by philosophers.

8. When we revolve in our minds, first, the great changes, which we see naturally produced in animals after their nativity, as in the production of the butterfly with painted wings from the crawling caterpillar; or of the respiring frog from the subnatant tadpole; from the feminine boy to the bearded man, and from the infant girl to the lactescent woman; both which changes may be prevented by certain mutilations of the glands necessary to reproduction.

Secondly, when we think over the great changes introduced into various animals by artificial or accidental cultivation, as in horses, which we have exercised for the different purposes of strength or swiftness, in carrying burthens or in running races; or in dogs, which have been cultivated for strength and courage, as the bull-dog; or for acuteness of his sense or smell, as the hound and spaniel; or for the swiftness of his foot, as the greyhound; or for his swimming in the water, or for drawing snow-sledges, as the rough-haired dogs of the north; or lastly, as a play-dog for children, as the lap-dog; with the changes of the forms of the cattle, which have been domesticated from the greatest antiquity, as camels, and sheep; which have undergone so total a transformation, that we are now ignorant from what species of wild animals they had their origin. Add to these the great changes of shape and colour, which we daily see produced in smaller animals from our domestication of them, as rabbits, or pigeons; or from the difference of climates and even of seasons; thus the sheep of warm climates are covered with hair instead of wool; and the hares and partridges of the latitudes, which are long buried in snow, become white during the winter months; add to these the various changes produced in the forms of mankind, by their early modes of exertion; or by the diseases occasioned by their habits of life; both of which became hereditary, and that through many generations. Those who labour at the anvil, the oar, or the loom, as well as those who carry sedan-chairs, or who have been educated to dance upon the rope, are distinguishable by the shape of their limbs; and the diseases occasioned by intoxication deform the countenance with leprous eruptions, or the body with tumid viscera, or the joints with knots and distortions.

Thirdly, when we enumerate the great changes produced in the species of animals before their nativity; these are such as resemble the form or colour of their parents, which have been altered by the cultivation or accidents above related, and are thus continued to their posterity. Or they are changes produced by the mixture of species as in mules; or changes produced probably by the exuberance of nourishment supplied to the fetus, as in monstrous births with additional limbs; many of these enormities of shape are propagated, and continued as a variety at least, if not as a new species of animal. I have seen a breed of cats with an additional claw on every foot; of poultry also with an additional claw, and with wings to their feet; and of others without rumps. Mr. Buffon mentions a breed of dogs without tails, which are common at Rome and at Naples, which he supposes to have been produced by a custom long established of cutting their tails close off. There are many kinds of pigeons, admired for their peculiarities, which are monsters thus produced and propagated. And to these must be added, the changes produced by the imagination of the male parent, as will be treated of more at large in No. VI. of this Section.

When we consider all these changes of animal form, and innumerable others, which may be collected from the books of natural history; we cannot but be convinced, that the fetus or embryon is formed by apposition of new parts, and not by the distention of a primordial nest of germs, included one within another, like the cups of a conjurer.

Fourthly, when we revolve in our minds the great similarity of structure, which obtains in all the warm-blooded animals, as well quadrupeds, birds, and amphibious animals, as in mankind; from the mouse and bat to the elephant and whale; one is led to conclude, that they have alike been produced from a similar living filament. In some this filament in its advance to maturity has acquired hands and fingers, with a fine sense of touch, as in mankind. In others it has acquired claws or talons, as in tygers and eagles. In others, toes with an intervening web, or membrane, as in seals and geese. In others it has acquired cloven hoofs, as in cows and swine; and whole hoofs in others, as in the horse. While in the bird kind this original living filament has put forth wings instead of arms or legs, and feathers instead of hair. In some it has protruded horns on the forehead instead of teeth in the fore part of the upper jaw; in others tushes instead of horns; and in others beaks instead of either. And all this exactly as is daily seen in the transmutations of the tadpole, which acquires legs and lungs, when he wants them; and loses his tail, when it is no longer of service to him.

Fifthly, from their first rudiment, or primordium, to the termination of their lives, all animals undergo perpetual transformations; which are in part produced by their own exertions in consequence of their desires and aversions, of their pleasures and their pains, or of irritations, or of associations; and many of these acquired forms or propensities are transmitted to their posterity. See Sect. XXXI. 1.

As air and water are supplied to animals in sufficient profusion, the three great objects of desire, which have changed the forms of many animals by their exertions to gratify them, are those of lust, hunger, and security. A great want of one part of the animal world has consisted in the desire of the exclusive possession of the females; and these have acquired weapons to combat each other for this purpose, as the very thick, shield-like, horny skin on the shoulder of the boar is a defence only against animals of his own species, who strike obliquely upwards, nor are his tushes for other purposes, except to defend himself, as he is not naturally a carnivorous animal. So the horns of the stag are sharp to offend his adversary, but are branched for the purpose of parrying or receiving the thrusts of horns similar to his own, and have therefore been formed for the purpose of combating other stags for the exclusive possession of the females; who are observed, like the ladies in the times of chivalry, to attend the car of the victor.

The birds, which do not carry food to their young, and do not therefore marry, are armed with spurs for the purpose of fighting for the exclusive possession of the females, as cocks and quails. It is certain that these weapons are not provided for their defence against other adversaries, because the females of these species are without this armour. The final cause of this contest amongst the males seems to be, that the strongest and most active animal should propagate the species, which should thence become improved.

Another great want consists in the means of procuring food, which has diversified the forms of all species of animals. Thus the nose of the swine has become hard for the purpose of turning up the soil in search of insects and of roots. The trunk of the elephant is an elongation of the nose for the purpose of pulling down the branches of trees for his food, and for taking up water without bending his knees. Beasts of prey have acquired strong jaws or talons. Cattle have acquired a rough tongue and a rough palate to pull off the blades of grass, as cows and sheep. Some birds have acquired harder beaks to crack nuts, as the parrot. Others have acquired beaks adapted to break the harder seeds, as sparrows. Others for the softer seeds of flowers, or the buds of trees, as the finches. Other birds have acquired long beaks to penetrate the moister soils in search of insects or roots, as woodcocks; and others broad ones to filtrate the water of lakes, and to retain aquatic insects. All which seem to have been gradually produced during many generations by the perpetual endeavour of the creatures to supply the want of food, and to have been delivered to their posterity with constant improvement of them for the purposes required.

The third great want amongst animals is that of security, which seems much to have diversified the forms of their bodies and the colour of them; these consist in the means of escaping other animals more powerful than themselves. Hence some animals have acquired wings instead of legs, as the smaller birds, for the purpose of escape. Others great length of fin, or of membrane, as the flying fish, and the bat. Others great swiftness of foot, as the hare. Others have acquired hard or armed shells, as the tortoise and the echinus marinus.

Mr. Osbeck, a pupil of Linnæus, mentions the American frog fish, Lophius Histrio, which inhabits the large floating islands of sea-weed about the Cape of Good Hope, and has fulcra resembling leaves, that the fishes of prey may mistake it for the sea-weed, which it inhabits. Voyage to China, p. 113.

The contrivances for the purposes of security extend even to vegetables, as is seen in the wonderful and various means of their concealing or defending their honey from insects, and their seeds from birds. On the other hand swiftness of wing has been acquired by hawks and swallows to pursue their prey; and a proboscis of admirable structure has been acquired by the bee, the moth, and the humming bird, for the purpose of plundering the nectaries of flowers. All which seem to have been formed by the original living filament, excited into action by the necessities of the creatures, which possess them, and on which their existence depends.

From thus meditating on the great similarity of the structure of the warm-blooded animals, and at the same time of the great changes they undergo both before and after their nativity; and by considering in how minute a portion of time many of the changes of animals above described have been produced; would it be too bold to imagine, that in the great length of time, since the earth began to exist, perhaps millions of ages before the commencement of the history of mankind, would it be too bold to imagine, that all warm-blooded animals have arisen from one living filament, which THE GREAT FIRST CAUSE endued with animality, with the power of acquiring new parts, attended with new propensities, directed by irritations, sensations, volitions, and associations; and thus possessing the faculty of continuing to improve by its own inherent activity, and of delivering down those improvements by generation to its posterity, world without end!

Sixthly, The cold-blooded animals, as the fish-tribes, which are furnished with but one ventricle of the heart, and with gills instead of lungs, and with fins instead of feet or wings, bear a great similarity to each other; but they differ, nevertheless, so much in their general structure from the warm-blooded animals, that it may not seem probable at first view, that the same living filament could have given origin to this kingdom of animals, as to the former. Yet are there some creatures, which unite or partake of both these orders of animation, as the whales and seals; and more particularly the frog, who changes from an aquatic animal furnished with gills to an aerial one furnished with lungs.

The numerous tribes of insects without wings, from the spider to the scorpion, from the flea to the lobster; or with wings, from the gnat and the ant to the wasp and the dragon-fly, differ so totally from each other, and from the red-blooded classes above described, both in the forms of their bodies, and their modes of life; besides the organ of sense, which they seem to possess in their antennæ or horns, to which it has been thought by some naturalists, that other creatures have nothing similar; that it can scarcely be supposed that this nation of animals could have been produced by the same kind of living filament, as the red-blooded classes above mentioned. And yet the changes which many of them undergo in their early state to that of their maturity, are as different, as one animal can be from another. As those of the gnat, which passes his early state in water, and then stretching out his new wings, and expanding his new lungs, rises in the air; as of the caterpillar, and bee-nymph, which feed on vegetable leaves or farina, and at length bursting from their self-formed graves, become beautiful winged inhabitants of the skies, journeying from flower to flower, and nourished by the ambrosial food of honey.

There is still another class of animals, which are termed vermes by Linnæus, which are without feet, or brain, and are hermaphrodites, as worms, leeches, snails, shell-fish, coralline insects, and sponges; which possess the simplest structure of all animals, and appear totally different from those already described. The simplicity of their structure, however, can afford no argument against their having been produced from a living filament as above contended.

Last of all the various tribes of vegetables are to be enumerated amongst the inferior orders of animals. Of these the anthers and stigmas have already been shewn to possess some organs of sense, to be nourished by honey, and to have the power of generation like insects, and have thence been announced amongst the animal kingdom in Sect. XIII. and to these must be added the buds and bulbs which constitute the viviparous offspring of vegetation. The former I suppose to be beholden to a single living filament for their seminal or amatorial procreation; and the latter to the same cause for their lateral or branching generation, which they possess in common with the polypus, tænia, and volvox; and the simplicity of which is an argument in favour of the similarity of its cause.

Linnæus supposes, in the Introduction to his Natural Orders, that very few vegetables were at first created, and that their numbers were increased by their intermarriages, and adds, suadent hæc Creatoris leges a simplicibus ad composita. Many other changes seem to have arisen in them by their perpetual contest for light and air above ground, and for food or moisture beneath the soil. As noted in Botanic Garden, Part II. Note on Cuscuta. Other changes of vegetables from climate, or other causes, are remarked in the Note on Curcuma in the same work. From these one might be led to imagine, that each plant at first consisted of a single bulb or flower to each root, as the gentianella and daisy; and that in the contest for air and light new buds grew on the old decaying flower stem, shooting down their elongated roots to the ground, and that in process of ages tall trees were thus formed, and an individual bulb became a swarm of vegetables. Other plants, which in this contest for light and air were too slender to rise by their own strength, learned by degrees to adhere to their neighbours, either by putting forth roots like the ivy, or by tendrils like the vine, or by spiral contortions like the honeysuckle; or by growing upon them like the misleto, and taking nourishment from their barks; or by only lodging or adhering on them, and deriving nourishment from the air, as tillandsia.

Shall we then say that the vegetable living filament was originally different from that of each tribe of animals above described? And that the productive living filament of each of those tribes was different originally from the other? Or, as the earth and ocean were probably peopled with vegetable productions long before the existence of animals; and many families of these animals long before other families of them, shall we conjecture that one and the same kind of living filaments is and has been the cause of all organic life?

This idea of the gradual formation and improvement of the animal world accords with the observations of some modern philosophers, who have supposed that the continent of America has been raised out of the ocean at a later period of time than the other three quarters of the globe, which they deduce from the greater comparative heights of its mountains, and the consequent greater coldness of its respective climates, and from the less size and strength of its animals, as the tygers and allegators compared with those of Asia or Africa. And lastly, from the less progress in the improvements of the mind of its inhabitants in respect to voluntary exertions.

This idea of the gradual formation and improvement of the animal world seems not to have been unknown to the ancient philosophers. Plato having probably observed the reciprocal generation of inferior animals, as snails and worms, was of opinion, that mankind with all other animals were originally hermaphrodites during the infancy of the world, and were in process of time separated into male and female. The breasts and teats of all male quadrupeds, to which no use can be now assigned, adds perhaps some shadow of probability to this opinion. Linnæus excepts the horse from the male quadrupeds, who have teats; which might have shewn the earlier origin of his exigence; but Mr. J. Hunter asserts, that he has discovered the vestiges of them on his sheath, and has at the same time enriched natural history with a very curious fact concerning the male pigeon; at the time of hatching the eggs both the male and female pigeon undergo a great change in their crops; which thicken and become corrugated, and secrete a kind of milky fluid, which coagulates, and with which alone they for a few days feed their young, and afterwards feed them with this coagulated fluid mixed with other food. How this resembles the breasts of female quadrupeds after the production of their young! and how extraordinary, that the male should at this time give milk as well as the female! See Botanic Garden, Part II. Note on Curcuma.

The late Mr. David Hume, in his posthumous works, places the powers of generation much above those of our boasted reason; and adds, that reason can only make a machine, as a clock or a ship, but the power of generation makes the maker of the machine; and probably from having observed, that the greatest part of the earth has been formed out of organic recrements; as the immense beds of limestone, chalk, marble, from the shells of fish; and the extensive provinces of clay, sandstone, ironstone, coals, from decomposed vegetables; all which have been first produced by generation, or by the secretions of organic life; he concludes that the world itself might have been generated, rather than created; that is, it might have been gradually produced from very small beginnings, increasing by the activity of its inherent principles, rather than by a sudden evolution of the whole by the Almighty fire.—What a magnificent idea of the infinite power of THE GREAT ARCHITECT! THE CAUSE OF CAUSES! PARENT OF PARENTS! ENS ENTIUM!

For if we may compare infinities, it would seem to require a greater infinity of power to cause the causes of effects, than to cause the effects themselves. This idea is analogous to the improving excellence observable in every part of the creation; such as in the progressive increase of the solid or habitable parts of the earth from water; and in the progressive increase of the wisdom and happiness of its inhabitants; and is consonant to the idea of our present situation being a state of probation, which by our exertions we may improve, and are consequently responsible for our actions.

V. 1. The efficient cause of the various colours of the eggs of birds, and of the air and feathers of animals, is a subject so curious, that I shall beg to introduce it in this place. The colours of many animals seem adapted to their purposes of concealing themselves either to avoid danger, or to spring upon their prey. Thus the snake and wild cat, and leopard, are so coloured as to resemble dark leaves and their lighter interstices; birds resemble the colour of the brown ground, or the green hedges, which they frequent; and moths and butterflies are coloured like the flowers which they rob of their honey. Many instances are mentioned of this kind in Botanic Garden, p. 2. Note on Rubia.

These colours have, however, in some instances another use, as the black diverging area from the eyes of the swan; which, as his eyes are placed less prominent than those of other birds, for the convenience of putting down his head under water, prevents the rays of light from being reflected into his eye, and thus dazzling his sight, both in air and beneath the water; which must have happened, if that surface had been white like the rest of his feathers.

There is a still more wonderful thing concerning these colours adapted to the purpose of concealment; which is, that the eggs of birds are so coloured as to resemble the colour of the adjacent objects and their interfaces. The eggs of hedge-birds are greenish with dark spots; those of crows and magpies, which are seen from beneath through wicker nests, are white with dark spots; and those of larks and partridges are russet or brown, like their nests or situations.

A thing still more astonishing is, that many animals in countries covered with snow become white in winter, and are said to change their colour again in the warmer months, as bears, hares, and partridges. Our domesticated animals lose their natural colours, and break into great variety, as horses, dogs, pigeons. The final cause of these colours is easily understood, as they serve some purposes of the animal, but the efficient cause would seem almost beyond conjecture.

First, the choroid coat of the eye, on which the semitransparent retina is expanded, is of different colour in different animals; in those which feed on grass it is green; from hence there would appear some connexion between the colour of the choroid coat and of that constantly painted on the retina by the green grass. Now, when the ground becomes covered with snow, it would seem, that that action of the retina, which is called whiteness, being constantly excited in the eye, may be gradually imitated by the extremities of the nerves of touch, or rete mucosum of the skin. And if it be supposed, that the action of the retina in producing the perception of any colour consists in so disposing its own fibres or surface, as to reflect those coloured rays only, and transmit the others like soap-bubbles; then that part of the retina, which gives us the perception of snow, must at that time be white; and that which gives us the perception of grass, must be green.

Then if by the laws of imitation, as explained in Section XII. 3. 3. and XXXIX. 6. the extremities of the nerves of touch in the rete mucosum be induced into similar action, the skin or feathers, or hair, may in like manner so dispose their extreme fibres, as to reflect white; for it is evident, that all these parts were originally obedient to irritative motions during their growth, and probably continue to be so; that those irritative motions are not liable in a healthy state to be succeeded by sensation; which however is no uncommon thing in their diseased state, or in their infant state, as in plica polonica, and in very young pen-feathers, which are still full of blood.

It was shewn in Section XV. on the Production of Ideas, that the moving organ of sense in some circumstances resembled the object which produced that motion. Hence it may be conceived, that the rete mucosum, which is the extremity of the nerves of touch, may by imitating the motions of the retina become coloured. And thus, like the fable of the camelion, all animals may possess a tendency to be coloured somewhat like the colours they most frequently inspect, and finally, that colours may be thus given to the egg-shell by the imagination of the female parent; which shell is previously a mucous membrane, indued with irritability, without which it could not circulate its fluids, and increase in its bulk. Nor is this more wonderful than that a single idea of imagination mould in an instant colour the whole surface of the body of a bright scarlet, as in the blush of shame, though by a very different process. In this intricate subject nothing but loose analogical conjectures can be had, which may however lead to future discoveries; but certain it is that both the change of the colour of animals to white in the winters of snowy countries, and the spots on birds eggs, must have some efficient cause; since the uniformity of their production shews it cannot arise from a fortuitous concurrence of circumstances; and how is this efficient cause to be detected, or explained, but from its analogy to other animal facts?

2. The nutriment supplied by the female parent in viviparous animals to their young progeny may be divided into three kinds, corresponding with the age of the new creature. 1. The nutriment contained in the ovum as previously prepared for the embryon in the ovary. 2. The liquor amnii prepared for the fetus in the uterus, and in which it swims; and lastly, the milk prepared in the pectoral glands for the new born-child. There is reason to conclude that variety of changes may be produced in the new animal from all these sources of nutriment, but particularly from the first of them..

The organs of digestion and of sanguification in adults, and afterwards those of secretion, prepare or separate the particles proper for nourishment from other combinations of matter, or recombine them into new kinds of matter, proper to excite into action the filaments, which absorb or attract them by animal appetency. In this process we must attend not only to the action of the living filament which receives a nutritive particle to its bosom, but also to the kind of particle, in respect to form, or size, or colour, or hardness, which is thus previously prepared for it by digestion, sanguification, and secretion. Now as the first filament of entity cannot be furnished with the preparative organs above mentioned, the nutritive particles, which are at first to be received by it, are prepared by the mother; and deposited in the ovum ready for its reception. These nutritive particles must be supposed to differ in some respects, when thus prepared by different animals. They may differ in size, solidity, colour, and form; and yet may be sufficiently congenial to the living filament, to which they are applied, as to excite its activity by their stimulus, and its animal appetency to receive them, and to combine them with itself into organization.

By this first nutriment thus prepared for the embryon is not meant the liquor amnii, which is produced afterwards, nor the larger exterior parts of the white of the egg; but the fluid prepared, I suppose, in the ovary of viviparous animals, and that which immediately surrounds the cicatricula of an impregnated egg, and is visible to the eye in a boiled one.

Now these ultimate particles of animal matter prepared by the glands of the mother may be supposed to resemble the similar ultimate particles, which were prepared for her own nourishment; that is, to the ultimate particles of which her own organization consists. And that hence when these become combined with a new embryon, which in its early state is not furnished with stomach, or glands, to alter them; that new embryon will bear some resemblance to the mother.

This seems to be the origin of the compound forms of mules, which evidently partake of both parents, but principally of the male parent. In this production of chimeras the antients seem to have indulged their fancies, whence the sphinxes, griffins, dragons, centaurs, and minotaurs, which are vanished from modern credulity.

It would seem, that in these unnatural conjunctions, when the nutriment deposited by the female was so ill adapted to stimulate the living filament derived from the male into action, and to be received; or embraced by it, and combined with it into organization, as not to produce the organs necessary to life, as the brain, or heart, or stomach, that no mule was produced. Where all the parts necessary to life in these compound animals were formed sufficiently perfect, except the parts of generation, those animals were produced which are now called mules.

The formation of the organs of sexual generation, in contradistinction to that by lateral buds, in vegetables, and in some animals, as the polypus, the tænia, and the volvox, seems the chef d'œuvre, the master-piece of nature; as appears from many flying insects, as in moths and butterflies, who seem to undergo a general change of their forms solely for the purpose of sexual reproduction, and in all other animals this organ is not complete till the maturity of the creature. Whence it happens that, in the copulation of animals of different species, the parts necessary to life are frequently completely formed; but those for the purpose of generation are defective, as requiring a nicer organization; or more exact coincidence of the particles of nutriment to the irritabilities or appetencies of the original living filament. Whereas those mules, where all the parts could be perfectly formed, may have been produced in early periods of time, and may have added to the numbers of our various species of animals, as before observed.

As this production of mules is a constant effect from the conjunction of different species of animals, those between the horse and the female ass always resembling the horse more than the ass; and those, on the contrary, between the male ass and the mare, always resembling the ass more than the mare; it cannot be ascribed to the imagination of the male animal which cannot be supposed to operate so uniformly; but to the form of the first nutritive particles, and to their peculiar stimulus exciting the living filament to select and combine them with itself. There is a similar uniformity of effect in respect to the colour of the progeny produced between a white man, and a black woman, which, if I am well informed, is always of the mulatto kind, or a mixture of the two; which may perhaps be imputed to the peculiar form of the particles of nutriment supplied to the embryon by the mother at the early period of its existence, and their peculiar stimulus; as this effect, like that of the mule progeny above treated of, is uniform and consistent, and cannot therefore be ascribed to the imagination of either of the parents.

Dr. Thunberg observes, in his Journey to the Cape of Good Hope, that there are some families, which have descended from blacks in the female line for three generations. The first generation proceeding from an European, who married a tawny slave, remains tawny, but approaches to a white complexion; but the children of the third generation, mixed with Europeans, become quite white, and are often remarkably beautiful. V. i. p. 112.

When the embryon has produced a placenta, and furnished itself with vessels for selection of nutritious particles, and for oxygenation of them, no great change in its form or colour is likely to be produced by the particles of sustenance it now takes from the fluid, in which it is immersed; because it has now acquired organs to alter or new combine them. Hence it continues to grow, whether this fluid, in which it swims, be formed by the uterus or by any other cavity of the body, as in extra-uterine gestation; and which would seem to be produced by the stimulus of the fetus on the sides of the cavity, where it is found, as mentioned before. And thirdly, there is still less reason to expect any unnatural change to happen to the child after its birth from the difference of the milk it now takes; because it has acquired a stomach, and lungs, and glands, of sufficient power to decompose and recombine the milk; and thus to prepare from it the various kinds of nutritious particles, which the appetencies of the various fibrils or nerves may require.

From all this reasoning I would conclude, that though the imagination of the female may be supposed to affect the embryon by producing a difference in its early nutriment; yet that no such power can affect it after it has obtained a placenta, and other organs; which may select or change the food, which is presented to it either in the liquor amnii, or in the milk. Now as the eggs in pullets, like the seeds in vegetables, are produced gradually, long before they are impregnated, it does not appear how any sudden effect of imagination of the mother at the time of impregnation can produce any considerable change in the nutriment already thus laid up for the expected or desired embryon. And that hence any changes of the embryon, except those uniform ones in the production of mules and mulattoes, more probably depend on the imagination of the male parent. At the same time it seems manifest, that those monstrous births, which consist in some deficiencies only, or some redundancies of parts, originate from the deficiency or redundance of the first nutriment prepared in the ovary, or in the part of the egg immediately surrounding the cicatricula, as described above; and which continues some time to excite the first living filament into action, after the simple animal is completed; or ceases to excite it, before the complete form is accomplished. The former of these circumstances is evinced by the eggs with double yolks, which frequently happen to our domesticated poultry, and which, I believe, are so formed before impregnation, but which would be well worth attending to, both before and after impregnation; as it is probable, something valuable on this subject might be learnt from them. The latter circumstance, or that of deficiency of original nutriment, may be deduced from reverse analogy.

There are, however, other kinds of monstrous births, which neither depend on deficiency of parts, or supernumerary ones; nor are owing to the conjunction of animals of different species; but which appear to be new conformations, or new dispositions of parts in respect to each other, and which, like the variation of colours and forms of our domesticated animals, and probably the sexual parts of all animals, may depend on the imagination of the male parent, which we now come to consider.

VI. 1. The nice actions of the extremities of our various glands are exhibited in their various productions, which are believed to be made by the gland, and not previously to exist as such in the blood.

Thus the glands, which constitute the liver, make bile; those of the stomach make gastric acid; those beneath the jaw, saliva; those of the ears, ear-wax; and the like. Every kind of gland must possess a peculiar irritability, and probably a sensibility, at the early state of its existence; and must be furnished with a nerve of sense, or of motion, to perceive, and to select, and to combine the particles, which compose the fluid it secretes. And this nerve of sense which perceives the different articles which compose the blood, must at least be conceived to be as fine and subtile an organ, as the optic or auditory nerve, which perceive light or sound. See Sect. XIV. 9.

But in nothing is this nice action of the extremities of the blood-vessels so wonderful, as in the production of contagious matter. A small drop of variolous contagion diffused in the blood, or perhaps only by being inserted beneath the cuticle, after a time, (as about a quarter of a lunation,) excites the extreme vessels of the skin into certain motions, which produce a similar contagious material, filling with it a thousand pustules. So that by irritation, or by sensation in consequence of irritation, or by association of motions, a material is formed by the extremities of certain cutaneous vessels, exactly similar to the stimulating material, which caused the irritation, or consequent sensation, or association.

Many glands of the body have their motions, and in consequence their secreted fluids, affected by pleasurable or painful ideas, since they are in many instances influenced by sensitive associations, as well as by the irritations of the particles of the passing blood. Thus the idea of meat, excited in the minds of hungry dogs, by their sense of vision, or of smell, increases the discharge of saliva, both in quantity and viscidity; as is seen in its hanging down in threads from their mouths, as they stand round a dinner-table. The sensations of pleasure, or of pain, of peculiar kinds, excite in the same manner a great discharge of tears; which appear also to be more saline at the time of their secretion, from their inflaming the eyes and eye-lids. The paleness from fear, and the blush of shame, and of joy, are other instances of the effects of painful, or pleasurable sensations, on the extremities of the arterial system.

It is probable, that the pleasurable sensation excited in the stomach by food, as well as its irritation, contributes to excite into action the gastric glands, and to produce a greater secretion of their fluids. The same probably occurs in the secretion of bile; that is, that the pleasurable sensation excited in the stomach, affects this secretion by sensitive association, as well as by irritative association.

And lastly it would seem, that all the glands in the body have their secreted fluids affected, in quantity and quality, by the pleasurable or painful sensations, which produce or accompany those secretions. And that the pleasurable sensations arising from these secretions may constitute the unnamed pleasure of exigence, which is contrary to what is meant by tedium vitæ, or ennui; and by which we sometimes feel ourselves happy, without being able to ascribe it to any mental cause, as after an agreeable meal, or in the beginning of intoxication.

Now it would appear, that no secretion or excretion of fluid is attended with so much agreeable sensation, as that of the semen; and it would thence follow, that the glands, which perform this secretion, are more likely to be much affected by their catenations with pleasurable sensations. This circumstance is certain, that much more of this fluid is produced in a given time, when the object of its exclusion is agreeable to the mind.

2. A forceable argument, which shews the necessity of pleasurable sensation to copulation, is, that the act cannot be performed without it;

it is easily interrupted by the pain of fear or bashfulness; and no efforts of volition or of irritation can effect this process, except such as induce pleasurable ideas or sensations. See Sect. XXXIII. 1. 1.

A curious analogical circumstance attending hermaphrodite insects, as snails and worms, still further illustrates this theory; if the snail or worm could have impregnated itself, there might have been a saving of a large male apparatus; but as this is not so ordered by nature, but each snail and worm reciprocally receives and gives impregnation, it appears, that a pleasurable excitation seems also to have been required.

This wonderful circumstance of many insects being hermaphrodites, and at the same time not having power to impregnate themselves, is attended to by Dr. Lister, in his Exercitationes Anatom. de Limacibus, p. 145; who, amongst many other final causes, which he adduces to account for it, adds, ut tam tristibus et frigidis animalibus majori cum voluptate perficiatur venus.

There is, however, another final cause, to which this circumstance may be imputed: it was observed above, that vegetable buds and bulbs, which are produced without a mother, are always exact resemblances of their parent; as appears in grafting fruit-trees, and in the flower-buds of the dioiceous plants, which are always of the same sex on the same tree; hence those hermaphrodite insects, if they could have produced young without a mother, would not have been, capable of that change or improvement, which is seen in all other animals, and in those vegetables, which are procreated by the male embryon received and nourished by the female. And it is hence probable, that if vegetables could only have been produced by buds and bulbs, and not by sexual generation, that there would not at this time have existed one thousandth part of their present number of species; which have probably been originally mule-productions; nor could any kind of improvement or change have happened to them, except by the difference of soil or climate.

3. I conclude, that the imagination of the male at the time of copulation, or at the time of the secretion of the semen, may so affect this secretion by irritative or sensitive association, as described in No. 5. 1. of this section, as to cause the production of similarity of form and of features, with the distinction of sex; as the motions of the chissel of the turner imitate or correspond with those of the ideas of the artist. It is not here to be understood, that the first living fibre, which is to form an animal, is produced with any similarity of form to the future animal; but with propensities, or appetences, which shall produce by accretion of parts the similarity of form, feature, or sex, corresponding to the imagination of the father.

Our ideas are movements of the nerves of sense, as of the optic nerve in recollecting visible ideas, suppose of a triangular piece of ivory. The fine moving fibres of the retina act in a manner to which I give the name of white; and this action is confined to a defined part of it; to which figure I give the name of triangle. And it is a preceding pleasurable sensation existing in my mind, which occasions me to produce this particular motion of the retina, when no triangle is present. Now it is probable, that the acting fibres of the ultimate terminations of the secreting apertures of the vessels of the testes, are as fine as those of the retina; and that they are liable to be thrown into that peculiar action, which marks the sex of the secreted embryon, by sympathy with the pleasurable motions of the nerves of vision or of touch; that is, with certain ideas of imagination. From hence it would appear, that the world has long been mistaken in ascribing great power to the imagination of the female, whereas from this account of it, the real power of imagination, in the act of generation, belongs solely to the male. See Sect. XII. 3. 3.

It may be objected to this theory, that a man may be supposed to have in his mind, the idea of the form and features of the female, rather than his own, and therefore there should be a greater number of female births. On the contrary, the general idea of our own form occurs to every one almost perpetually, and is termed consciousness of our existence, and thus may effect, that the number of males surpasses that of females. See Sect. XV. 3. 4. and XVIII. 13. And what further confirms this idea is, that the male children most frequently resemble the father in form, or feature, as well as in sex; and the female most frequently resemble the mother, in feature, and form, as well as in sex.

It may again be objected, if a female child sometimes resembles the father, and a male child the mother, the ideas of the father, at the time of procreation, must suddenly change from himself to the mother, at the very instant, when the embryon is secreted or formed. This difficulty ceases when we consider, that it is as easy to form an idea of feminine features with male organs of reproduction, or of male features with female ones, as the contrary; as we conceive the idea of a sphinx or mermaid as easily and as distinctly as of a woman. Add to this, that at the time of procreation the idea of the male organs, and of the female features, are often both excited at the same time, by contact, or by vision.

I ask, in my turn, is the sex of the embryon produced by accident? Certainly whatever is produced has a cause; but when this cause is too minute for our comprehension, the effect is said in common language to happen by chance, as in throwing a certain number on dice. Now what cause can occasionally produce the male or female character of the embryon, but the peculiar actions of those glands, which form the embryon? And what can influence or govern these actions of the gland, but its associations or catenations with other sensitive motions? Nor is this more extraordinary, than that the catenations of irritative motions with the apparent vibrations of objects at sea should produce sickness of the stomach; or that a nauseous story should occasion vomiting.

4. An argument, which evinces the effect of imagination on the first rudiment of the embryon, may be deduced from the production of some peculiar monsters. Such, for instance, as those which have two heads joined to one body, and those which have two bodies joined to one head; of which frequent examples occur amongst our domesticated quadrupeds, and poultry. It is absurd to suppose, that such forms could exist in primordial germs, as explained in No. IV. 4. of this section. Nor is it possible, that such deformities could be produced by the growth of two embryons, or living filaments; which should afterwards adhere together; as the head and tail part of different polypi are said to do (Blumenbach on Generation, Cadel, London); since in that case one embryon, or living filament, must have begun to form one part first, and the other another part first. But such monstrous conformations become less difficult to comprehend, when they are considered as an effect of the imagination, as before explained, on the living filament at the time of its secretion; and that such duplicature of limbs were produced by accretion of new parts, in consequence of propensities, or animal appetencies thus acquired from the male parent.

For instance, I can conceive, if a turkey-cock should behold a rabbit, or a frog, at the time of procreation, that it might happen, that a forcible or even a pleasurable idea of the form of a quadruped might so occupy his imagination, as to cause a tendency in the nascent filament to resemble such a form, by the apposition of a duplicature of limbs. Experiments on the production of mules and monsters would be worthy the attention of a Spallanzani, and might throw much light upon this subject, which at present must be explained by conjectural analogies.

The wonderful effect of imagination, both in the male and female parent, is shewn in the production of a kind of milk in the crops both of the male and female pigeons after the birth of their young, as observed by Mr. Hunter, and mentioned before. To this should be added, that there are some instances of men having had milk secreted in their breasts, and who have given suck to children, as recorded by Mr. Buffon. This effect of imagination, of both the male and female parent, seems to have been attended to in very early times; Jacob is said not only to have placed rods of trees, in part stripped of their bark, so as to appear spotted, but also to have placed spotted lambs before the flocks, at the time of their copulation. Genesis, chap. xxx. verse 40.

5. In respect to the imagination of the mother, it is difficult to comprehend, how this can produce any alteration in the fetus, except by affecting the nutriment laid up for its first reception, as described in No. V. 2. of this section, or by affecting the nourishment or oxygenation with which she supplies it afterwards. Perpetual anxiety may probably affect the secretion of the liquor amnii into the uterus, as it enfeebles the whole system; and sudden fear is a frequent cause of miscarriage; for fear, contrary to joy, decreases for a time the action of the extremities of the arterial system; hence sudden paleness succeeds, and a shrinking or contraction of the vessels of the skin, and other membranes. By this circumstance, I imagine, the terminations of the placental vessels are detached from their adhesions, or insertions, into the membrane of the uterus; and the death of the child succeeds, and consequent miscarriage.

Of this I recollect a remarkable instance, which could be ascribed to no other cause, and which I shall therefore relate in few words. A healthy young woman, about twenty years of age, had been about five months pregnant, and going down into her cellar to draw some beer, was frighted by a servant boy starting up from behind the barrel, where he had concealed himself with design to alarm the maid-servant, for whom he mistook his mistress. She came with difficulty up stairs, began to flood immediately, and miscarried in a few hours. She has since borne several children, nor ever had any tendency to miscarry of any of them.

6. In respect to the power of the imagination of the male over the form, colour, and sex of the progeny, the following instances have fallen under my observation, and may perhaps be found not very unfrequent, if they were more attended to. I am acquainted with a gentleman, who has one child with dark hair and eyes, though his lady and himself have light hair and eyes; and their other four children are like their parents. On observing this dissimilarity of one child to the others he assured me, that he believed it was his own imagination, that produced the difference; and related to me the following story. He said, that when his lady lay in of her third child, he became attached to a daughter of one of his inferior tenants, and offered her a bribe for her favours in vain; and afterwards a greater bribe, and was equally unsuccessful; that the form of this girl dwelt much in his mind for some weeks, and that the next child, which was the dark-ey'd young lady above mentioned, was exceedingly like, in both features and colour, to the young woman who refused his addresses.

To this instance I must add, that I have known two families, in which, on account of an intailed estate in expectation, a male heir was most eagerly desired by the father; and on the contrary, girls were produced to the seventh in one, and to the ninth in another; and then they had each of them a son. I conclude, that the great desire of a male heir by the father produced rather a disagreeable than an agreeable sensation; and that his ideas dwelt more on the fear of generating a female, than on the pleasurable sensations or ideas of his own male form or organs at the time of copulation, or of the secretion of the semen; and that hence the idea of the female character was more present to his mind than that of the male one; till at length in despair of generating a male these ideas ceased, and those of the male character presided at the genial hour.

7. Hence I conclude, that the act of generation cannot exist without being accompanied with ideas, and that a man must have at that time either a general idea of his own male form, or of the form of his male organs; or an idea of the female form, or of her organs; and that this marks the sex, and the peculiar resemblances of the child to either parent. From whence it would appear, that the phalli, which were hung round the necks of the Roman ladies, or worn in their hair, might have effect in producing a greater proportion of male children; and that the calipædia, or art of begetting beautiful children, and of procreating either males or females, may be taught by affecting the imagination of the male-parent; that is, by the fine extremities of the seminal glands, imitating the actions of the organs of sense either of sight or touch. But the manner of accomplishing this cannot be unfolded with sufficient delicacy for the public eye; but may be worth the attention of those, who are seriously interested in the procreation of a male or female child.

Recapitulation.

VII. 1. A certain quantity of nutritive particles are produced by the female parent before impregnation, which require no further digestion, secretion, or oxygenation. Such are seen in the unimpregnated eggs of birds, and in the unimpregnated seed-vessels of vegetables.

2. A living filament is produced by the male, which being inserted amidst these first nutritive particles, is stimulated into action by them; and in consequence of this action, some of the nutritive particles are embraced, and added to the original living filament; in the same manner as common nutrition is performed in the adult animal.

3. Then this new organization, or additional part, becomes stimulated by the nutritive particles in its vicinity, and sensation is now superadded to irritation; and other particles are in consequence embraced, and added to the living filament; as is seen in the new granulations of flesh in ulcers.

By the power of association, or by irritation, the parts already produced continue their motions, and new ones are added by sensation, as above mentioned; and lastly by volition, which last sensorial power is proved to exist in the fetus in its maturer age, because it has evidently periods of activity and of sleeping; which last is another word for a temporary suspension of volition.

The original living filament may be conceived to possess a power of repulsing the particles applied to certain parts of it, as well as of embracing others, which stimulate other parts of it; as these powers exist in different parts of the mature animal; thus the mouth of every gland embraces the particles or fluid, which suits its appetency; and its excretory duct repulses those particles, which are disagreeable to it.

4. Thus the outline or miniature of the new animal is produced gradually, but in no great length of time; because the original nutritive particles require no previous preparation by digestion, secretion, and oxygenation: but require simply the selection and apposition, which is performed by the living filament. Mr. Blumenbach says, that he possesses a human fetus of only five weeks old, which is the size of a common bee, and has all the features of the face, every finger, and every toe, complete; and in which the organs of generation are distinctly seen. P. 76. In another fetus, whose head was not larger than a pea, the whole of the basis of the skull with all its depressions, apertures, and processes, were marked in the most sharp and distinct manner, though without any ossification. Ib.

5. In some cases by the nutriment originally deposited by the mother the filament acquires parts not exactly similar to those of the father, as in the production of mules and mulattoes. In other cases, the deficiency of this original nutriment causes deficiencies of the extreme parts of the fetus, which are last formed, as the fingers, toes, lips. In other cases, a duplicature of limbs are caused by the superabundance of this original nutritive fluid, as in the double yolks of eggs, and the chickens from them with four legs and four wings. But the production of other monsters, as those with two heads, or with parts placed in wrong situations, seems to arise from the imagination of the father being in some manner imitated by the extreme vessels of the seminal glands; as the colours of the spots on eggs, and the change of the colour of the hair and feathers of animals by domestication, may be caused in the same manner by the imagination of the mother.

6. The living filament is a part of the father, and has therefore certain propensities, or appetencies, which belong to him; which may have been gradually acquired during a million of generations, even from the infancy of the habitable earth; and which now possesses such properties, as would render, by the apposition of nutritious particles, the new fetus exactly similar to the father; as occurs in the buds and bulbs of vegetables, and in the polypus, and tænia or tape-worm. But as the first nutriment is supplied by the mother, and therefore resembles such nutritive particles, as have

been used for her own nutriment or growth, the progeny takes in part of the likeness of the mother.

Other similarity of the excitability, or of the form of the male parent, such as the broad or narrow shoulders, or such as constitute certain hereditary diseases, as scrophula, epilepsy, insanity, have their origin produced in one or perhaps two generations; as in the progeny of those who drink much vinous spirits; and those hereditary propensities cease again, as I have observed, if one or two sober generations succeed; otherwise the family becomes extinct.

This living filament from the father is also liable to have its propensities, or appetencies, altered at the time of its production by the imagination of the male parent; the extremities of the seminal glands imitating the motions of the organs of sense; and thus the sex of the embryon is produced; which may be thus made a male or a female by affecting the imagination of the father at the time of impregnation. See Sect. XXXIX. 6. 3. and 7.

7. After the fetus is thus completely formed together with its umbilical vessels and placenta, it is now supplied with a different kind of food, as appears by the difference of consistency of the different parts of the white of the egg, and of the liquor amnii, for it has now acquired organs for digestion or secretion, and for oxygenation, though they are as yet feeble; which can in some degree change, as well as select, the nutritive particles, which are now presented to it. But may yet be affected by the deficiency of the quantity of nutrition supplied by the mother, or by the degree of oxygenation supplied to its placenta by the maternal blood.

The augmentation of the complete fetus by additional particles of nutriment is not accomplished by distention only, but by apposition to every part both external and internal; each of which acquires by animal appetencies the new addition of the particles which it wants. And hence the enlarged parts are kept similar to their prototypes, and may be said to be extended; but their extension must be conceived only as a necessary consequence of the enlargement of all their parts by apposition of new particles.

Hence the new apposition of parts is not produced by capillary attraction, because the whole is extended; whereas capillary attraction would rather tend to bring the sides of flexible tubes together, and not to distend them. Nor is it produced by chemical affinities, for then a solution of continuity would succeed, as when sugar is dissolved in water; but it is produced by an animal process, which is the consequence of irritation, or sensation; and which may be termed animal appetency.

This is further evinced from experiments, which have been instituted to shew, that a living muscle of an animal body requires greater force to break it, than a similar muscle of a dead body. Which evinces, that besides the attraction of cohesion, which all matter possesses, and besides the chemical attractions of affinities, which hold many bodies together, there is an animal adhesion, which adds vigour to these common laws of the inanimate world.

8. At the nativity of the child it deposits the placenta or gills, and by expanding its lungs acquires more plentiful oxygenation from the currents of air, which it must now continue perpetually to respire to the end of its life; as it now quits the liquid element, in which it was produced, and like the tadpole, when it changes into a frog, becomes an aerial animal.

9. As the habitable parts of the earth have been, and continue to be, perpetually increasing by the production of sea-shells and corallines, and by the recrements of other animals, and vegetables; so from the beginning of the existence of this terraqueous globe, the animals, which inhabit it, have constantly improved, and are still in a state of progressive improvement.

This idea of the gradual generation of all things seems to have been as familiar to the ancient philosophers as to the modern ones; and to have given rise to the beautiful hieroglyphic figure of the προτον ωον, or first great egg, produced by NIGHT, that is, whose origin is involved in obscurity, and animated by ερος, that is, by DIVINE LOVE; from whence proceeded all things which exist.

Conclusion.

VIII. 1. Cause and effect may be considered as the progression, or successive motions, of the parts of the great system of Nature. The state of things at this moment is the effect of the state of things, which existed in the preceding moment; and the cause of the state of things, which shall exist in the next moment.

These causes and effects may be more easily comprehended, if motion be considered as a change of the figure of a group of bodies, as proposed in Sect. XIV. 2. 2. inasmuch as our ideas of visible or tangible objects are more distinct, than our abstracted ideas of their motions. Now the change of the configuration of the system of nature at this moment must be an effect of the preceding configuration, for a change of configuration cannot exist without a previous configuration; and the proximate cause of every effect must immediately precede that effect. For example, a moving ivory ball could not proceed onwards, unless it had previously began to proceed; or unless an impulse had been previously given it; which previous motion or impulse constitutes a part of the last situation of things.

As the effects produced in this moment of time become causes in the next, we may consider the progressive motions of objects as a chain of causes only; whose first link proceeded from the great Creator, and which have existed from the beginning of the created universe, and are perpetually proceeding.

2. These causes may be conveniently divided into two kinds, efficient and inert causes, according with the two kinds of entity supposed to exist in the natural world, which may be termed matter and spirit, as proposed in Sect. I. and further treated of in Sect. XIV. The efficient causes of motion, or new configuration, consist either of the principle of general gravitation, which actuates the sun and planets; or of the principle of particular gravitation, as in electricity, magnetism, heat; or of the principle of chemical affinity, as in combustion, fermentation, combination; or of the principle of organic life, as in the contraction of vegetable and animal fibres. The inert causes of motion, or new configuration, consist of the parts of matter, which are introduced within the spheres of activity of the principles above described. Thus, when an apple falls on the ground, the principle of gravitation is the efficient cause, and the matter of the apple the inert cause. If a bar of iron be approximated to a magnet, it may be termed the inert cause of the motion, which brings these two bodies into contact; while the magnetic principle may be termed the efficient cause. In the same manner the fibres, which constitute the retina, may be called the inert cause of the motions of that organ in vision, while the sensorial power may be termed the efficient cause.

3. Another more common distribution of the perpetual chain of causes and effects, which constitute the motions, or changing configurations, of the natural world, is into active and passive. Thus, if a ball in motion impinges against another ball at rest, and communicates its motion to it, the former ball is said to act, and the latter to be acted upon. In this sense of the words a magnet is said to attract iron; and the prick of a spur to stimulate a horse into exertion; so that in this view of the works of nature all things may be said either simply to exist, or to exist as causes, or to exist as effects; that is, to exist either in an active or passive state.

This distribution of objects, and their motions, or changes of position, has been found so convenient for the purposes of common life, that on this foundation rests the whole construction or theory of language. The names of the things themselves are termed by grammarians Nouns, and their modes of existence are termed Verbs. The nouns are divided into substantives, which denote the principal things spoken of; and into adjectives, which denote some circumstances, or less kinds of things, belonging to the former. The verbs are divided into three kinds, such as denote the existence of things simply, as, to be; or their existence in an active state, as, to eat; or their existence in a passive state, as, to be eaten. Whence it appears, that all languages consist only of nouns and verbs, with their abbreviations for the greater expedition of communicating our thoughts; as explained in the ingenious work of Mr. Horne Tooke, who has unfolded by a single flash of light the whole theory of language, which had so long lain buried beneath the learned lumber of the schools. Diversions of Purley. Johnson. London.

4. A third division of causes has been into proximate and remote; these have been much spoken of by the writers on medical subjects, but without sufficient precision. If to proximate and remote causes we add proximate and remote effects, we shall include four links of the perpetual chain of causation; which will be more convenient for the discussion of many philosophical subjects.

Thus if a particle of chyle be applied to the mouth of a lacteal vessel, it may be termed the remote cause of the motions of the fibres, which compose the mouth of that lacteal vessel; the sensorial power is the proximate cause; the contraction of the fibres of the mouth of the vessel is the proximate effect; and their embracing the particle of chyle is the remote effect; and these four links of causation constitute absorption.

Thus when we attend to the rising sun, first the yellow rays of light stimulate the sensorial power residing in the extremities of the optic nerve, this is the remote cause. 2. The sensorial power is excited into a state of activity, this is the proximate cause. 3. The fibrous extremities of the optic nerve are contracted, this is the proximate effect. 4. A pleasurable or painful sensation is produced in consequence of the contraction of these fibres of the optic nerve, this is the remote effect; and these four links of the chain of causation constitute the sensitive idea, or what is commonly termed the sensation of the rising sun.

5. Other causes have been announced by medical writers under the names of causa procatarctica, and causa proegumina, and causa sine quâ non. All which are links more or less distant of the chain of remote causes.

To these must be added the final cause, so called by many authors, which means the motive, for the accomplishment of which the preceding chain of causes was put into action. The idea of a final cause, therefore, includes that of a rational mind, which employs means to effect its purposes; thus the desire of preserving himself from the pain of cold, which he has frequently experienced, induces the savage to construct his hut; the fixing stakes into the ground for walls, branches of trees for rafters, and turf for a cover, are a series of successive voluntary exertions; which are so many means to produce a certain effect. This effect of preserving himself from cold, is termed the final cause; the construction of the hut is the remote effect; the action of the muscular fibres of the man, is the proximate effect; the volition, or activity of desire to preserve himself from cold, is the proximate cause; and the pain of cold, which excited that desire, is the remote cause.

6. This perpetual chain of causes and effects, whose first link is rivetted to the throne of GOD, divides itself into innumerable diverging branches, which, like the nerves arising from the brain, permeate the most minute and most remote extremities of the system, diffusing motion and sensation to the whole. As every cause is superior in power to the effect, which it has produced, so our idea of the power of the Almighty Creator becomes more elevated and sublime, as we trace the operations of nature from cause to cause, climbing up the links of these chains of being, till we ascend to the Great Source of all things.

Hence the modern discoveries in chemistry and in geology, by having traced the causes of the combinations of bodies to remoter origins, as well as those in astronomy, which dignify the present age, contribute to enlarge and amplify our ideas of the power of the Great First Cause. And had those ancient philosophers, who contended that the world was formed from atoms, ascribed their combinations to certain immutable properties received from the hand of the Creator, such as general gravitation, chemical affinity, or animal appetency, instead of ascribing them to a blind chance; the doctrine of atoms, as constituting or composing the material world by the variety of their combinations, so far from leading the mind to atheism, would strengthen the demonstration of the existence of a Deity, as the first cause of all things; because the analogy resulting from our perpetual experience of cause and effect would have thus been exemplified through universal nature.

The heavens declare the glory of GOD, and the firmament sheweth his handywork! One day telleth another, and one night certifieth another; they have neither speech nor language, yet their voice is gone forth into all lands, and their words into the ends of the world. Manifold are thy works, O LORD! in wisdom hast thou made them all. Psal. xix. civ.

SECT. XL

On the OCULAR SPECTRA of Light and Colours, by Dr. R. W. Darwin, of Shrewsbury. Reprinted, by Permission, from the Philosophical Transactions, Vol. LXXVI. p. 313.

Spectra of four kinds. 1. Activity of the retina in vision. 2. Spectra from defect of sensibility. 3. Spectra from excess of sensibility. 4. Of direct ocular spectra. 5. Greater stimulus excites the retina into spasmodic action. 6. Of reverse ocular spectra. 7. Greater stimulus excites the retina into various successive spasmodic actions. 8. Into fixed spasmodic action. 9. Into temporary paralysis. 10. Miscellaneous remarks; 1. Direct and reverse spectra at the same time. A spectral halo. Rule to predetermine the colours of spectra. 2. Variation of spectra from extraneous light. 3. Variation of spectra in number, figure, and remission. 4. Circulation of the blood in the eye is visible. 5. A new way of magnifying objects. Conclusion.

When any one has long and attentively looked at a bright object, as at the setting sun, on closing his eyes, or removing them, an image, which resembles in form the object he was attending to, continues some time to be visible; this appearance in the eye we shall call the ocular spectrum of that object.

These ocular spectra are of four kinds: 1st, Such as are owing to a less sensibility of a defined part of the retina; or *spectra from defect of sensibility.* 2d, Such as are owing to a greater sensibility of a defined part of the retina; or *spectra from excess of sensibility.* 3d, Such as resemble their object in its colour as well as form; which may be termed *direct ocular spectra.* 4th, Such as are of a colour contrary to that of their object; which may be termed *reverse ocular spectra.*

The laws of light have been most successfully explained by the great Newton, and the perception of visible objects has been ably investigated by the ingenious Dr. Berkeley and M. Malebranche; but these minute phenomena of vision have yet been thought reducible to no theory, though many philosophers have employed a considerable degree of attention upon them: among these are Dr. Jurin, at the end of Dr. Smith's Optics;

M. Æpinus, in the Nov. Com. Petropol. V. 10.; M. Beguelin, in the Berlin Memoires, V. II. 1771; M. d'Arcy, in the Histoire de l'Acad. des Scienc. 1765; M. de la Hire; and, lastly, the celebrated M. de Buffon, in the Memoires de l'Acad. des Scien. who has termed them accidental colours, as if subjected to no established laws, Ac. Par. 1743. M. p. 215.

I must here apprize the reader, that it is very difficult for different people to give the same names to various shades of colours; whence, in the following pages, something must be allowed, if on repeating the experiments the colours here mentioned should not accurately correspond with his own names of them.

I. *Activity of the Retina in Vision.*

From the subsequent experiments it appears, that the retina is in an active not in a passive state during the existence of these ocular spectra; and it is thence to be concluded, that all vision is owing to the activity of this organ.

1. Place a piece of red silk, about an inch in diameter, as in plate 1, at Sect. III. 1., on a sheet of white paper, in a strong light; look steadily upon it from about the distance of half a yard for a minute; then closing your eyelids cover them with your hands, and a green spectrum will be seen in your eyes, resembling in form the piece of red silk: after some time, this spectrum will disappear and shortly reappear; and this alternately three or four times, if the experiment is well made, till at length it vanishes entirely.

2. Place on a sheet of white paper a circular piece of blue silk, about four inches in diameter, in the sunshine; cover the center of this with a circular piece of yellow silk, about three inches in diameter; and the center of the yellow silk with a circle of pink silk, about two inches in diameter; and the center of the pink silk with a circle of green silk, about one inch in diameter; and the centre of this with a circle of indigo, about half an inch in diameter; make a small speck with ink in the very center of the whole, as in plate 3, at Sect. III. 3. 6.; look steadily for a minute on this central spot, and then closing your eyes, and applying your hand at about an inch distance before them, so as to prevent too much or too little light from passing through the eyelids, you will see the most beautiful circles of colours that imagination can conceive, which are most resembled by the colours occasioned by pouring a drop or two of oil on a still lake in a bright day; but these circular irises of colours are not only different from the colours of the silks above mentioned, but are at the same time perpetually changing as long as they exist.

3. When any one in the dark presses either corner of his eye with his finger, and turns his eye away from his finger, he will see a circle of colours

like those in a peacock's tail: and a sudden flash of light is excited in the eye by a stroke on it. (Newton's Opt. Q. 16.)

4. When any one turns round rapidly on one foot, till he becomes dizzy, and falls upon the ground, the spectra of the ambient objects continue to present themselves in rotation, or appear to librate, and he seems to behold them for some time still in motion.

From all these experiments it appears, that the spectra in the eye are not owing to the mechanical impulse of light impressed on the retina, nor to its chemical combination with that organ, nor to the absorption and emission of light, as is observed in many bodies; for in all these cases the spectra must either remain uniformly, or gradually diminish; and neither their alternate pretence and evanescence as in the first experiment, nor the perpetual changes of their colours as in the second, nor the flash of light or colours in the pressed eye as in the third, nor the rotation or libration of the spectra as in the fourth, could exist.

It is not absurd to conceive, that the retina may be stimulated into motion, as well as the red and white muscles which form our limbs and vessels; since it consists of fibres, like those, intermixed with its medullary substance. To evince this structure, the retina of an ox's eye was suspended in a glass of warm water, and forcibly torn in a few places; the edges of these parts appeared jagged and hairy, and did not contract, and become smooth like simple mucus, when it is distended till it breaks; which shews that it consists of fibres; and that its fibrous construction became still more distinct to the sight, by adding some caustic alkali to the water, as the adhering mucus was first eroded, and the hair-like fibres remained floating in the vessel. Nor does the degree of transparency of the retina invalidate the evidence of its fibrous structure, since Leeuwenhoek has shewn that the crystalline humour itself consists of fibres. (Arcana Naturæ, V. 1. p. 70.)

Hence it appears, that as the muscles have larger fibres intermixed with a smaller quantity of nervous medulla, the organ of vision has a greater quantity of nervous medulla intermixed with smaller fibres; and it is probable that the locomotive muscles, as well as the vascular ones, of microscopic animals have much greater tenuity than these of the retina.

And besides the similar laws, which will be shewn in this paper to govern alike the actions of the retina and of the muscles, there are many other analogies which exist between them. They are both originally excited into action by irritations, both are nearly in the same quantity of time, are alike strengthened or fatigued by exertion, are alike painful if excited into action when they are in an inflamed state, are alike liable to paralysis, and to the torpor of old age.

II. OF SPECTRA FROM DEFECT OF SENSIBILITY.

The retina is not so easily excited into action by less irritation after having been lately subjected to greater.

1. When any one passes from the bright daylight into a darkened room, the irises of his eyes expand themselves to their utmost extent in a few seconds of time; but it is very long before the optic nerve, after having been stimulated by the greater light of the day, becomes sensible of the less degree of it in the room; and, if the room is not too obscure, the irises will again contract themselves in some degree, as the sensibility of the retina returns.

2. Place about half an inch square of white paper on a black hat, and looking steadily on the center of it for a minute, remove your eyes to a sheet of white paper; and after a second or two a dark square will be seen on the white paper, which will continue some time. A similar dark square will be seen in the closed eye, if light be admitted through the eyelids.

So after looking at any luminous object of a small size, as at the sun, for a short time, so as not much to fatigue the eyes, this part of the retina becomes less sensible to smaller quantities of light; hence, when the eyes are turned on other less luminous parts of the sky, a dark spot is seen resembling the shape of the sun, or other luminous object which we last beheld. This is the source of one kind of the dark-coloured *muscæ volitantes*. If this dark spot lies above the center of the eye, we turn our eyes that way, expecting to bring it into the center of the eye, that we may view it more distinctly; and in this case the dark spectrum seems to move upwards. If the dark spectrum is found beneath the centre of the eye, we pursue it from the same motive, and it seems to move downwards. This has given rise to various conjectures of something floating in the aqueous humours of the eyes; but whoever, in attending to these spots, keeps his eyes unmoved by looking steadily at the corner of a cloud, at the same time that he observes the dark spectra, will be thoroughly convinced, that they have no motion but what is given to them by the movement of our eyes in pursuit of them. Sometimes the form of the spectrum, when it has been received from a circular luminous body, will become oblong; and sometimes it will be divided into two circular spectra, which is not owing to our changing the angle made by the two optic axises, according to the distance of the clouds or other bodies to which the spectrum is supposed to be contiguous, but to other causes mentioned in No. X. 3. of this section. The apparent size of it will also be variable according to its supposed distance.

As these spectra are more easily observable when our eyes are a little weakened by fatigue, it has frequently happened, that people of delicate constitutions have been much alarmed at them, fearing a beginning decay of their sight, and have thence fallen into the hands of ignorant oculists; but I believe they never are a prelude to any other disease of the eye, and that it is from habit alone, and our want of attention to them, that we do not see them on all objects every hour of our lives. But as the nerves of very weak people lose their sensibility, in the same manner as their muscles lose their activity, by a small time of exertion, it frequently happens, that sick people in the extreme debility of fevers are perpetually employed in picking something from the bed-clothes, occasioned by their mistaking the appearance of these *muscæ volitantes* in their eyes. Benvenuto Celini, an Italian artist, a man of strong abilities, relates, that having passed the whole night on a distant mountain with some companions and a conjurer, and performed many ceremonies to raise the devil, on their return in the morning to Rome, and looking up when the sun began to rise, they saw numerous devils run on the tops of the houses, as they passed along; so much were the spectra of their weakened eyes magnified by fear, and made subservient to the purposes of fraud or superstition. (Life of Ben. Celini.)

3. Place a square inch of white paper on a large piece of straw-coloured silk; look steadily some time on the white paper, and then move the centre of your eyes on the silk, and a spectrum of the form of the paper will appear on the silk, of a deeper yellow than the other part of it: for the central part of the retina, having been some time exposed to the stimulus of a greater quantity of white light, is become less sensible to a smaller quantity of it, and therefore sees only the yellow rays in that part of the straw-coloured silk.

Facts similar to these are observable in other parts of our system: thus, if one hand be made warm, and the other exposed to the cold, and then both of them immersed in subtepid water, the water is perceived warm to one hand, and cold to the other; and we are not able to hear weak sounds for some time after we have been exposed to loud ones; and we feel a chilliness on coming into an atmosphere of temperate warmth, after having been some time confined in a very warm room: and hence the stomach, and other organs of digestion, of those who have been habituated to the greater stimulus of spirituous liquor, are not excited into their due action by the less stimulus of common food alone; of which the immediate consequence is indigestion and hypochondriacism.

III. OF SPECTRA FROM EXCESS OF SENSIBILITY.

The retina is more easily excited into action by greater irritation after having been lately subjected to less.

1. If the eyes are closed, and covered perfectly with a hat, for a minute or two, in a bright day; on removing the hat a red or crimson light is seen through the eyelids. In this experiment the retina, after being some time kept in the dark, becomes so sensible to a small quantity of light, as to perceive distinctly the greater quantity of red rays than of others which pass through the eyelids. A similar coloured light is seen to pass through the edges of the fingers, when the open hand is opposed to the flame of a candle.

2. If you look for some minutes steadily on a window in the beginning of the evening twilight, or in a dark day, and then move your eyes a little, so that those parts of the retina, on which the dark frame-work of the window was delineated, may now fall on the glass part of it, many luminous lines, representing the frame-work, will appear to lie across the glass panes: for those parts of the retina, which were before least stimulated by the dark frame-work, are now more sensible to light than the other parts of the retina which were exposed to the more luminous parts of the window,

3. Make with ink on white paper a very black spot, about half an inch in diameter, with a tail about an inch in length, so as to represent a tadpole, as in plate 2, at Sect. III. 3. 3.; look steadily for a minute on this spot, and, on moving the eye a little, the figure of the tadpole will be seen on the white part of the paper, which figure of the tadpole will appear whiter or more luminous than the other parts of the white paper; for the part of the retina on which the tadpole was delineated, is now more sensible to light, than the other parts of it, which were exposed to the white paper. This experiment is mentioned by Dr. Irwin, but is not by him ascribed to the true cause, namely, the greater sensibility of that part of the retina which has been exposed to the black spot, than of the other parts which had received the white field of paper, which is put beyond a doubt by the next experiment.

4. On closing the eyes after viewing the black spot on the white paper, as in the foregoing experiment, a red spot is seen of the form of the black spot: for that part of the retina, on which the black spot was delineated, being now more sensible to light than the other parts of it, which were exposed to the white paper, is capable of perceiving the red rays which penetrate the eyelids. If this experiment be made by the light of a tallow candle, the spot will be yellow instead of red; for tallow candles abound much with yellow light, which passes in greater quantity and force through the eyelids than blue light; hence the difficulty of distinguishing blue and green by this kind of candle light. The colour of the spectrum may possibly vary in the daylight, according to the different colour of the meridian or the morning or evening light.

M. Beguelin, in the Berlin Memoires, V. II. 1771, observes, that, when he held a book so that the sun shone upon his half-closed eyelids, the black letters, which he had long inspected, became red, which must have been thus occasioned. Those parts of the retina which had received for some time the black letters, were so much more sensible than those parts which had been opposed to the white paper, that to the former the red light, which passed through the eyelids, was perceptible. There is a similar story told, I think, in de Voltaire's Historical Works, of a Duke of Tuscany, who was playing at dice with the general of a foreign army, and, believing he saw bloody spots upon the dice, portended dreadful events, and retired in confusion. The observer, after looking for a minute on the black spots of a die, and carelessly closing his eyes, on a bright day; would see the image of a die with red spots upon it, as above explained.

5. On emerging from a dark cavern, where we have long continued, the light of a bright day becomes intolerable to the eye for a considerable time, owing to the excess of sensibility existing in the eye, after having been long exposed to little or no stimulus. This occasions us immediately to contract the iris to its smallest aperture, which becomes again gradually dilated, as the retina becomes accustomed to the greater stimulus of the daylight.

The twinkling of a bright star, or of a distant candle in the night, is perhaps owing to the same cause. While we continue to look upon these luminous objects, their central parts gradually appear paler, owing to the decreasing sensibility of the part of the retina exposed to their light; whilst, at the same time, by the unsteadiness of the eye, the edges of them are perpetually falling on parts of the retina that were just before exposed to the darkness of the night, and therefore tenfold more sensible to light than the part on which the star or candle had been for some time delineated. This pains the eye in a similar manner as when we come suddenly from a dark room into bright daylight, and gives the appearance of bright scintillations. Hence the stars twinkle most when the night is darkest, and do not twinkle through telescopes, as observed by Musschenbroeck; and it will afterwards be seen why this twinkling is sometimes of different colours when the object is very bright, as Mr. Melvill observed in looking at Sirius. For the opinions of others on this subject, see Dr. Priestley's valuable History of Light and Colours, p. 494.

Many facts observable in the animal system are similar to these; as the hot glow occasioned by the usual warmth of the air, or our clothes, on coming out of a cold bath; the pain of the fingers on approaching the fire after having handled snow; and the inflamed heels from walking in snow. Hence those who have been exposed to much cold have died on being brought to a fire, or their limbs have become so much inflamed as to mortify. Hence much food or wine given suddenly to those who have almost perished by hunger has destroyed them; for all the organs of the famished body are now become so much more irritable to the stimulus of food and wine, which they have long been deprived of, that inflammation is excited, which terminates in gangrene or fever.

IV. OF DIRECT OCULAR SPECTRA.

*A quantity of stimulus somewhat greater than natural excites the
retina into spasmodic action, which ceases in a few seconds.*

A certain duration and energy of the stimulus of light and colours excites the perfect action of the retina in vision; for very quick motions are imperceptible to us, as well as very slow ones, as the whirling of a top, or the shadow on a sun-dial. So perfect darkness does not affect the eye at all; and excess of light produces pain, not vision.

1. When a fire-coal is whirled round in the dark, a lucid circle remains a considerable time in the eye; and that with so much vivacity of light, that it is mistaken for a continuance of the irritation of the object. In the same manner, when a fiery meteor shoots across the night, it appears to leave a long lucid train behind it, part of which, and perhaps sometimes the whole, is owing to the continuance of the action of the retina after having been thus vividly excited. This is beautifully illustrated by the following experiment: fix a paper sail, three or four inches in diameter, and made like that of a smoke jack, on a tube of pasteboard; on looking through the tube at a distant prospect, some disjointed parts of it will be seen through the narrow intervals between the sails; but as the fly begins to revolve, these intervals appear larger; and when it revolves quicker, the whole prospect is seen quite as distinct as if nothing intervened, though less luminous.

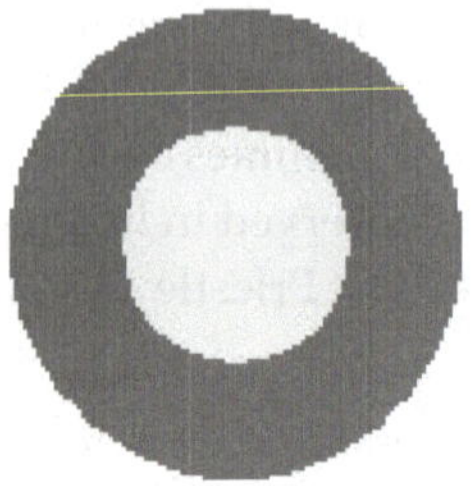

Fig. 3.

2. Look through a dark tube, about half a yard long, at the area of a yellow circle of half an inch diameter, lying upon a blue area of double

that diameter, for half a minute; and on closing your eyes the colours of the spectrum will appear similar to the two areas, as in fig. 3.; but if the eye is kept too long upon them, the colours of the spectrum will be the reverse of those upon the paper, that is, the internal circle will become blue, and the external area yellow; hence some attention is required in making this experiment.

3. Place the bright flame of a spermaceti candle before a black object in the night; look steadily at it for a short time, till it is observed to become somewhat paler; and on closing the eyes, and covering them carefully, but not so as to compress them, the image of the blazing candle will continue distinctly to be visible.

4. Look steadily, for a short time, at a window in a dark day, as in Exp. 2. Sect. III. and then closing your eyes, and covering them with your hands, an exact delineation of the window remains for some time visible in the eye. This experiment requires a little practice to make it succeed well; since, if the eyes are fatigued by looking too long on the window, or the day be too bright, the luminous parts of the window will appear dark in the spectrum, and the dark parts of the frame-work will appear luminous, as in Exp. 2. Sect. III. And it is even difficult for many, who first try this experiment, to perceive the spectrum at all; for any hurry of mind, or even too great attention to the spectrum itself, will disappoint them, till they have had a little experience in attending to such small sensations.

The spectra described in this section, termed direct ocular spectra, are produced without much fatigue of the eye; the irritation of the luminous object being soon withdrawn, or its quantity of light being not so great as to produce any degree of uneasiness in the organ of vision; which distinguishes them from the next class of ocular spectra, which are the consequence of fatigue. These direct spectra are best observed in such circumstances that no light, but what comes from the object, can fall upon the eye; as in looking through a tube, of half a yard long, and an inch wide, at a yellow paper on the side of a room, the direct spectrum was easily produced on closing the eye without taking it from the tube; but if the lateral light is admitted through the eyelids, or by throwing the spectrum on white paper, it becomes a reverse spectrum, as will be explained below.

The other senses also retain for a time the impressions that have been made upon them, or the actions they have been excited into. So if a hard body is pressed upon the palm of the hand, as is practised in tricks of legerdemain, it is not easy to distinguish for a few seconds whether it remains or is removed; and tastes continue long to exist vividly in the mouth, as the smoke of tobacco, or the taste of gentian, after the sapid material is withdrawn.

V. A quantity of stimulus somewhat greater than the last mentioned excites the retina into spasmodic action, which ceases and recurs alternately.

1. On looking for a time on the setting sun, so as not greatly to fatigue the sight, a yellow spectrum is seen when the eyes are closed and covered, which continues for a time, and then disappears and recurs repeatedly before it entirely vanishes. This yellow spectrum of the sun when the eyelids are opened becomes blue; and if it is made to fall on the green grass, or on other coloured objects, it varies its own colour by an intermixture of theirs, as will be explained in another place.

2. Place a lighted spermaceti candle in the night about one foot from your eye, and look steadily on the centre of the flame, till your eye becomes much more fatigued than in Sect. IV. Exp. 3.; and on closing your eyes a reddish spectrum will be perceived, which will cease and return alternately.

The action of vomiting in like manner ceases, and is renewed by intervals, although the emetic drug is thrown up with the first effort: so after-pains continue some time after parturition; and the alternate pulsations of the heart of a viper are renewed for some time after it is cleared from its blood.

VI. OF REVERSE OCULAR SPECTRA.

The retina, after having been excited into action by a stimulus somewhat greater them the last mentioned falls into opposite spasmodic action.

The actions of every part of animal bodies may be advantageously compared with each other. This strict analogy contributes much to the investigation of truth; while those looser analogies, which compare the phenomena of animal life with those of chemistry or mechanics, only serve to mislead our inquiries.

When any of our larger muscles have been in long or in violent action, and their antagonists have been at the same time extended, as soon as the action of the former ceases, the limb is stretched the contrary way for our ease, and a pandiculation or yawning takes place.

By the following observations it appears, that a similar circumstance obtains in the organ of vision; after it has been fatigued by one kind of action, it spontaneously falls into the opposite kind.

1. Place a piece of coloured silk, about an inch in diameter, on a sheet of white paper, about half a yard from your eyes; look steadily upon it for a minute; then remove your eyes upon another part of the white paper, and a

spectrum will be seen of the form of the silk thus inspected, but of a colour opposite to it. A spectrum nearly similar will appear if the eyes are closed, and the eyelids shaded by approaching the hand near them, so as to permit some, but to prevent too much light falling on them.

Red silk produced a green spectrum.

Green produced a red one.

Orange produced blue.

Blue produced orange.

Yellow produced violet.

Violet produced yellow.

That in these experiments the colours of the spectra are the reverse of the colours which occasioned them, may be seen by examining the third figure in Sir Isaac Newton's Optics, L. II. p. 1, where those thin laminæ of air, which reflected yellow, transmitted violet; those which reflected red, transmitted a blue green; and so of the rest, agreeing with the experiments above related.

2. These reverse spectra are similar to a colour, formed by a combination of all the primary colours except that with which the eye has been fatigued in making the experiment: thus the reverse spectrum of red must be such a green as would be produced by a combination of all the other prismatic colours. To evince this fact the following satisfactory experiment was made. The prismatic colours were laid on a circular pasteboard wheel, about four inches in diameter, in the proportions described in Dr. Priestley's History of Light and Colours, pl. 12. fig. 83. except that the red compartment was entirely left out, and the others proportionably extended so as to complete the circle. Then, as the orange is a mixture of red and yellow, and as the violet is a mixture of red and indigo, it became necessary to put yellow on the wheel instead of orange, and indigo instead of violet, that the experiment might more exactly quadrate with the theory it was designed to establish or confute; because in gaining a green spectrum from a red object, the eye is supposed to have become insensible to red light. This wheel, by means of an axis, was made to whirl like a top; and on its being put in motion, a green colour was produced, corresponding with great exactness to the reverse spectrum of red.

3. In contemplating any one or these reverse spectra in the closed and covered eye, it disappears and re-appears several times successively, till at length it entirely vanishes, like the direct spectra in Sect. V.; but with this additional circumstance, that when the spectrum becomes faint or

evanescent, it is instantly revived by removing the hand from before the eyelids, so as to admit more light: because then not only the fatigued part of the retina is inclined spontaneously to fall into motions of a contrary direction, but being still sensible to all other rays of light, except that with which it was lately fatigued, is by these rays at the same time stimulated into those motions which form the reverse spectrum.

From these experiments there is reason to conclude, that the fatigued part of the retina throws itself into a contrary mode of action, like oscitation or pandiculation, as soon as the stimulus which has fatigued it is withdrawn; and that it still remains sensible, that is, liable to be excited into action by any other colours at the same time, except the colour with which it has been fatigued.

VII. *The retina after having been excited into action by a stimulus somewhat greater than the last mentioned falls into various successive spasmodic actions.*

1. On looking at the meridian sun as long as the eyes can well bear its brightness, the disk first becomes pale, with a luminous crescent, which seems to librate from one edge of it to the other, owing to the unsteadiness of the eye; then the whole phasis of the sun becomes blue, surrounded with a white halo; and on closing the eyes, and covering them with the hands, a yellow spectrum is seen, which in a little time changes into a blue one.

M. de la Hire observed, after looking at the bright sun, that the impression in his eye first assumed a yellow appearance, and then green, and then blue; and wishes to ascribe these appearances to some affection of the nerves. (Porterfield on the Eye, Vol. I. p. 313.)

2. After looking steadily on about an inch square of pink silk, placed on white paper, in a bright sunshine, at the distance of a foot from my eyes, and closing and covering my eyelids, the spectrum of the silk was at first a dark green, and the spectrum of the white paper became of a pink. The spectra then both disappeared; and then the internal spectrum was blue; and then, after a second disappearance, became yellow, and lastly pink, whilst the spectrum of the field varied into red and green.

These successions of different coloured spectra were not exactly the same in the different experiments, though observed, as near as could be, with the same quantity of light, and other similar circumstances; owing, I suppose, to trying too many experiments at a time; so that the eye was not quite free from the spectra of the colours which were previously attended to.

The alternate exertions of the retina in the preceding section resembled the oscitation or pandiculation of the muscles, as they were performed in directions contrary to each other, and were the consequence of fatigue rather than of pain. And in this they differ from the successive dissimilar exertions of the retina, mentioned in this section, which resemble in miniature the more violent agitations of the limbs in convulsive diseases, as epilepsy, chorea S. Viti, and opisthotonos; all which diseases are perhaps, at first, the consequence of pain, and have their periods afterwards established by habit.

VIII. *The retina, after having been excited into action by a stimulus somewhat greater than the last mentioned falls into a fixed spasmodic action, which continues for some days.*

1. After having looked long at the meridian sun, in making some of the preceding experiments, till the disks faded into a pale blue, I frequently observed a bright blue spectrum of the sun on other objects all the next and the succeeding day, which constantly occurred when I attended to it, and frequently when I did not previously attend to it. When I closed and covered my eyes, this appeared of a dull yellow; and at other times mixed with the colours of other objects on which it was thrown. It may be imagined, that this part of the retina was become insensible to white light, and thence a bluish spectrum became visible on all luminous objects; but as a yellowish spectrum was also seen in the closed and covered eye, there can remain no doubt of this being the spectrum of the sun. A similar appearance was observed by M. Æpinus, which he acknowledges he could give no account of. (Nov. Com. Petrop. V. 10. p. 2. and 6.)

The locked jaw, and some cataleptic spasms, are resembled by this phenomenon; and from hence we may learn the danger to the eye by inspecting very luminous objects too long a time.

IX. *A quantity of stimulus greater than the preceding induces a temporary paralysis of the organ of vision.*

1. Place a circular piece of bright red silk, about half an inch in diameter, on the middle of a sheet of white paper; lay them on the floor in a bright sunshine, and fixing your eyes steadily on the center of the red circle, for three or four minutes, at the distance of four or six feet from the object, the red silk will gradually become paler, and finally cease to appear red at all.

2. Similar to these are many other animal facts; as purges, opiates, and even poisons, and contagious matter, cease to stimulate our system, after we have been habituated to their use. So some people sleep undisturbed by a clock, or even by a forge hammer in their neighbourhood: and not only continued irritations, but violent exertions of any kind, are succeeded by temporary paralysis. The arm drops down after violent action, and

continues for a time useless; and it is probable, that those who have perished suddenly in swimming, or in scating on the ice, have owed their deaths to the paralysis, or extreme fatigue, which succeeds every violent and continued exertion.

X. MISCELLANEOUS REMARKS.

There were some circumstances occurred in making these experiments, which were liable to alter the results of them, and which I shall here mention for the assistance of others, who may wish to repeat them.

1. *Of direct and inverse spectra existing at the same time; of reciprocal direct spectra; of a combination of direct and inverse spectra; of a spectral halo; rules to pre-determine the colours of spectra.*

a. When an area, about six inches square, of bright pink Indian paper, had been viewed on an area, about a foot square, of white writing paper, the internal spectrum in the closed eye was green, being the reverse spectrum of the pink paper; and the external spectrum was pink, being the direct spectrum of the pink paper. The same circumstance happened when the internal area was white, and external one pink; that is, the internal spectrum was pink, and the external one green. All the same appearances occurred when the pink paper was laid on a black hat.

b. When six inches square of deep violet polished paper was viewed on a foot square of white writing paper, the internal spectrum was yellow, being the reverse spectrum of the violet paper, and the external one was violet, being the direct spectrum of the violet paper.

c. When six inches square of pink paper was viewed on a foot square of blue paper, the internal spectrum was blue, and the external spectrum was pink; that is, the internal one was the direct spectrum of the external object, and the external one was the direct spectrum of the internal object, instead of their being each the reverse spectrum of the objects they belonged to.

d. When six inches square of blue paper were viewed on a foot square of yellow paper, the interior spectrum became a brilliant yellow, and the exterior one a brilliant blue. The vivacity of the spectra was owing to their being excited both by the stimulus of the interior and exterior objects; so that the interior yellow spectrum was both the reverse spectrum of the blue paper, and the direct one of the yellow paper; and the exterior blue spectrum was both the reverse spectrum of the yellow paper, and the direct one of the blue paper.

e. When the internal area was only a square half-inch of red paper, laid on a square foot of dark violet paper, the internal spectrum was green, with a reddish-blue halo. When the red internal paper was two inches square, the internal spectrum was a deeper green, and the external one redder. When the internal paper was six inches square, the spectrum of it became blue, and the spectrum of the external paper was red.

f. When a square half-inch of blue paper was laid on a six-inch square of yellow paper, the spectrum of the central paper in the closed eye was yellow, incircled with a blue halo. On looking long on the meridian sun, the disk fades into a pale blue surrounded with a whitish halo.

These circumstances, though they very much perplexed the experiments till they were investigated, admit of a satisfactory explanation; for while the rays from the bright internal object in exp. *a.* fall with their full force on the center of the retina, and, by fatiguing that part of it, induce the reverse spectrum, many scattered rays, from the same internal pink paper, fall on the more external parts of the retina, but not in such quantity as to occasion much fatigue, and hence induce the direct spectrum of the pink colour in those parts of the eye. The same reverse and direct spectra occur from the violet paper in exp. *b.*: and in exp. *c.* the scattered rays from the central pink paper produce a direct spectrum of this colour on the external parts of the eye, while the scattered rays from the external blue paper produce a direct spectrum of that colour on the central part of the eye, instead of these parts of the retina falling reciprocally into their reverse spectra. In exp. *d.* the colours being the reverse of each other, the scattered rays from the exterior object falling on the central parts of the eye, and there exciting their direct spectrum, at the same time that the retina was excited into a reverse spectrum by the central object, and this direct and reverse spectrum being of similar colour, the superior brilliancy of this spectrum was produced. In exp. *e.* the effect of various quantities of stimulus on the retina, from the different respective sizes of the internal and external areas, induced a spectrum of the internal area in the center of the eye, combined of the reverse spectrum of that internal area and the direct one of the external area, in various shades of colour, from a pale green to a deep blue, with similar changes in the spectrum of the external area. For the same reasons, when an internal bright object was small, as in exp. *f.* instead of the whole of the spectrum of the external object being reverse to the colour of the internal object, only a kind of halo, or radiation of colour, similar to that of the internal object, was spread a little way on the external spectrum. For this internal blue area being so small, the scattered rays from it extended but a little way on the image of the external area of yellow paper, and could therefore produce only a blue halo round the yellow spectrum in the center.

If any one should suspect that the scattered rays from the exterior coloured object do not intermix with the rays from the interior coloured object, and thus affect the central part of the eye, let him look through an opake tube, about two feet in length, and an inch in diameter, at a coloured wall of a room with one eye, and with the other eye naked; and he will find, that by shutting out the lateral light, the area of the wall seen through a tube appears as if illuminated by the sunshine, compared with the other parts of it; from whence arises the advantage of looking through a dark tube at distant paintings.

Hence we may safely deduce the following rules to determine before-hand the colours of all spectra. 1. The direct spectrum without any lateral light is an evanescent representation of its object in the unfatigued eye. 2. With some lateral light it becomes of a colour combined of the direct spectrum of the central object, and of the circumjacent objects, in proportion to their respective quantity and brilliancy. 3. The reverse spectrum without lateral light is a representation in the fatigued eye of the form of its objects, with such a colour as would be produced by all the primary colours, except that of the object. 4. With lateral light the colour is compounded of the reverse spectrum of the central object, and the direct spectrum of the circumjacent objects, in proportion to their respective quantity and brilliancy.

2. Variation and vivacity of the spectra
occasioned by extraneous light.

The reverse spectrum, as has been before explained, is similar to a colour, formed by a combination of all the primary colours, except that with which the eye has been fatigued in making the experiment: so the reverse spectrum of red is such a green as would be produced by a combination of all the other prismatic colours. Now it must be observed, that this reverse spectrum of red is therefore the direct spectrum of a combination of all the other prismatic colours, except the red; whence, on removing the eye from a piece of red silk to a sheet of white paper, the green spectrum, which is perceived, may either be called the reverse spectrum of the red silk, or the direct spectrum of all the rays from the white paper, except the red; for in truth it is both. Hence we see the reason why it is not easy to gain a direct spectrum of any coloured object in the day-time, where there is much lateral light, except of very bright objects, as of the setting sun, or by looking through an opake tube; because the lateral external light falling also on the central part of the retina, contributes to induce the reverse spectrum, which is at the same time the direct spectrum of that lateral light, deducting only the colour of the central object which we have been viewing. And for the same reason, it is difficult to gain the reverse spectrum, where there is no lateral light to contribute to its formation. Thus, in looking through an opake tube on a yellow wall, and closing my eye, without admitting any lateral light, the spectra were all at first yellow; but at length changed into blue. And on looking in the same manner on red paper, I did at length get a green spectrum; but they were all at first red ones: and the same after looking at a candle in the night.

The reverse spectrum was formed with greater facility when the eye was thrown from the object on a sheet of white paper, or when light was admitted through the closed eyelids; because not only the fatigued part of the retina was inclined spontaneously to fall into motions of a contrary direction; but being still sensible to all other rays of light except that with which it was lately fatigued, was by these rays stimulated at the same time into those motions which form the reverse spectrum. Hence, when, the reverse spectrum of any colour became faint, it was wonderfully revived by admitting more light through the eyelids, by removing the hand from before them: and hence, on covering the closed eyelids, the spectrum would often cease for a time, till the retina became sensible to the stimulus of the smaller quantity of light, and then it recurred. Nor was the spectrum only changed in vivacity, or in degree, by this admission of light through the eyelids; but it frequently happened, after having viewed bright objects, that the spectrum in the closed and covered eye was changed into a third spectrum, when light was admitted through the eyelids: which third spectrum was composed of such colours as could pass through the eyelids, except those of the object. Thus, when an area of half an inch diameter of pink paper was viewed on a sheet of white paper in the sunshine, the spectrum with closed and covered eyes was green; but on removing the hands from before the closed eyelids, the spectrum became yellow, and returned instantly again to green, as often as the hands were applied to cover the eyelids, or removed from them: for the retina being now insensible to red light, the yellow rays passing through the eyelids in greater quantity than the other colours, induced a yellow spectrum; whereas if the spectrum was thrown on white paper, with the eyes open, it became only a lighter green.

Though a certain quantity of light facilitates the formation of the reverse spectrum, a greater quantity prevents its formation, as the more powerful stimulus excites even the fatigued parts of the eye into action; otherwise we should see the spectrum of the last viewed object as often as we turn our eyes. Hence the reverse spectra are best seen by gradually approaching the hand near the closed eyelids to a certain distance only, which must be varied with the brightness of the day, or the energy of the spectrum. Add to this, that all dark spectra, as black, blue, or green, if light be admitted through the eyelids, after they have been some time covered, give reddish spectra, for the reasons given in Sect. III. Exp. 1.

From these circumstances of the extraneous light coinciding with the spontaneous efforts of the fatigued retina to produce a reverse spectrum, as was observed before, it is not easy to gain a direct spectrum, except of objects brighter than the ambient light; such as a candle in the night, the setting sun, or viewing a bright object through an opake tube; and then the reverse spectrum is instantaneously produced by the admission of some external light; and is as instantly converted again to the direct spectrum by the exclusion of it. Thus, on looking at the setting sun, on closing the eyes, and covering them, a yellow spectrum is seen, which is the direct spectrum of the setting sun; but on opening the eyes on the sky, the yellow spectrum is immediately changed into a blue one, which is the reverse spectrum of the yellow sun, or the direct spectrum of the blue sky, or a combination of both. And this is again transformed into a yellow one on closing the eyes, and so reciprocally, as quick as the motions of the opening and closing eyelids. Hence, when Mr. Melvill observed the scintillations of the star Sirius to be sometimes coloured, these were probably the direct spectrum of the blue sky on the parts of the retina fatigued by the white light of the star. (Essays Physical and Literary, p. 81. V. 2.)

When a direct spectrum is thrown on colours darker than itself, it mixes with them; as the yellow spectrum of the setting sun, thrown on the green grass, becomes a greener yellow. But when a direct spectrum is thrown on colours brighter than itself, it becomes instantly changed into the reverse spectrum, which mixes with those brighter colours. So the yellow spectrum of the setting sun thrown on the luminous sky becomes blue, and changes with the colour or brightness of the clouds on which it appears. But the reverse spectrum mixes with every kind of colour on which it is thrown, whether brighter than itself or not; thus the reverse spectrum, obtained by viewing a piece of yellow silk, when thrown on white paper, was a lucid blue green; when thrown on black Turkey leather, becomes a deep violet. And the spectrum of blue silk, thrown on white paper, was a light yellow; on black silk was an obscure orange; and, the blue spectrum, obtained from orange-coloured silk, thrown on yellow, became a green.

In these cases the retina is thrown into activity or sensation by the stimulus of external colours, at the same time that it continues the activity or sensation which forms the spectra; in the same manner as the prismatic colours, painted on a whirling top, are seen to mix together. When these colours of external objects are brighter than the direct spectrum which is thrown upon them, they change it into the reverse spectrum, like the admission of external light on a direct spectrum, as explained above. When they are darker than the direct spectrum, they mix with it, their weaker stimulus being inefficient to induce the reverse spectrum.

3. Variation of spectra in respect to
number, and figure, and remission.

Fig. 4.

When we look long and attentively at any object, the eye cannot always be kept entirely motionless; hence, on inspecting a circular area of red silk placed on white paper, a lucid crescent or edge is seen to librate on one side or other of the red circle: for the exterior parts of the retina sometimes falling on the edge of the central silk, and sometimes on the white paper, are less fatigued with red light than the central part of the retina, which is constantly, exposed to it; and therefore, when they fall on the edge of the red silk, they perceive it more vividly. Afterwards, when the eye becomes fatigued, a green spectrum in the form of a crescent is seen to librate on one side or other of the central circle, as by the unsteadiness of the eye a part of the fatigued retina falls on the white paper; and as by the increasing fatigue of the eye the central part of the silk appears paler, the edge on which the unfatigued part of the retina occasionally falls will appear of a deeper red than the original silk, because it is compared with the pale internal part of it. M. de Buffon in making this experiment observed, that the red edge of the silk was not only deeper coloured than the original silk; but, on his retreating a little from it, it became oblong, and at length divided into two, which must have been owing to his observing it either before or behind the point of intersection of the two optic axises. Thus, if a pen is held up before a distant candle, when we look intensely at the pen two candles are seen behind it; when we look intensely at the candle two pens are seen. If the sight be unsteady at the time of beholding the sun, even though one eye only be used, many images of the sun will appear, or luminous lines, when the eye is closed. And as some parts of these will be more vivid than others, and some parts of them will be produced nearer the center of the eye than others, these will disappear sooner than the others; and hence the number and shape of these spectra of the sun will continually vary, as long as they exist. The cause of some being more vivid than others, is the unsteadiness of the eye of the beholder, so that some parts of the retina have been longer exposed to the sunbeams. That some parts of a complicated spectrum fade and return before other parts of it, the following experiment evinces. Draw three concentric circles; the external one an inch and a half in diameter, the middle one an inch, and the internal one half an inch; colour the external and internal areas blue, and the remaining one yellow, as in Fig. 4.; after having looked about a minute on the center of these circles, in a bright light, the spectrum of the external area appears first in the closed eye, then the middle area, and lastly the central one; and then the central one disappears, and the others in inverted order. If concentric circles of more colours are added, it produces the beautiful ever changing spectrum in Sect. I. Exp. 2.

From hence it would seem, that the center of the eye produces quicker remissions of spectra, owing perhaps to its greater sensibility; that is, to its more energetic exertions. These remissions of spectra bear some analogy to the tremors of the hands, and palpitations of the heart, of weak people: and perhaps a criterion of the strength of any muscle or nerve may be taken from the time it can be continued in exertion.

4. *Variation of spectra in respect to brilliancy; the visibility of the circulation of the blood in the eye.*

1. The meridian or evening light makes a difference in the colours of some spectra; for as the sun descends, the red rays, which are less refrangible by the convex atmosphere, abound in great quantity. Whence the spectrum of the light parts of a window at this time, or early in the morning, is red; and becomes blue either a little later or earlier; and white in the meridian day; and is also variable from the colour of the clouds or sky which are opposed to the window.

2. All these experiments are liable to be confounded, if they are made too soon after each other, as the remaining spectrum will mix with the new ones. This is a very troublesome circumstance to painters, who are obliged to look long upon the same colour; and in particular to those whose eyes, from natural debility, cannot long, continue the same kind of exertion. For the same reason, in making these experiments, the result becomes much varied if the eyes, after viewing any object, are removed on other objects for but an instant of time, before we close them to view the spectrum; for the light from the object, of which we had only a transient view, in the very time of closing our eyes acts as a stimulus on the fatigued retina; and for a time prevents the defined spectrum from appearing, or mixes its own spectrum with it. Whence, after the eyelids are closed, either a dark field, or some unexpected colours, are beheld for a few seconds, before the desired spectrum becomes distinctly visible.

3. The length of time taken up in viewing an object, of which we are to observe the spectrum, makes a great difference in the appearance of the spectrum, not only in its vivacity, but in its colour; as the direct spectrum of the central object, or of the circumjacent ones, and also the reverse spectra of both, with their various combinations, as well as the time of their duration in the eye, and of their remissions or alternations, depend upon the degree of fatigue the retina is subjected to. The Chevalier d'Arcy constructed a machine by which a coal of fire was whirled round in the dark, and found, that when a luminous body made a revolution in eight thirds of time, it presented to the eye a complete circle of fire; from whence he concludes, that the impression continues on the organ about the seventh part of a second. (Mem. de l'Acad. des Sc. 1765.) This, however, is only to be considered as the shortest time of the duration of these direct spectra; since in the fatigued eye both the direct and reverse spectra, with their intermissions, appear to take up many seconds of time, and seem very variable in proportion to the circumstances of fatigue or energy.

4. It sometimes happens, if the eyeballs have been rubbed hard with the fingers, that lucid sparks are seen in quick motion amidst the spectrum we are attending to. This is similar to the flashes of fire from a stroke on the eye in fighting, and is resembled by the warmth and glow, which appears upon the skin after friction, and is probably owing to an acceleration of the arterial blood into the vessels emptied by the previous pressure. By being accustomed to observe such small sensations in the eye, it is easy to see the circulation of the blood in this organ. I have attended to this frequently, when I have observed my eyes more than commonly sensible to other spectra. The circulation may be seen either in both eyes at a time, or only in one of them; for as a certain quantity of light is necessary to produce this curious phenomenon, if one hand be brought nearer the closed eyelids than the other, the circulation in that eye will for a time disappear. For the easier viewing the circulation, it is sometimes necessary to rub the eyes with a certain degree of force after they are closed, and to hold the breath rather longer than is agreeable, which, by accumulating more blood in the eye, facilitates the experiment; but in general it may be seen distinctly after having examined other spectra with your back to the light, till the eyes become weary; then having covered your closed eyelids for half a minute, till the spectrum is faded away which you were examining, turn your face to the light, and removing your hands from the eyelids, by and by again shade them a little, and the circulation becomes curiously distinct. The streams of blood are however generally seen to unite, which shews it to be the venous

circulation, owing, I suppose, to the greater opacity of the colour of the blood in these vessels; for this venous circulation is also much more easily seen by the microscope in the tail of a tadpole.

5. Variation of spectra in respect to distinctness and size; with a new way of magnifying objects.

1. It was before observed, that when the two colours viewed together were opposite to each other, as yellow and blue, red and green, &c. according to the table of reflections and transmissions of light in Sir Isaac Newton's Optics, B. II. Fig. 3. the spectra of those colours were of all others the most brilliant, and best defined; because they were combined of the reverse spectrum of one colour, and of the direct spectrum of the other. Hence, in books printed with small types, or in the minute graduation of thermometers, or of clock-faces, which are to be seen at a distance, if the letters or figures are coloured with orange, and the ground with indigo; or the letters with red, and the ground with green; or any other lucid colour is used for the letters, the spectrum of which is similar to the colour of the ground; such letters will be seen much more distinctly, and with less confusion, than in black or white: for as the spectrum of the letter is the same colour with the ground on which they are seen, the unsteadiness of the eye in long attending to them will not produce coloured lines by the edges of the letters, which is the principal cause of their confusion. The beauty of colours lying in vicinity to each other, whose spectra are thus reciprocally similar to each colour, is owing to this greater ease that the eye experiences in beholding them distinctly; and it is probable, in the organ of hearing, a similar circumstance may constitute the pleasure of melody. Sir Isaac Newton observes, that gold and indigo were agreeable when viewed together; and thinks there may be some analogy between the sensations of light and sound. (Optics, Qu. 14.)

In viewing the spectra of bright objects, as of an area of red silk of half an inch diameter on white paper, it is easy to magnify it to tenfold its size: for if, when the spectrum is formed, you still keep your eye fixed on the silk area, and remove it a few inches further from you, a green circle is seen round the red silk: for the angle now subtended by the silk is less than it was when the spectrum was formed, but that of the spectrum continues the same, and our imagination places them at the same distance. Thus when you view a spectrum on a sheet of white paper, if you approach the paper to the eye, you may diminish it to a point; and if the paper is made to recede from the eye, the spectrum will appear magnified in proportion to the distance.

Fig. 5.

I was surprised, and agreeably amused, with the following experiment. I covered a paper about four inches square with yellow, and with a pen filled with a blue colour wrote upon the middle of it the word BANKS in capitals, as in Fig. 5, and sitting with my back to the sun, fixed my eyes for a minute exactly on the center of the letter N in the middle of the word; after closing my eyes, and shading them somewhat with my hand, the word was distinctly seen in the spectrum in yellow letters on a blue field; and then, on opening my eyes on a yellowish wall at twenty feet distance, the magnified name of BANKS appeared written on the wall in golden characters.

Conclusion.

It was observed by the learned M. Sauvage (Nosol. Method. Cl. VIII. Ord. i.) that the pulsations of the optic artery might be perceived by looking attentively on a white wall well illuminated. A kind of net-work, darker than the other parts of the wall, appears and vanishes alternately with every pulsation. This change of the colour of the wall he well ascribes to the compression of the retina by the diastole of the artery. The various colours produced in the eye by the pressure of the finger, or by a stroke on it, as mentioned by Sir Isaac Newton, seem likewise to originate from the unequal pressure on various parts of the retina. Now as Sir Isaac Newton has shewn, that all the different colours are reflected or transmitted by the laminæ of soap bubbles, or of air, according to their different thickness or thinness, is it not probable, that the effect of the activity of the retina may be to alter its thickness or thinness, so as better to adapt it to reflect or transmit the colours which stimulate it into action? May not muscular fibres exist in the retina for this purpose, which may be less minute than the locomotive muscles of microscopic animals? May not these muscular actions of the retina constitute the sensation of light and colours; and the voluntary repetitions of them, when the object is withdrawn, constitute our memory of them? And lastly, may not the laws of the sensations of light, here investigated, be applicable to all our other senses, and much contribute to elucidate many phenomena of animal bodies both in their healthy and diseased state; and thus render this investigation well worthy the attention of the physician, the metaphysician, and the natural philosopher?

November 1, 1785.

> Dum, Liber! astra petis volitans trepidantibus alis, Irruis immemori, parvula gutta, mari.

> Me quoque, me currente rotâ revolubilis ætas Volverit in tenebras,—i, Liber, ipse sequor.

INDEX TO THE SECTIONS OF PART FIRST

A

B

Balance ourselves by vision

Bandage increases absorption

Barrenness

Battement of sounds

Bath, cold. See Cold Bath

Beauty, sense of

Bile-ducts

—— stones

—— regurgitates into the blood

—— vomiting of

Birds of passage

—— nests of

—— colour of their eggs

Biting in pain

—— of mad animals

Black spots on dice appear red

Bladder, communication of with the intestines

—— of fish

Blood, transfusion of in nervous fevers

—— deficiency of, . and

—— from the vena portarum into the intestines

—— its momentum

—— momentum increased by venesection

—— drawn in nervous pains

—— its oxygenation

Breasts of men

Breathing, how learnt

Brutes differ from men

Brutes. See Animals

Buxton bath, why it feels warm

C

D

G

Gall-stone

—— See Bile-stones

Generation

Gills of fish

Glands

—— conglobate glands

—— have their peculiar stimulus

—— their senses

—— invert their motions

—— increase their motions

Golden rule for exhibiting wine

—— for leaving off wine

Gout from inflamed liver

—— in the stomach

—— why it returns after evacuations

—— owing to vinous spirit only

—— periods of

Grinning in pain

Gyration on one foot, . and

H

Habit defined

Hæmorrhages, periods of

—— from paralysis of veins, . and

Hair and nails

—— colour of

Harmony

Head-achs

Hearing

Heat, sense of

—— produced by the glands

Identity

Iliac passion

Imagination

— — of the male forms the sex

Imitation, origin of

Immaterial beings

Impediment of speech

Infection. See Contagion

Inflammation

— — great vascular exertion in

— — not from pains from defect of stimulus

— — of parts previously insensible

— — often distant from its cause

— — observes solar days

— — of the eye

— — of the bowels prevented by their continued action in sleep

Inoculation with blood

Insane people, their great strength

Insanity (see Madness) pleasurable one

Insects, their knowledge, . and

— — in the heads of calves

— — class of

Instinctive actions defined

Intestines

Intoxication relieves pain, why

— — from food after fatigue

— — diseases from it

— — See Drunkenness

Intuitive analogy

Invention

Irritability increases during sleep

—— in increasing doses

Organs of sense, . and

Organs when destroyed cease to produce ideas

Organic particles of Buffon

Organ-pipes

Oxygenation of the blood

P

Pain from excess and defect of motion

—— not felt during exertion

—— from greater contraction of fibres

—— from accumulation of sensorial power

—— from light, pressure, heat, caustics

—— in epilepsy

—— distant from its cause

—— from stone in the bladder

—— of head and back from defect

—— from a gall-stone

—— of the stomach in gout

—— of shoulder in hepatitis

—— produces volition

Paleness in cold fit

Palsies explained

Paralytic limbs stretch from irritation

—— patients move their sound limb much

Paralysis from great exertion

—— from less exertion

—— of the lacteals

—— of the liver

—— of the right arm, why

—— of the veins

Particles of matter will not approach

Passions

—— connate

Pecking of chickens

Perception defined

Periods of agues, how formed

—— of diseases

—— of natural actions and of diseased actions

Perspiration in fever-fits, . See Sweat

Petechiæ

Pigeons secrete milk in their stomachs

Piles

Placenta a pulmonary organ

Pleasure of life

—— from greater fibrous contractions

—— what kind causes laughter

—— what kind causes sleep

Pleurisy, periods of

—— cause of

Prometheus, story of

Prostration of strength in fevers

Pupils of the eyes large

Pulse quick in fevers with debility

—— in fevers with strength

—— from defect of blood

—— weak from emetics

Q

Quack advertisements injurious. Preface

Quadrupeds have no sanguiferous lochia

—— have nothing similar to the yolk of egg

R

Rhaphania, periods of

Reason

Reasoning

Recollection

Relaxation and bracing

Repetition, why agreeable

Respiration affected by attention

Restlessness in fevers

Retrograde motions

—— of the stomach

—— of the skin

—— of fluids, how distinguished

—— how caused

—— vegetable motions

Retina is fibrous

—— is active in vision

—— excited into spasmodic motions

—— is sensible during sleep

Reverie

—— case of a sleep-walker

—— is an epileptic disease

Rhymes in poetry, why agreeable

Rheumatism, three kinds of

Rocking young children

Ruminating animals

S

Saliva produced by mercury

—— by food

—— not much, accumulated in sleep

—— powers, accumulation of

—— exhaustion of

—— wasted below natural in hot fits

—— less exertion of produces pain

—— less quantity of it

Sensual motions distinguished from muscular

Sex owing to the imagination of the father

Shingles from inflamed kidney

Shoulders broad

Shuddering from cold, . and

Sight, its accuracy in men

Skin, skurf on it

Sleep suspends volition

—— defined

—— remote causes

—— sensation continues in it

—— from food

—— from rocking, uniform sounds

—— from wine and opium

—— why it invigorates

—— pulse slower and fuller

—— interrupted

—— from breathing less oxygene

—— from being whirled on a millstone

—— from application of cold

—— induced by regular hours

Sleeping animals

Sleep-walkers. See Reverie

Small-pox

Suggestion defined

Sun and moon, their influence

Surprise

Suspicion attends madness

Swallowing, act of

Sweat, cold

—— in hot fit of fever

—— in a morning, why

Sweaty hands cured by lime

Swinging and rocking, why agreeable

Sympathy

Syncope

T

Tædium vitæ. See Ennui

Tape-worm

Taste, sense of

Tears, secretion of

—— from grief

—— from tender pleasure

—— from stimulus of nasal duct

—— by volition

Teeth decaying cause headachs

Temperaments

Theory of medicine, wanted. Preface

Thirst, sense of

—— why in dropsies

Tickle themselves, children cannot

Tickling

Time

—— stopped by quicksilver

—— weakens the pulse

W

Waking, how

Walking, how learnt

Warmth in sleep, why

Weakness defined

—— cure of

—— See Debility

Wit producing laughter

World generated